Learn R

Learning a computer language like R can be either frustrating, fun or boring. Having fun requires challenges that wake up the learner's curiosity but also provide an emotional reward for overcoming them. The book is designed so that it includes smaller and bigger challenges, in what I call playgrounds, in the hope that all readers will enjoy their path to R fluency. Fluency in the use of a language is a skill that is acquired through practice and exploration. For students and professionals in the biological sciences, humanities and many applied fields, recognizing the parallels between R and natural languages should help them feel at home with R. The approach I use is similar to that of a travel guide, encouraging exploration and describing the available alternatives and how to reach them. The intention is to guide the reader through the R landscape of 2024 and beyond.

What is new in the second edition?

- Text expanded by more than 25% to include additional R features and gentler and more detailed explanations
- Contains 24 new diagrams and flowcharts, seven new tables, and revised text and code examples for clarity
- All three indexes were expanded, and answers to 28 frequently asked questions added

What will you find in this book?

- Programming concepts explained as they apply to current R
- Emphasis on the role of abstractions in programming
- Few prescriptive rules—mostly the author's preferences together with alternatives
- Presentation of the R language emphasizing the "R way of doing things"
- Tutoring for "programming in the small" using scripts for data analysis
- Explanation of the differences between R proper and extensions for data wrangling
- The grammar of graphics is described as a language for the construction of data visualisations
- Examples of data exchange between R and the foreign world using common file formats
- Coaching to become an independent R user, capable of writing original scripts and solving future challenges

Pedro J. Aphalo is a PhD graduate from the University of Edinburgh, currently a senior lecturer at the University of Helsinki. He is a plant biologist and agriculture scientist with a passion for data, electronics, computers, and photography, in addition to plants. He has been a user of R for 28 years, who first organized an R course for MSc students 21 years ago and is the author of 14 R packages currently in CRAN.

Chapman & Hall/CRC
The R Series

Recently Published Titles

Engineering Production-Grade Shiny Apps
Colin Fay, Sébastien Rochette, Vincent Guyader, and Cervan Girard

Javascript for R
John Coene

Advanced R Solutions
Malte Grosser, Henning Bumann, and Hadley Wickham

Event History Analysis with R, Second Edition
Göran Broström

Behavior Analysis with Machine Learning Using R
Enrique Garcia Ceja

Rasch Measurement Theory Analysis in R: Illustrations and Practical Guidance for Researchers and Practitioners
Stefanie Wind and Cheng Hua

Spatial Sampling with R
Dick R. Brus

Crime by the Numbers: A Criminologist's Guide to R
Jacob Kaplan

Analyzing US Census Data: Methods, Maps, and Models in R
Kyle Walker

ANOVA and Mixed Models: A Short Introduction Using R
Lukas Meier

Tidy Finance with R
Stefan Voigt, Patrick Weiss and Christoph Scheuch

Deep Learning and Scientific Computing with R torch
Sigrid Keydana

Model-Based Clustering, Classification, and Density Estimation Using mclust in R
Lucca Scrucca, Chris Fraley, T. Brendan Murphy, and Adrian E. Raftery

Spatial Data Science: With Applications in R
Edzer Pebesma and Roger Bivand

Modern Data Visualization with R
Robert Kabacoff

Learn R: As a Language, Second Edition
Pedro J. Aphalo

For more information about this series, please visit: https://www.crcpress.com/Chapman--HallCRC-The-R-Series/book-series/CRCTHERSER

Learn R

As a Language

Second Edition

Pedro J. Aphalo

CRC Press is an imprint of the
Taylor & Francis Group, an **informa** business
A CHAPMAN & HALL BOOK

Second edition published 2024
by CRC Press
2385 NW Executive Center Drive, Suite 320, Boca Raton, FL 33431

and by CRC Press
4 Park Square, Milton Park, Abingdon, Oxon, OX14 4RN

CRC Press is an imprint of Taylor & Francis Group, LLC

First edition published by CRC Press 2020

ISBN: 978-1-032-51843-5 (hbk)
ISBN: 978-1-032-51699-8 (pbk)
ISBN: 978-1-003-40418-7 (ebk)

Typeset in LucidaBrightOT-Identity-H font
by KnowledgeWorks Global Ltd.

DOI: 10.1201/9781003404187

Publisher's note: This book has been prepared from camera-ready copy provided by the author.

Contents

List of Figures

List of Tables

Preface

> Suppose that you want to teach the 'cat' concept to a very young child. Do you explain that a cat is a relatively small, primarily carnivorous mammal with retractible claws, a distinctive sonic output, etc.? I'll bet not. You probably show the kid a lot of different cats, saying 'kitty' each time, until it gets the idea. To put it more generally, generalisations are best made by abstraction from experience.
>
> R. P. Boas
> *Can we make mathematics intelligible?*, 1981

Why did I choose "*Learn R: As a Language*" as the title? This book is based on exploration and practice that aims at teaching how to express various operations on data using the R language. It focuses on the language, rather than on specific types of data analysis, exposes the reader to current usage, and does not spare the quirks of the language. When we use our native language in everyday life, we do not think about grammar rules or sentence structure, except for the trickier or unfamiliar situations. My aim is for this book to help readers learn to use R in this same way, i.e., to become fluent in R. The book is structured around the elements of natural languages like English with chapter titles that highlight the parallels between them and the R language.

Learn R: As a Language is different from other books about R in that it emphasises the learning of the language itself, rather than how to use it to address specific data analysis tasks. My aim has been to enable readers to use R to implement original solutions to the data analysis and data visualisation tasks they encounter. The use of quantitative methods and data analysis has become more frequent in fields with limited long-term tradition in their use, like humanities, or, the complexity of the methods used has dramatically increased, like in Biology. Such trends can be expected to continue in the future.

Currently, many students of biological and environmental sciences learn to use R in courses about statistics or data analysis. However, this is frequently not in enough depth to effectively use R in scripts for automating data analyses or to ensure their reproducibility. There are also many researchers in various fields who are already familiar with statistical principles and willing to switch from other software to R. *Learn R: As a Language* is written with these readers in mind to serve both as a textbook and as a reference.

A language is a system of communication. Basic concepts and operations are based on abstractions that are shared across programming languages and relevant to programs of all sizes and complexities; these abstractions are explained in the book together with their implementation in the R language. Other abstractions and programming concepts, outside the scope of this book, are relevant to large and complex pieces of software meant to be widely distributed. In other words, *Learn R: As a Language* aims at teaching and supporting *programming in the small*: the use of R to automate the drudgery of data manipulation, including the different steps spanning from data input and exploration to the production of publication-ready illustrations and reproducible data-based reports.

Using a language actively is the most efficient way of learning it. By using it, I mean actually reading, writing, and running scripts or programs. *Learn R: As a Language* supports learning the R language in a way comparable to how children learn to speak: they work out what the rules are, simply by listening to people speak and trying to utter what they want to tell their parents. Of course, small children also receive guidance through feedback, but they are not taught a prescriptive set of rules like when learning a second language at school. Instead of listening, readers will read and run code, and instead of speaking, readers will write and try to run R-language code on a computer. I do provide explanations and guidance, as understanding how R works greatly helps with its use. However, the approach I encourage in this book is for readers to play with the numerous examples and to create variations upon them, to find out by themselves the patterns behind the R language. Instead of parents being the sounding board for the first utterances of readers new to R, the computer will play this role. Although working through the examples in *Learn R: As a Language* in a group of peers or in class is beneficial, the book is designed to be useful also in the absence of such support.

Changes in the second edition:

I edited the text from the first edition to correct all errors and outdated examples or explanations known to me. This revised second edition reflects changes in R and the contributed packages used in the book. Very little of the code from the first edition had stopped working but deprecations meant that a few examples triggered messages or warnings, and will eventually fail. Recent (> 4.0.0) versions of R have significant enhancements, including the new pipe operator described and used in this second edition. Packages have also evolved, acquiring new features like a new approach to flipping plots in 'ggplot2'.

I have aimed at making the book more accessible to readers with no previous experience in computer programming. Feedback from readers and reviewers highlighted a few gaps in the content and some difficult-to-follow explanations. I revised the text, in some cases changing the sequence of presentation. I added diagrams to illustrate the structure of different types of objects and flowcharts to describe how program constructs work. I added tables listing groups of related functions. New sections cover character string operations, and details of data wrangling in R. Some of the most frequently asked questions about R are answered in the text and separately indexed. All exercises or "playgrounds" are numbered to facilitate their use as class work and the sharing of model answers. As the first edition has been frequently found useful as a reference, I expanded the already thorough indexing and added more cross-references connecting related sections across the whole book.

An additional change is in my view about packages 'dplyr' and 'tidyr', part of the 'tidyverse'. I have come to think that the rate of development of these two packages can make them difficult for users for whom data analysis is just one aspect of their occupation. As these packages are widely used, I emphasise more than in the first edition the differences between functions and classes from packages 'dplyr' and 'tidyr' and equivalent ones from base R. I added a section on working with dates and times using the 'lubridate' package. I updated and reorganised the chapter describing package 'ggplot2' and some of its extensions.

In numbers, the page count has increased by 27%, the number of figures from eight to twenty-six plus nine in-text diagrams, and tables from none to seven. As for the design, text boxes have been replaced by call-outs marked with marginal bars. In addition, starting from version 2.0.0, the 'learnrbook' package supports the first and second editions of the book. It contains data, scripts, and all the code examples from both editions. It also helps with the installation of all the packages used in the book. The website at `https://www.learnr-book.info/` provides updated open-access content.

Acknowledgements

I thank Jaakko Heinonen for introducing me in the late 1990s to the then new R. Along the way many experts have answered my questions in usenet and more recently in StackOverflow. I wish to warmly thank members of my research group, students, collaborators, authors of books, and people I have met online or at conferences. They have made it possible for me to write this book. I am specially indebted to Dan Yavorsky, Tarja Lehto, Titta Kotilainen, Tautvydas Zalnierius, Fang Wang, Yan Yan, Neha Rai, Markus Laurel, Brett Cooper, Viivi Lindholm, Matěj Rzehulka, Zuzana Svarna, colleagues, students, and anonymous reviewers for many very helpful comments on the draft manuscript and/or the first edition. Rob Calver, editor of both editions, provided advice and encouragement with great patience. Paul Boyd, Shashi Kumar, Ashraf Reza, Vaishali Singh, Lara Spieker, and Sherry Thomas efficiently helped with different aspects of this project.

The writing of this second edition was helped by a six-month sabbatical granted by the Faculty of Biological and Environmental Sciences of the University of Helsinki, Finland. I thank Prof. Kurt Fagerstedt for his support.

In many ways this text owes more to people who are not authors than to myself. However, as I am the one who has written *Learn R: As a Language* and decided what to include and exclude, I take full responsibility for any errors and inaccuracies.

Pedro J. Aphalo
Helsinki, 5th March 2024

1

Using the Book to Learn R

> The important part of becoming a programmer is learning to think like a programmer. You don't need to know the details of a programming language by heart, you can just look that stuff up.
>
> The treasure is in the structure, not the nails.
>
> P. Burns
> *Tao Te Programming*, 2012

1.1 Aims of This Chapter

In this chapter, I describe how I imagine the book can be used most effectively to learn the R language. Learning R and remembering what one has previously learnt and forgotten makes it also necessary to use this book and other sources as references. Learning to use R effectively also involves learning how to search for information and how to ask questions from other users, for example through on-line forums. Thus, I also give advice on how to find answers to R-related questions and how to use the available documentation.

1.2 Approach and Structure

Depending on previous experience, reading *Learn R: As a Language* will be about exploring a new world or revisiting a familiar one. In both cases this book aims to be a travel guide, neither a traveller's account nor a cookbook of R recipes. It can be used as a course book, supplementary reading or for self-instruction, and also as a reference. My hope is that as a guide to the use of R, this book will remain useful to readers as they gain experience and develop skills.

I encourage readers to approach R like a child approaches his or her mother tongue when learning to speak: do not struggle, just play, and fool around with R! If the

DOI: 10.1201/9781003404187-1

going gets difficult and frustrating, take a break! If you get a new insight, take a break to enjoy the victory!

In R, like in most "rich" languages, there are multiple ways of coding the same operations. I have included code examples that aim to strike a balance between execution speed and readability. One could write equivalent R books using substantially different code examples. Keep this in mind when reading the book and using R. Keep also in mind that it is impossible to remember everything about R, and as a user you will frequently need to consult the documentation, even while doing the exercises in this book. The R language, in a broad sense, is vast because it can be expanded with independently developed packages. Learning to use R mainly consists of learning the basics plus developing the skill of finding your way in R, its documentation and on-line question-and-answer forums.

Readers should not aim to remember all the details presented in the book. This is impossible for most of us. Later use of this and other books, and documentation effectively as references, depends on a good grasp of a broad picture of how R works and on learning how to navigate the documentation; i.e., it is more important to remember abstractions and in what situations they are used, and function names, than the details of how to use them. Developing a sense of when one needs to be careful not to fall into a "language trap" is also important.

The book can be used both as a textbook for learning R and as a reference. It starts with simple concepts and language elements progressing towards more complex language structures and uses. Along the way readers will find, in each chapter, descriptions and examples of the common (usual) cases and the exceptions. Some books hide the exceptions and counterintuitive features from learners to make the learning easier; I instead have included these but marked them using icons and marginal bars. There are two reasons for choosing this approach. First, the boundary between boringly easy and frustratingly challenging is different for each of us, and varies depending on the subject dealt with. So, I hope the marks will help readers predict what to expect, how much effort to put into each section, and even what to read and what to skip. Second, if I had hidden the tricky bits of the R language, I would have made later use of R by the readers more difficult. It would have also made the book less useful as a reference.

The book contains many code examples as well as exercises. I expect readers will run code examples and try as many variations of them as needed to develop an understanding of the "rules" of the R language, e.g., how the function or feature exemplified works. This is what long-time users of R do when facing an unfamiliar feature or a gap in their understanding.

Readers who are new to R should read at least chapters 2 to 6 sequentially. Possibly, skipping parts of the text and exercises marked as advanced. However, I expect to be most useful to these readers not to completely skip the description of unusual features and special cases but rather to skim enough from them so as to get an idea of what special situations they may face as R users. Exercises should not be skipped, as they are a key component of the didactic approach used.

Readers already familiar with R will be able to read the chapters in the book in any order, as the need arises. Marginal bars and icons, and the back and forward cross-references among sections, make possible for readers to *find a good path* within the book both when learning R and when using the book as a reference.

I expect *Learn R: As a Language* to remain useful as a reference to those readers who use it to learn R. It will also be useful as a reference to readers already familiar with R. To support the use of the book as a reference, I have been thorough with indexing, including many carefully chosen terms, their synonyms, and the names of all R objects and constructs discussed, collecting them in three alphabetical indexes: *General index*, *Index of R names by category*, and *Alphabetic index of R names* starting at pages 429, 446 and 438, respectively. I have also included back and forward cross-references linking related sections throughout the whole book.

1.3 Typographic and Naming Conventions

1.3.1 Call-outs

Marginal bars and icons are used in the book to inform about what content is advanced or included with a specific aim. The following icons and colours are used.

Signals in-depth explanations of specific R features or general programming concepts. Several of these explanations make reference to programming concepts or features of the R language that are explained later in the book. Readers new to R and computer programming can safely skip these call-outs on the first reading of the book. To become proficient in the use of R these readers are expected to return at a later time without hurry, preferably with a cup of coffee or tea to these call-outs. Readers with more experience, like those possibly reading individual chapters or using the book as a reference, will find these in-depth explanations useful.

Signals important bits of information that must be remembered when using R—i.e., explanations of some unusual, but important, feature of the language or concepts that in my experience are easily missed by those new to R.

Frequently asked question
Signals my answer to a question that I expect to be useful to readers based on the popularity of similar or related questions posted in online forums. When reading through the book, they highlight things that are worth special attention. When using the book as a reference, they help find solutions to frequently encountered difficulties. Index on page 446.

1.1 Signals a *playground* containing open-ended exercises—ideas and pieces of R code to play with at the R console. I expect readers to run these examples both as is and after creating variations by editing the code, studying the output, or diagnosis messages, returned by R in each case. Numbered by chapter for easy reference.

1.2 Signals an *advanced playground* that requires more time to play with before grasping concepts than regular *playgrounds*. Numbered by chapter together with other playgrounds.

1.3.2 Code conventions and syntax highlighting

Small sections of program code interspersed within the main text, receive the name of *code chunks*. In this book R code chunks are typeset in a typewriter font, using colour to highlight the different elements of the syntax, such as variables, functions, constant values, etc. The command line prompts (> and +) are not displayed in the chunks. R code elements embedded in the text are similarly typeset but always black. For example, in the code chunk below, `mean()` and `print()` are functions; 1, 5, and 3 are constant numeric values, and `z` is the name of a variable where the result of the computation done in the first line of code is stored. The line starting with `##` shows what is printed or shown when executing the second statement: `[1] 1`. In the book, `##` is used as a marker to signal output from R, it is not part of the output. As `#` is the marker for comments in the R language, prepending `#` to the output makes it possible to copy and paste into the R console the whole contents of the code chunks as they appear in the book.

```
z <- mean(1, 5, 3)
print(z)
## [1] 1
```

When explaining general concepts I use short abstract names, while for real-life examples I use descriptive names. Although not required, for clarity, I use abstract names that hint at the structure of objects stored, such as `mat1` for a matrix, `vct4` for a vector and `df3` for a data frame. This convention resembles that followed by the base R documentation.

Code in playgrounds either works in isolation, or when it depends on objects created in the examples in the main text, this is mentioned within the playground. In playgrounds I use names in capital letters so that they are distinct. The code outside playgrounds does reuse objects created earlier in the same section, and occasionally in earlier sections of the same chapter.

1.3.3 Diagrams

To describe data objects, I use diagrams similar to Joseph N. Hall's PEGS (Perl Graphical Structures) (Hall and Schwartz 1997). I use colour fill to highlight the type of the stored objects. I use the "signal" sign for the names of whole objects and of their component members, the former with a thicker border. Below is an example from chapter 3.

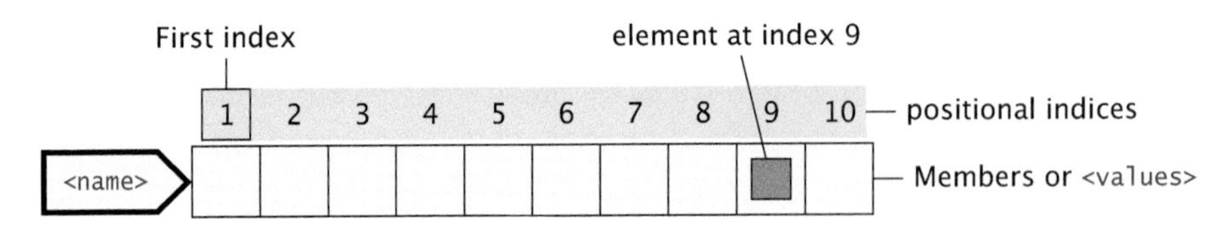

To describe code structure, I use diagrams based on boxes and arrows, while to describe the flow of code execution I use traditional flow charts.

In the different diagrams, I use the notation `<value>`, `<statement>`, `<name>`, etc., as generic placeholders indicating *any valid value, any valid R statement, any valid R name*, etc.

1.4 Finding Answers to Problems

1.4.1 What are the options?

First of all, do not panic! Every programmer, even those with decades of experience, gets stuck with problems from time to time and can run out of ideas for a while. This is normal and happens to all of us.

It is important to learn how to find answers as part of the routine of using R. We should start by reading the documentation of the function or object that we are trying to use, which in many cases also includes examples. R's help pages tell how to use individual functions or objects. In contrast, R's manual *An Introduction to R*, and other books describe what functions or overall approaches to use for different tasks.

Reading the documentation and books not always helps. Sometimes one can become blind to the obvious, by being too familiar with a piece of code, as it also happens when writing in a natural language like English. A second useful step is, thus, looking at the code with "different eyes", those of a friend or workmate, or your own eyes a day or a week later.

One can also seek help in specialised online forums or from peers or "local experts". If searching in forums for existing questions and answers fails to yield a useful answer, one can write a new question in a forum.

When searching for answers, asking for an advice, or reading books, one can be confronted with different ways of approaching the same tasks. Do not allow this to overwhelm you; in most cases, it will not matter which approach you use as many computations can be done in R, as in any computer language, in several different ways, still obtaining the same result. Use the alternative that you find easier to understand.

1.4.2 R's built-in help

Every object available in base R or exported by an R extension package (functions, methods, classes, and data) is documented in R's help system. Sometimes a single help page documents several R objects. Not only help pages are always available, but they are structured consistently with a title, short description, and frequently also a detailed description. In the case of functions, parameter names, their purpose, and expected arguments are always described, as well as the returned value. Usually at the bottom of help pages, several examples of the use of the objects or functions are given. How to access R help is described in section 2.3 on page 12.

In addition to help pages, R's distribution includes useful manuals as PDF or HTML files. These manuals are also available at `https://rstudio.github.io/r-manuals/` restyled for easier reading in web browsers. In addition to help pages, many packages, contain *vignettes* such as user guides or articles describing the algorithms used and/or containing use case examples. In the case of some packages, a web site with documentation in HTML format is also available. Package documentation can be also found in repositories like the *Comprehensive R Archive Network*, better known as CRAN. From CRAN it is possible to download R and many extensions to it. The DESCRIPTION file of each R package provides contact information for the maintainer, links to web sites, and instructions on how to report bugs. Similar information plus a short description are frequently also available in a README file.

Error messages tend to be terse in R, and may require some lateral thinking and/or "experimentation" to understand the real cause behind problems. Learning to interpret error messages is necessary to become a proficient user of R, so forcing errors and warnings with purposely written "bad" code is a useful exercise.

1.4.3 Online forums

Netiquette

When posting requests for help, one needs to abide by what is usually described as "netiquette", which in many respects also applies to asking in person or by e-mail for help from a peer or local expert. Preference among sources of information depends on what one finds easier to use. Consideration towards others' time is necessary but has to be balanced against wasting too much of one's own time.

In most internet forums, a certain behaviour is expected from those asking and answering questions. Some types of misbehaviour, such as the use of offensive or inappropriate language, will usually result in the user losing writing rights in a forum. Occasional minor misbehaviour usually results in the original question not being answered and, instead, the problem highlighted in a comment. In general, following the steps listed below will greatly increase your chances of getting a detailed and useful answer.

- Do your homework: first search for existing answers to your question, both online and in the documentation. (Do mention that you attempted this without success when you post your question.)
- Provide a clear explanation of the problem, and all the relevant information. The version of R, operating system, and any packages loaded and their versions can be important.
- If at all possible, provide a simplified and short, but self-contained, code example that reproduces the problem (sometimes called a *reprex*).
- Be polite.
- Contribute to the forum by answering other users' questions when you know the answer.

Carefully preparing a reproducible example ("reprex") is crucial. A *reprex* is a self-contained and as simple as possible piece of computer code that triggers (and so demonstrates) a problem. If possible, when data are needed, a data set included in base R or artificial data generated within the reprex code should be used. If the problem can only be reproduced with one's own data, then one needs to provide a minimal subset of it that still triggers the problem.

While preparing a *reprex* one has to simplify the code, and sometimes this step makes clear the nature of the problem. Always, before posting a reprex online, check it with the latest versions of R and any package being used. If sharing data, be careful about confidential information and either remove or mangle it.

I must say that about two out of three times I prepare a *reprex*, it allows me to find the root of the problem and a solution or a work-around on my own. Preparing a *reprex* takes some effort but it is worthwhile even if it ends up not being posted online.

R package 'reprex' and its RStudio add-in simplify the creation of reproducible code examples, by creating and copying to the clipboard a reprex encoded in Markdown and ready to be pasted into a question at StackOverflow or an issue at GitHub. See `https://reprex.tidyverse.org/` for details.

StackOverflow

Nowadays, StackOverflow (`http://stackoverflow.com/`) is the best question-and-answer (Q&A) support site for R. Within the StackOverflow site there is an R collective. In most cases, searching for existing questions and their answers will be all that you need to do. If asking a question, make sure that it is really a new question. If there is some question that looks similar, make clear how your question is different.

StackOverflow has a user-rights system based on reputation, and questions and answers can be up- and down-voted. Questions with the most up-votes are listed at the top of searches, and the most-voted answers to each question are also displayed first. Those who ask a question are expected to accept correct answers to help future readers. If the questions or answers one writes are up-voted or if answers are accepted one gains reputation (expressed as a number). As one accumulates reputation, one gets badges and additional rights, such as editing other users' questions and answers or later on, even deleting wrong answers or off-topic questions from the system. This sounds complicated, but works extremely well at ensuring that the base of questions and answers is relevant and correct, without relying heavily on nominated *moderators*. When using StackOverflow, do contribute by accepting correct answers, up-voting questions and answers that you find useful, down-voting those you consider poor, and flagging or correcting errors you may discover.

Being careful in the preparation of a reproducible example is important in two situations: 1) when asking a question at StackOverflow or other online forums and 2) when reporting a bug to the maintainer of any piece of software. For the question to be reliably answered or the problem to be fixed, the person answering a question, needs to be able to reproduce the problem, and after modifying the code, needs to be able to test if the problem has been solved or not. However, even if you are facing a problem caused by your misunderstanding of how R works, the simpler the example, the more likely that someone will quickly realise what your intention was when writing the code that produces a result different from what you expected. Even when it is not possible to create a reprex, one needs to ask clearly only one thing per question.

The code of conduct (`https://stackoverflow.com/conduct`) and help that explains expected behaviour (`https://stackoverflow.com/help`) are available at the site and worthwhile reading before using the site actively for the first time.

Contacting the author

The best way to get in contact with me about this book is by raising an issue at `https://github.com/aphalo/learnr-book-crc/issues`. Issues can be used both to ask for support questions related to the book, report mistakes and suggest changes to the text, diagrams and/or example code. Edits to the manuscript of this book can be submitted as a pull request.

Issues are raised by filling-in an online form, on a web page that also contains brief instructions. Git issues are a very efficient way of keeping track of corrections that need to be done. As support questions usually reveal unclear explanations or other problems, raising issues to ask them facilitates the tasks of improving and keeping the book up-to-date.

1.5 Further Reading

To understand what programming as an activity is, you can read *Tao Te Programming* (Burns 2012). It will make easier the learning of programming in R, both practically and emotionally. In Burns's words "This is a book about what goes on in the minds of programmers".

2

R: The Language and the Program

> In a world of … relentless pressure for more of everything, one can lose sight of the basic principles—simplicity, clarity, generality—that form the bedrock of good software.
>
> Brian W. Kernighan and Rob Pike
> *The Practice of Programming*, 1999

2.1 Aims of This Chapter

I share some facts about the history and design of the R language so that you can gain a good vantage point from which to grasp the logic behind R's features, making it easier to understand and remember them. You will learn the distinction between the R program itself and the front-end programs, like RStudio, frequently used together with R.

You will also learn how to interact with R when sitting at a computer. You will learn the difference between typing commands interactively and reading each partial result from R on the screen as you enter them, versus using R scripts containing multiple commands stored in a file to execute or run a "job" that saves results to another file for later inspection.

I describe the steps taken in a typical scientific or technical study, including the data analysis workflow and the roles that R can play in it. I share my views on the advantages and disadvantages of textual command languages such as R compared to menu-driven user interfaces, frequently used in other statistics software. I discuss the role of textual languages and *literate programming* in the very important question of the reproducibility of data analyses and mention how I have used them while writing and typesetting this book.

DOI: 10.1201/9781003404187-2

2.2 What is R?

2.2.1 R as a language

R is a computer language designed for data analysis and data visualisation, however, in contrast to some other scripting languages, it is, from the point of view of computer programming, a complete language—it is not missing any important feature. In other words, no fundamental operations or data types are lacking (Chambers 2016). I attribute much of its success to the fact that its design achieves a very good balance between simplicity, clarity, and generality. R excels at generality, thanks to its extensibility at the cost of only a moderate loss of simplicity, while clarity is ensured by enforced documentation of extensions and support for both object-oriented and functional approaches to programming. The same three principles can be also easily followed by user code written in R.

In the case of languages like C++, C, Pascal, and FORTRAN, multiple software implementations exist (different compilers and interpreters, i.e., pieces of software that translate programs encoded in these languages into *machine code* instructions for computer processors to run). So in addition to different flavours of each language stemming from different definitions, e.g., versions of international standards, different implementations of the same standard may have, usually small, unintentional and intentional differences.

Most people think of R as a computer program, similar to SAS or SPSS. R is indeed a computer program—a piece of software—but it is also a computer language, implemented in the R program. At the moment, this difference is not as important as for other languages because the R program is the only widely used implementation of the R language.

R started as a partial implementation of the then relatively new S language (Becker and Chambers 1984; Becker et al. 1988). When designed, S, developed at Bell Labs in the U.S.A., provided a novel way of carrying out data analyses. S evolved into S-Plus (Becker et al. 1988). S-Plus was available as a commercial program, most recently from TIBCO, U.S. R started as a poor man's home-brewed implementation of S, for use in teaching, developed by Robert Gentleman and Ross Ihaka at the University of Auckland, in New Zealand (Ihaka and Gentleman 1996). Initially, R, the program, implemented a subset of the S language. The R program evolved until only relatively few differences between S and R remained. These remaining differences are intentional—thought of as significant improvements. In more recent times, R overtook S-Plus in popularity. The R language is not standardised, and no formal definition of its grammar exists. Consequently, the R language is defined by the behaviour of its implementation in the R program.

What makes R different from SPSS, SAS, etc., is that S was designed from the start as a computer programming language. This may look unimportant for someone not actually needing or willing to write software for data analysis. However, in reality, it makes a huge difference because R is easily extensible, both using the R language for implementation and by calling from R functions and routines written in other computer programming languages such as C, C++, FORTRAN, Python, or

Java. This flexibility means that new functionality can be easily added, and easily shared with a consistent R-based user interface. In other words, instead of having to switch between different pieces of software to do different types of analyses or plots, one can usually find a package that will make new tools seamlessly available within R.

The name "base R" is used to distinguish R itself, as in the R executable included in the R distribution and its default packages, from R in a broader sense, which includes contributed packages. A few packages are included in the R distribution, but most R packages are independently developed extensions and separately distributed. The number of freely available open-source R packages available is huge, in the order of 20 000.

The most important advantage of using a language like R is that instructions to the computer are given as text. This makes it easy to repeat or *reproduce* a data analysis. Textual instructions serve to communicate to other people what has been done in a way that is unambiguous. Sharing the instructions themselves avoids a translation from a set of instructions to the computer into text readable to humans—for example, the materials and methods section of a paper.

Readers with programming experience will notice that some features of R differ from those in other programming languages. R does not have the strict type checks of Pascal or C++. It has operators that can take vectors and matrices as operands. Reliable and fast R code tends to rely on different *idioms* than well-written Pascal or C++ code.

2.2.2 R as a computer program

The R program itself is open-source, i.e., its source code is available for anybody to inspect, modify, and use. A very small fraction of users will directly contribute improvements to the R program itself. However, those contributions and bug reports are important in making R extremely reliable. The executable R program we actually use can be built for different operating systems and computer hardware. The members of the R developing team aim to keep the results obtained from calculations done on all the different builds and computer architectures as consistent as possible. The idea is to ensure that computations return consistent results not only across updates to R but also across different operating systems, like Linux, Unix (including OS X) and MS-Windows, or computer hardware, like that based on ARM and x86 processors.

The R program does not have a full-fledged graphical user interface (GUI), or menus from which to start different types of analyses. Instead, the user types the commands at the R console and the result is displayed starting on the next line (Figure 2.1). The same textual commands can also be saved into a text file, line by line, and such a file, called a "script" can substitute for the direct typing of the same sequence of commands at the console (writing and use of R scripts are explained in chapter 5 on page 125). When we work at the console, typing-in commands one by one, we use R *interactively*. When we run a script, we may say that we run a "batch job". The two approaches described above are available in the R program itself.

```
R Console
> print("hello")
[1] "hello"
>
```

Figure 2.1
The R console. This is where the user can type textual commands line by line. Here a user has typed `print("Hello")` and *entered* it by ending the line of text by pressing the "enter" key. The result of running the command is displayed below the command. The character at the head of the input line, a ">" in this case, is called the command prompt, signalling where a command can be typed in. Commands entered by the user are displayed in red, while results returned by R are displayed in blue. "`[1]`" can be ignored here, its meaning is explained on page 28. The console as displayed in R GUI under MS-Windows is shown.

As R is essentially a command-line application, it can be used on what nowadays are frugal computing resources, equivalent to a personal computer of three decades ago. R can run even on the Raspberry Pi, a micro-controller board with the processing power of a modest smartphone (see `https://r4pi.org/`). At the other end of the spectrum, on really powerful servers, R can be used for the analysis of big data sets with millions of observations. How powerful a computer is needed for a given data analysis task depends on the size of the data sets, on how patient one is, on the ability to select efficient algorithms and on writing "good" code.

2.3 Using R

2.3.1 Editors and IDEs

Integrated Development Environments (IDEs) are normally used when developing computer programs. IDEs provide a centralised user interface from within which the different tools used to create and test a computer program can be accessed and used in coordination. Most IDEs include a dedicated editor capable of syntax highlighting (automatically colouring "code words" based on their role in the programming language), and even able to report some mistakes in advance of running the code. One could describe such an editor as the equivalent of a word processor with spelling and grammar checking that can alert about spelling and syntax errors for a computer language like R instead of a natural language like English. IDEs frequently add other features that help navigation of the programme source code and give easy access to documentation.

Nowadays, it is very common to use an IDE as a front-end or middleman between the user and the R program. Computations are still done in the R program, which

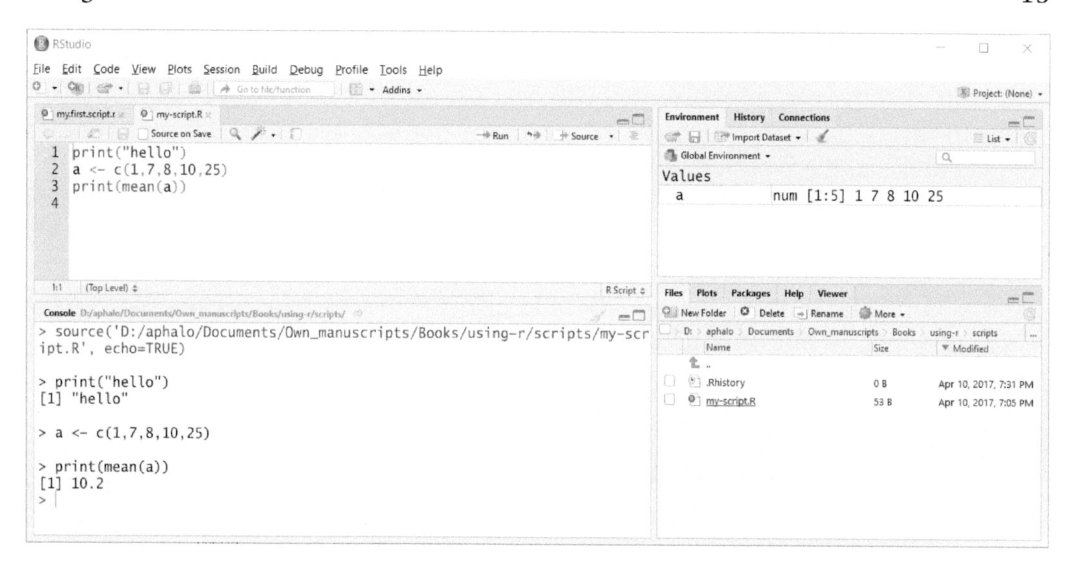

Figure 2.2
The RStudio interface after running the script that is visible in tab `my-script.R` of the editor pane (top left). Here I used the “Source” button to run the script and R printed the results to the R console in the lower left pane. The lower right pane shows a list of files, including the script open in the editor. The upper right pane displays a list of the objects currently visible in the user workspace, object a, which was created by the code in the second line of the R script.

is *not* built-in in the IDEs. Of the available IDEs for R, RStudio is currently the most popular by a wide margin. Recent versions of RStudio support Python in addition to R.

Readers with programming experience may be already familiar with Microsoft’s free Visual Studio Code or the open-source Eclipse IDEs for which plugins supporting R are available.

The main window of IDEs is in most cases divided into windows or panes, possibly with tabs. In RStudio one has access to the R console, a text editor, a file-system browser, a pane for graphical output, and access to several additional tools such as for installing and updating extension packages. Although RStudio supports very well the development of large scripts and packages, it is currently, in my opinion, also the best possible way of using R at the console as it has the R help system very well integrated both in the editor and R console. Figure 2.2 shows the main window displayed by RStudio after running the same script as shown at the R console (Figure 2.5) and at the operating system command prompt (Figure 2.6). By comparing these three figures, it is clear that RStudio is really only a software layer between the user and an unmodified R executable. In RStudio, the script was sourced by pressing the “Source” button at the top of the editor panel. RStudio, in response to this, generated the code needed to source the file and “entered” it at the console (2.2, lower left screen panel, text in purple), the same console where we can directly type this same R command if we wish.

When a script is run, if an error is triggered, RStudio automatically finds the location of the error, a feature you will find useful when running code from exercises in this book. Other features are beyond what one needs for simple everyday data analysis and are aimed at package development and report generation. Tools for debugging, code profiling, benchmarking of code and unit tests, make it possible to analyse and improve performance as well as help with quality assurance and certification of R packages and exceed what you will need for this book's exercises and simple data analysis. RStudio also integrates support for file version control, which is not only useful for package development but also for keeping track of the progress or concurrent work with collaborators in the analysis of data.

The "desktop" version of RStudio that one installs and uses locally, runs on most modern operating systems, such as Linux, Unix, OS X, and MS-Windows. There is also a server version that runs on Linux, as well as a cloud service (`https://posit.cloud/`) providing shared access to such a server. The RStudio server is used remotely through a web browser. The user interface is almost the same in all cases. Desktop and server versions are both distributed as unsupported free software and as supported commercial software.

RStudio and other IDEs support saving of their state and some settings per working folder under the name of *project*, so that work on a data analysis can be interrupted and later continued, even on a different computer. As mentioned in section 2.3.2 on page 14, when working with R we keep related files in a folder.

In this book, I provide only a minimum of guidance on the use of RStudio, and no guidance for other IDEs. To learn more about RStudio, please, read the documentation available through RStudio's help menu and keep at hand a printed copy of the RStudio cheat sheet while learning how to use it. This and other useful R-related cheatsheets can be downloaded at `https://posit.co/resources/cheatsheets/`. Additional instructions on the use of RStudio, including a video, are available through the Resources menu entry of the book's website at `https://www.learnr-book.info/`.

2.3.2 R sessions and workspaces

We use *session* to describe the interactive execution from start to finish of one running instance of the R program. We use *workspace* to name the imaginary space were all objects currently available in an R session are stored. In R, the whole workspace can be stored in a single file on disk at the end or during a session and restored later into another session, possibly on a different computer. Usually, when working with R, we dedicate a folder in disk storage to store all files from a given data analysis project. We normally keep in this folder files with data to read in, scripts, a file storing the whole contents of the workspace, named by default `.Rdata` and a text file with the history of commands entered interactively, named by default `.Rhistory`. The user's files within this folder can be located in nested folders. There are no strict rules on how the files should be organised or on their number. The recommended practice is to avoid crowded folders and folders containing unrelated files. It is a good idea to keep in a given folder and workspace the work in progress for a single data analysis project or experiment, so that the workspace can be saved and restored easily between sessions and work continued from where one left it independently of work done in other workspaces. The folder where files are currently read and saved is in R documentation called the *current working directory*. When opening an `.Rdata` file the current working directory is automatically set to the location where the `.Rdata` file was read from.

RStudio projects are implemented as a folder with a name ending in `.Rprj`, located under the same folder where scripts, data, `.Rdata`, and `.Rhistory` are stored. This folder is managed by RStudio and should be not modified or deleted by the user. Only in the very rare case of its corruption, it should be deleted, and the RStudio project created again from scratch. Files `.Rdata` and `.Rhistory` should not be deleted by the user, except to reset the R workspace. However, this is unnecessary as it can be also easily achieved from within R.

2.3.3 Using R interactively

Decades ago, users communicated with computers through a physical terminal (keyboard plus text-only screen) that was frequently called a *console*. A text-only interface to a computer program, in most cases a window or a pane within a graphical user interface, is still called a console. In our case, the R console (Figure 2.1). This is the native user interface of R.

Typing commands at the R console is useful when one is playing around, rather aimlessly exploring things, or trying to understand how an R function or operator we are not familiar with works. Once we want to keep track of what we are doing, there are better ways of using R, which allow us to keep a record of how an analysis has been carried out. The different ways of using R are not exclusive of each other, so most users will use the R console to test individual commands and plot data during the first stages of exploration. As soon as we decide how we want to plot or analyse the data, it is best to start using scripts. This is not enforced in any way by R, but scripts are what really brings to light the most important advantages of using a programming language for data analysis. In Figure 2.1, we can see how the R console looks. The text in red has been typed in by the user, except for the prompt >, and the text in blue is what R has displayed in response. It is essentially a dialogue between user and R. The console can *look* different when displayed within an IDE like RStudio, but the only difference is in the appearance of the text rather than in the text itself (cf. Figures 2.1 and 2.3).

The two previous figures showed the result of entering a single command. Figure 2.4 shows how the console looks after the user has entered several commands, each as a separate line of text.

The examples in this book require only the console window for user input. Menu-driven programs are not necessarily bad, they are just unsuitable when there is a need to set very many options and choose from many different actions. They are also difficult to maintain when extensibility is desired, and when independently developed modules of very different characteristics need to be integrated. Textual languages also have the advantage, to be addressed in later chapters, that command sequences can be stored in human- and computer-readable text files. Such

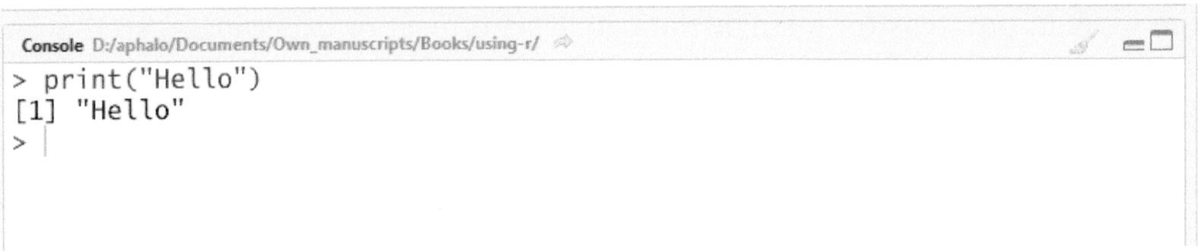

```
Console D:/aphalo/Documents/Own_manuscripts/Books/using-r/
> print("Hello")
[1] "Hello"
>
```

Figure 2.3
The R console embedded in RStudio. The same commands have been typed in as in Figure 2.1. Commands entered by the user are displayed in purple, while results returned by R are displayed in black.

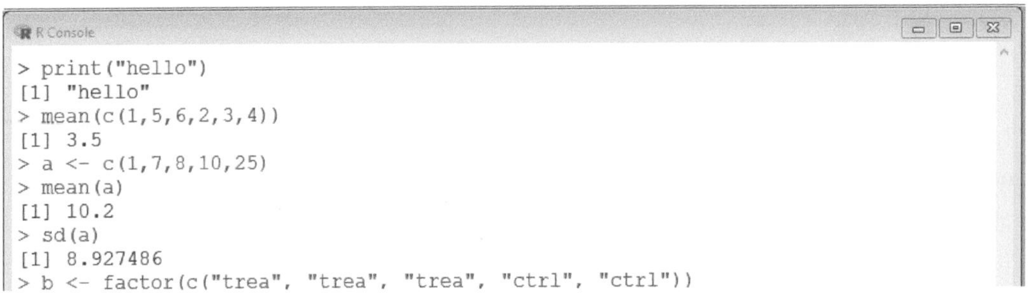

```
R Console
> print("hello")
[1] "hello"
> mean(c(1,5,6,2,3,4))
[1] 3.5
> a <- c(1,7,8,10,25)
> mean(a)
[1] 10.2
> sd(a)
[1] 8.927486
> b <- factor(c("trea", "trea", "trea", "ctrl", "ctrl"))
```

Figure 2.4
The R console after several commands have been entered. Commands entered by the user are displayed in red, while results returned by R are displayed in blue.

files constitute a record of all the steps used, and in most cases, make it trivial to manually reproduce the same steps at a later time. Scripts are a very simple and handy way of communicating to other users how a given data analysis has been done or can be done.

In the console one types commands at the > prompt. When one ends a line by pressing the return or enter key, if the line can be interpreted as an R command, the result will be printed at the console, followed by a new > prompt. If the command is incomplete, a + continuation prompt will be shown, and you will be able to type in the rest of the command. For example, if the whole calculation that you would like to do is $1 + 2 + 3$, if you enter in the console `1 + 2 +` in one line, you will get a continuation prompt where you will be able to type `3`. However, if you type `1 + 2`, the result will be calculated, and printed.

For example, one can search for a help page at the R console. Below are the first code example and the first playground in the book. This first example is for illustration only, and you can return to them later as only on page 20 I discuss how to install or get access to the R program.

```
help("sum")
?sum
```

2.1 Look at help for some other functions like `mean()`, `var()`, `plot()` and, why not, `help()` itself!

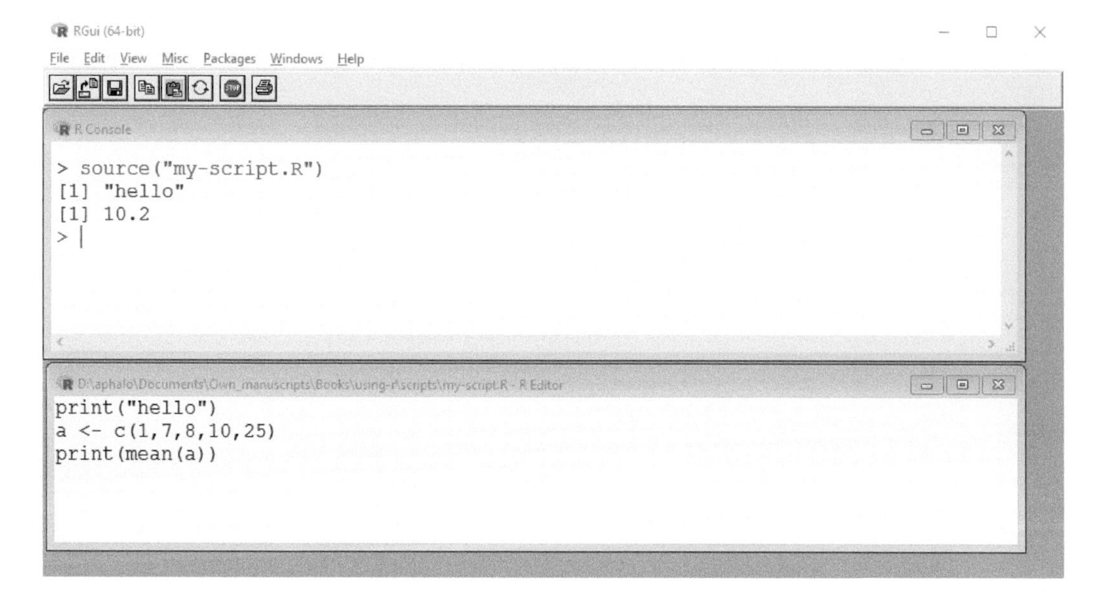

Figure 2.5
Screen capture of the R console and editor just after running a script. The upper pane shows the R console, and the lower pane, the script file in an editor.

```
help(help)
```

When trying to access help related to R extension packages through R's built in help, make sure the package is loaded into the current R session, as described on page 180, before calling `help()`.

When using RStudio, there are easier ways of navigating to a help page than calling function `help()` by typing its name, for example, with the cursor on the name of a function in the editor or console, pressing the F1 key opens the corresponding help page in the help pane. Letting the cursor hover for a few seconds over the name of a function at the R console will open "bubble help" for it. If the function is defined in a script or another file that is open in the editor pane, one can directly navigate from the line where the function is called to where it is defined. In RStudio one can also search for help through the graphical interface. The R manuals can also be accessed most easily through the Help menu in RStudio or RGUI.

2.3.4 Using R in a "batch job"

To run a script, we need first to prepare a script in a text editor. Figure 2.5 shows the console immediately after running the script file shown in the text editor. As before, red text, the command `source("my-script.R")`, was typed by the user, and the blue text in the console is what was displayed by R as a result of this action. The title bar of the console, shows "R-console", while the title bar of the editor shows the *path* to the script file that is open and ready to be edited followed by "R-editor".

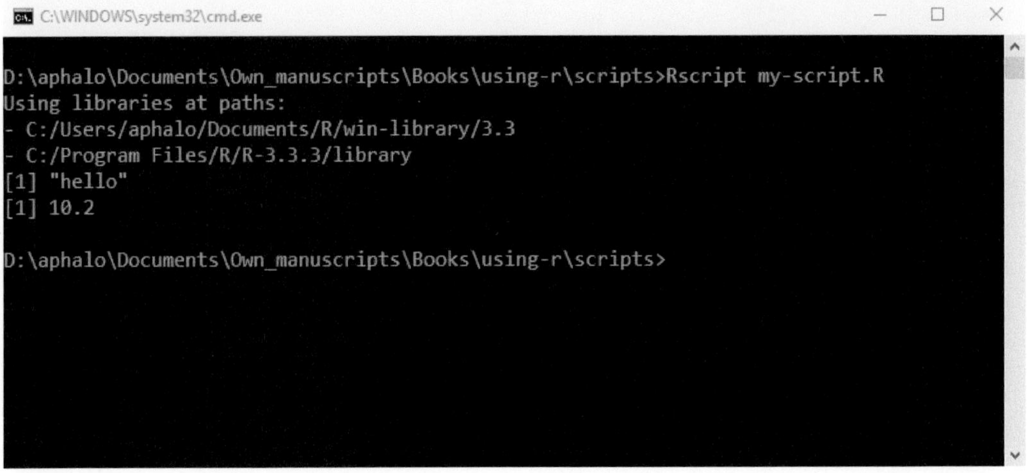

```
C:\WINDOWS\system32\cmd.exe

D:\aphalo\Documents\Own_manuscripts\Books\using-r\scripts>Rscript my-script.R
Using libraries at paths:
- C:/Users/aphalo/Documents/R/win-library/3.3
- C:/Program Files/R/R-3.3.3/library
[1] "hello"
[1] 10.2

D:\aphalo\Documents\Own_manuscripts\Books\using-r\scripts>
```

Figure 2.6
Screen capture of the MS-Windows command console just after running the same script. Here we use `Rscript` to run the script; the exact syntax will depend on the operating system in use. In this case, R prints the results at the operating system console or shell, rather than in its own R console.

When working at the command prompt, most results are printed by default. However, within scripts one needs to use function `print()` explicitly when a result is to be displayed.

A true "batch job" is not run at the R console but at the operating system command prompt, or shell. The shell is the console of the operating system—Linux, Unix, OS X, or MS-Windows. Figure 2.6 shows how running a script at the Windows command prompt looks. A script can be run at the operating system prompt to do time-consuming calculations with the output saved to a file. One may use this approach on a server, say, to leave a large data analysis job running overnight or even for several days.

Within RStudio desktop it is possible to access the operating system shell through the tab named "Terminal" and through the menu. It is also possible to run jobs in the background in the tab "Background jobs", i.e., while simultaneously using the R console. This is made possible by concurrently running two or more instances of the R program.

2.4 Reproducible Data Analysis with R

Statistical concepts and procedures are not only important after data are collected but also crucial at the design stage of any data-based study. Rather frequently, we deal with pre-existing data already at the planning stage of an experiment or survey. Statistics provides the foundation for the design of experiments and surveys, data analysis, and data visualisation. This is similar to the role played by grammar and vocabulary in communication in a natural language like English. Statistics makes possible decision-making based on partial evidence (or samples), but it is also a means of communication. Data visualisation also plays a key role in the written and oral communication of study conclusions. R is useful throughout all stages of the research process, from the design of studies to the communication of the results.

During recent years, the lack of reproducibility in scientific research, frequently described as a *reproducibility crisis*, has been broadly discussed and analysed (Gandrud 2015). One of the problems faced when attempting to reproduce scientific and technical studies is reproducing the data analysis. More generally, under any situation where accountability is important, from scientific research to decision making in commercial enterprises, industrial quality control and safety, and environmental impact assessments, being able to reproduce a data analysis reaching the same conclusions from the same data is crucial. Thus, an unambiguous description of the steps taken for an analysis is a requirement. Currently, most approaches to reproducible data analysis are based on automating report generation and including, as part of the report, all the computer commands that were used.

A reliable record of what commands have been run on which data is especially difficult to keep when issuing commands through menus and dialogue boxes in a graphical user interface or by interactively typing commands as text at a console. Even working interactively at the R console using copy and paste to include commands and results in a report typed in a word processor is error prone, and laborious. The use and archiving of R scripts alleviate this difficulty.

However, a further requirement to achieve reproducibility is the consistency between the saved and reported output and the R commands reported as having been used to produce them, saved separately when using scripts. This creates an error-prone step between data analysis and reporting. To solve this problem an approach to data analysis derived from what is called *literate programming* (Knuth 1984) was developed: running an especially formatted script that produces a document that includes the R code used for the analysis; the results of running this code and any explanatory text needed to describe the methodology used and interpret the results of the analysis.

Although a system capable of producing such reports with R, called 'Sweave' (Leisch 2002), has been available for a couple of decades, it was rather limited and not supported by an IDE, making its use rather tedious. Package 'knitr' (Xie 2013) further developed the approach and together with its integration into RStudio made the use of this type of report much easier. Less sophisticated reports, called R *notebooks*, formatted as HTML files can be created directly from ordinary R scripts containing no special formatting. Notebooks are HTML files that show as text the code used interspersed with the results, and can contain embedded the actual source script used to generate them.

Package 'knitr' supports the writing of reports with the textual explanations encoded using either Markdown or LaTeX as markup for text-formatting instructions. While Markdown (`https://daringfireball.net/projects/markdown/`) is an easy-to-learn and use text markup approach, LaTeX (Lamport 1994) is based on TeX (Knuth 1987), the most powerful typesetting engine freely available. There are different flavours of Markdown, including R markdown (see `https://rmarkdown.rstudio.com/`) with special support for R code. Quarto (see `https://quarto.org/`) was recently released as an enhancement of R markdown (see `https://rmarkdown.rstudio.com/`), improving typesetting and styling, and providing a single system capable of generating a broad selection of outputs. When used together with R, Quarto relies on package 'knitr' for the key step in the conversion, so in a strict sense Quarto does not replace it.

Because of the availability of these approaches to the generation of reports, the R language is extremely useful when reproducibility is important. Both 'knitr' and Quarto are powerful and flexible enough to write whole books, such as this very book you are now reading, produced with R, 'knitr' and LaTeX. All pages in the book were typeset directly, with plots and other R output generated on-the-fly by R and inserted automatically. All diagrams were generated by LaTeX during the typesetting step. The only exceptions are the figures in this chapter that have been manually captured from the computer screen. Why am I using this approach? First, because I want to make sure that every bit of code, as you will see printed, runs without error. In addition, I want to make sure that the output displayed below every line or chunk of R language code is exactly what R returns. Furthermore, it saves a lot of work for me as an author, as I can just update R and all the packages used to their latest version, and build the book again, after any changes needed to keep it up to date and free of errors. By using these tools and markup in plain text files, the indices, cross-references, citations, and list of references are all generated automatically.

Although the use of these tools is very important, they are outside the scope of this book and well described in other books dedicated to them (Gandrud 2015; Xie 2013). When using R in this way, a good command of R as a language for communication with both humans and computers is very useful.

2.5 Getting Ready to Use R

As the book is designed with the expectation that readers will run code examples as they read the text, you have to ensure access to the R before reading the next chapter. It is likely that your school, employer or teacher has already enabled access to R. If not, or if you are reading the book on your own, you should install R or secure access to an online service. Using RStudio or another IDE can facilitate the use of R, but all the code in the remaining chapters makes only use of R and packages available through CRAN.

I have written an R package, named 'learnrbook', containing original data and computer-readable listings for all code examples and exercises in the book. It also contains code and data that makes it easier to install the packages used in later chapters. Its name is 'learnrbook' and is available through CRAN. **It is not neces-**

sary for you to install this or any other packages until section 6.4.2 on page 180, where I explain how to install and use R packages.

Are there any resources to support the *Learn R: As a Language* book?
Please, visit https://www.learnr-book.info/ to find additional material related to this book, including additional free chapters. Up-to-date instructions for software installation are provided online at this and other sites, as these instructions are likely to change after the publication of the book.

How to install the R program in my computer?
Installation of R varies depending on the operating system and computer hardware, and is in general similar to that of other software under a given operating system distribution. For most types of computer hardware, the current version of R is available through the Comprehensive R Archive Network (CRAN) at https://cran.r-project.org/. Especially in the case of Linux distributions, R can frequently be installed as a component of the operating system distribution. There are some exceptions, such as the *R4Pi* distribution of R for the Raspberry Pi, which is maintained independently (https://r4pi.org/).

Installers for Linux, Windows and MacOS are available through CRAN (https://cran.r-project.org/) together with brief but up-to-date installation instructions.

How to install the RStudio IDE in my computer?
RStudio installers are available at Posit's web site (https://posit.co/products/open-source/rstudio/) of which the free version is suitable for running the code examples and exercises in the book. In many cases, the IT staff at your employer or school will install them, or they may be already included in the default computer setup.

How to get access to RStudio as a cloud service?
An alternative, that is very well suited for courses or learning as part of a group is the RStudio cloud service, recently renamed Posit cloud (https://posit.co/products/cloud/cloud/). For individual use, a free account is in many cases enough, and for groups that qualify for the discounted price, a low-cost teacher's account works very well.

2.6 Further Reading

Suggestions for further reading are dependent on how you plan to use R. If you envision yourself running batch jobs under Linux or Unix, you would profit from learning to write shell scripts. Because bash is widely used nowadays, *Learning the bash Shell* (Newham and Rosenblatt 2005) can be recommended. If you aim at writing R code that is going to be reused, and have some familiarity with C, C++ or Java, reading *The Practice of Programming* (Kernighan and Pike 1999) will provide a mostly language-independent view of programming as an activity and help you master the all-important tricks of the trade. The history of R, and its relation or

S, is best told by those who were involved at the early stages of its development, Chambers (2016, chapter 2), and Ihaka (1998).

3

Base R: "Words" and "Sentences"

> The desire to economise time and mental effort in arithmetical computations, and to eliminate human liability to error, is probably as old as the science of arithmetic itself.
>
> Howard Aiken
> *Proposed automatic calculating machine*, 1937; reprinted 1964

3.1 Aims of This Chapter

In my experience, for those who are not familiar with computer programming languages, the best first step in learning the R language is to use it interactively by typing textual commands at the R *console*. This teaches not only the syntax and grammar rules, but also gives a glimpse at the advantages and flexibility of this approach to data analysis. In this chapter, I focus on the different simple values or items that can be stored and manipulated in R, as well as the role of computer program statements, the equivalent of "sentences" in natural languages.

In the first part of the chapter, you will use R to do everyday calculations that should be so easy and familiar that you will not need to think about the operations themselves. This easy start will give you a chance to focus on learning how to issue textual commands at the command prompt.

Later in the chapter, you will gradually need to focus more on the R language and its grammar and less on how commands are entered. By the end of the chapter, you will be familiar with most of the kinds of simple "words" used in the R language and you will be able to read and write simple R statements.

Throughout the chapter, I will occasionally show the equivalent of the R code in mathematical notation. If you are not familiar with the mathematical notation, you can safely ignore the mathematics, as long as you understand the diagrams and the R code.

DOI: 10.1201/9781003404187-3

3.2 Natural and Computer Languages

Computer languages have strict rules, and the interpreters and compilers that translate these languages into machine code are unforgiving about errors. They will issue error messages, but in contrast to human readers or listeners, will not guess your intentions and continue. However, computer languages have a much smaller set of words than natural languages, such as English. If you are new to computer programming, understanding the parallels between computer and natural languages may be useful.

One can think of constant values and variables (values stored under a name) as nouns and of operators and functions as verbs. A complete command, or statement, is the equivalent of a natural language sentence: "a comprehensible utterance". The simple statement `a + 1` has three components: `a`, a variable, `+`, an operator and `1` a constant. The statement `sqrt(4)` has two components, a function `sqrt()` and a numerical constant `4`. We say that "to compute $\sqrt{4}$ we *call* `sqrt()` with `4` as its *argument*".

Although all values manipulated in a digital computer are stored as *bits* in memory, multiple interpretations are possible. Numbers, letters, logical values, etc., can be encoded into bits and decoded as long as their type or `mode` is known. The concept of `class` is not directly related to how values are encoded when stored in computer memory, but instead how they are interpretated as part of a computer program. We can have, for example, RGB colour values, stored as three numbers such as `0, 0, 255`, as hexadecimal numbers stored as characters #0000FF, or even use fancy names stored as character strings like `"blue"`. We could create a `class` for colours using any of these representations, based on two different modes: `numeric` and `character`.

3.3 Numeric Values and Arithmetic

When working in `R` with arithmetic expressions, the normal mathematical precedence rules are followed and parentheses can be used to alter this order. Parentheses can be nested, but in contrast to the usual practice in mathematics, the same parenthesis symbol is used at all nesting levels.

Both in mathematics and programming languages *operator precedence rules* determine which subexpressions are evaluated first and which later. Contrary to primitive electronic calculators, `R` evaluates numeric expressions containing operators according to the rules of mathematics. In the expression $1 + 2 \times 3$, the product 2×3 has precedence over the addition, and is evaluated first, yielding as the result of the whole expression, 7. Similar rules apply to other operators, even those taking as operands non-numeric values.

The equivalent of the math expression

$$\frac{3 + e^2}{\cos \pi}$$

is, in R, written as follows:

```
(3 + exp(2)) / cos(pi)
## [1] -10.38906
```

Where constant `pi` ($\pi = 3.1415\ldots$) and function `cos()` (cosine) are defined in base R. Many trigonometric and mathematical functions are available in addition to operators like +, −, *, /, and ^.

In R, angles are expressed in radians, thus $\cos(\pi) = 1$ and $\sin(\pi) = 0$, according to trigonometry. Degrees can be converted into radians taking into account that the circle corresponds to $2 \times \pi$ when expressed in radians and to 360° when expressed in degrees. Thus the cosine of an angle of 45° can be computed as follows.

```
sin(45/180 * pi)
## [1] 0.7071068
```

One thing to remember when translating fractions into R code is that in arithmetic expressions the bar of the fraction generates a grouping that alters the normal precedence of operations. In contrast, in R expressions this grouping must be explicitly signalled with additional parentheses.

If you are in doubt about how precedence rules work, you can add parentheses to make sure the order of computations is the one you intend. Redundant parentheses have no effect.

```
1 + 2 * 3
## [1] 7
1 + (2 * 3)
## [1] 7
(1 + 2) * 3
## [1] 9
```

The number of opening (left side) and closing (right side) parentheses must be balanced, and they must be located so that each enclosed term is a valid mathematical expression, i.e., code that can be evaluated to return a value, a value that can be inserted in place of the expression enclosed in parenthesis before evaluating the remaining of the expression. For example, `(1 + 2) * 3` after evaluating `(1 + 2)` becomes `3 * 3` yielding `9`. In contrast, `(1 +) 2 * 3` is a syntax error as `1 +` is incomplete and does not yield a number.

3.1 In *playgrounds* the output from running the code in R is not shown, as these are exercises for you to enter at the R console and run. In general, you should not skip them as in most cases playgrounds aim to teach or demonstrate concepts or features that I have *not* included in full-detail in the main text. You are strongly encouraged to *play*, in other words, to create new variations of the examples and execute them to explore how R works.

```
1 + 1
2 * 2
2 + 10 / 5
(2 + 10) / 5
10^2 + 1
sqrt(9)

pi
sin(pi)
log(100)
log10(100)
log2(8)
exp(1)
```

Variables are used to store values. After we *assign* a value to a variable, we can use in our code the name of the variable in place of the stored value. The "usual" assignment operator is <-. In R, all names, including variable names, are case sensitive. Variables a and A are two different variables. Variable names can be long in R, although it is not a good idea to use very long names. Here I am using very short names, something that is usually also a very bad idea. However, in the examples in this chapter, where the stored values have no connection to the real world, simple names emphasise their abstract nature. In the chunk below, `vct1` and `vct2` are arbitrarily chosen variable names; I should have used names like `height.cm` or `outside.temperature.C` if they had been useful to convey information.

In the book, I use variable names that help recognise the kind of object stored, as this is most relevant when learning R. Here I use `vct1` because in R, as we will see on page 28, numeric objects are always vectors, even when of length one.

```
vct1 <- 1
vct1 + 1
## [1] 2
vct1
## [1] 1
vct2 <- 10
vct2 <- vct1 + vct2
vct2
## [1] 11
```

Entering the name of a variable *at the R console* implicitly calls function `print()` displaying the stored value on the console. The same applies to any other statement entered *at the R console*: `print()` is implicitly called with the result of executing the statement as its argument.

```
vct1
## [1] 1
print(vct1)
## [1] 1
vct1 + 1
## [1] 2
print(vct1 + 1)
## [1] 2
```

3.2 There are some syntactically legal assignment statements that are not very frequently used, but you should be aware that they are valid, as they will not trigger error messages and may surprise you. The most important thing is to write code consistently. The "backwards" assignment operator -> and resulting code like `1 -> vct1` are valid but less frequently used. The use of the equals sign (=) for assignment in place of <- although valid is discouraged. Chaining assignments as in the first statement below can be used to signal to the human reader that `vct1`, `vct2` and `vct3` are being assigned the same value.

```
VCT1 <- VCT2 <- VCT3 <- 0
VCT1
VCT2
VCT3
1 -> VCT1
VCT1
VCT1 = 3
VCT1
remove(VCT1, VCT2, VCT3) # cleanup
```

In R, all numbers belong to mode `numeric` (we will discuss the concepts of *mode* and *class* in section 3.8 on page 59). We can query if the mode of an object is `numeric` with function `is.numeric()`. The returned values are either `TRUE` or `FALSE`. These are logical values that will be discussed in section 3.5 on page 49.

```
mode(1)
## [1] "numeric"
vct1 <- 1
is.numeric(vct1)
## [1] TRUE
```

Because numbers can be stored in computer memory in different formats, most computing languages, including R, implement multiple types of numerical values. In most cases, R's `numeric` values can be used everywhere that a number is expected. However, in some cases, explicitly using class `integer` to indicate that we will store or operate on whole numbers, can be advantageous, e.g., `integer` constants are identified by a trailing capital "L", as in `32L`.

```
is.numeric(1L)
## [1] TRUE
is.integer(1L)
## [1] TRUE
is.double(1L)
## [1] FALSE
```

Real numbers are a mathematical abstraction, and do not have an exact equivalent in computers. Instead of Real numbers, computers store and operate on numbers that are restricted to a broad but finite range of values and have a finite resolution. They are called, *floats* (or *floating-point* numbers); in R they go by the name of `double` and can be created with the constructor `double()`.

```
is.numeric(1)
## [1] TRUE
```

```
is.integer(1)
## [1] FALSE
is.double(1)
## [1] TRUE
```

Vectors are one-dimensional in structure, of varying length and used to store similar values, e.g., numbers. They are different from the vectors, commonly used in Physics when describing directional forces, which are symbolised with an arrow as an “accent”, such as $\vec{\mathbf{F}}$. In R numeric values and other atomic values are always `vector` s that can contain zero, one or more elements. The diagram below exemplifies a vector containing ten elements, also called members. These elements can be extracted using integer numbers as positional indices, and manipulated as described in more detail in section 3.10 on page 64.

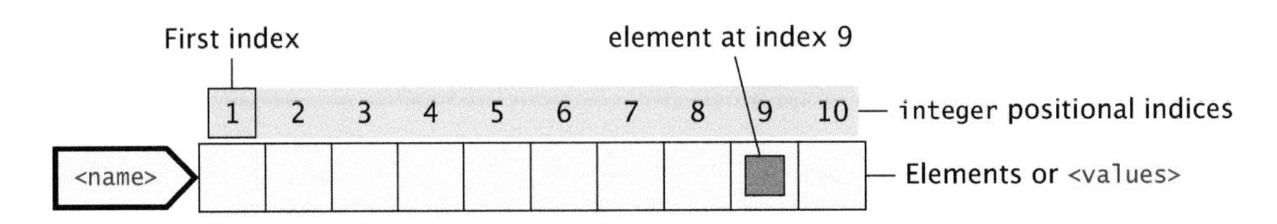

Vectors, in mathematical notation, are similarly represented using positional indexes as subscripts,

$$a_{1...n} = a_1, a_2, \cdots a_i, \cdots, a_n, \tag{3.1}$$

where $a_{1...n}$ is the whole vector and a_1 its first member. The length of $a_{1...n}$ is n as it contains n members. In the diagram above $n = 10$.

As you have seen above, the results of calculations were printed preceded with `[1]`. This is the index or position in the vector of the first number (or other value) displayed at the head of the current line. As in R single values are vectors of length one, when they are printed, they are also preceded with `[1]`.

One can use function `c()` “concatenate” to create a vector from other vectors, including vectors of length 1, or even vectors of length 0, such as the `numeric` constants in the statements below. The first example shows an anonymous vector created, printed, and then automatically discarded.

```
c(3, 1, 2)
## [1] 3 1 2
```

To be able to reuse the vector, we assign it to a variable, giving a name to it. The length of a vector can be queried with function `length()`. Below, R code is followed by diagrams depicting the structure of the vectors created.

```
vct4 <- c(3, 1, 2)
length(vct4)
## [1] 3
vct4
## [1] 3 1 2
```

1 2 3 — integer positional indices
vct4 3 1 2 — numeric values

```
vct5 <- c(4, 5, 0)
vct5
## [1] 4 5 0
```

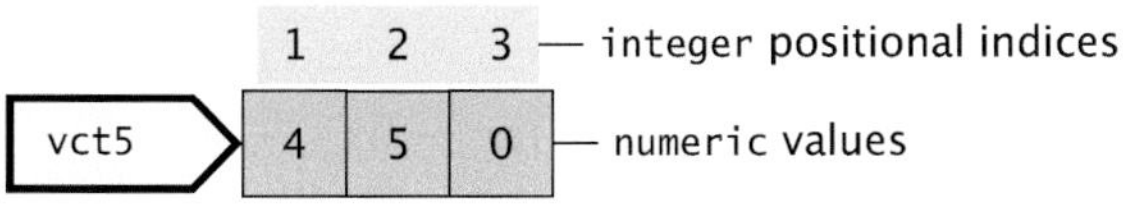

1 2 3 — integer positional indices
vct5 4 5 0 — numeric values

```
vct6 <- c(vct4, vct5)
vct6
## [1] 3 1 2 4 5 0
```

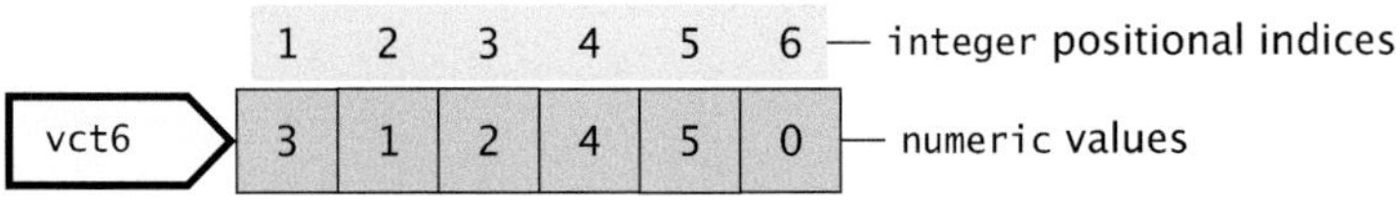

1 2 3 4 5 6 — integer positional indices
vct6 3 1 2 4 5 0 — numeric values

```
vct7 <- c(vct5, vct4)
vct7
## [1] 4 5 0 3 1 2
```

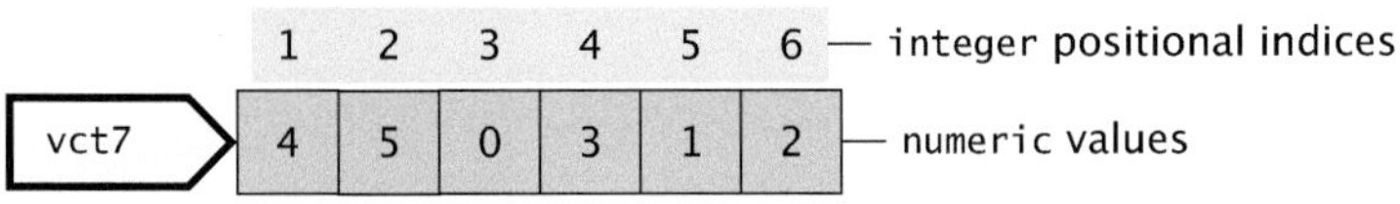

1 2 3 4 5 6 — integer positional indices
vct7 4 5 0 3 1 2 — numeric values

One or more member values of a vector can be extracted using the positional indexes and the extraction operator `[ ]`. The returned value is a new vector. Member extraction is discussed in detail in section 3.10 on page 64.

```
vct7[3]
## [1] 0
vct7[c(6, 2)]
## [1] 2 5
```

How to create an empty vector?

```
numeric()
## numeric(0)
```

Next, I show concatenation of two vectors of the same class, the second of them of length zero.

```
c(vct7, numeric())
## [1] 4 5 0 3 1 2
```

Function `c()` accepts as arguments two or more vectors and concatenates them, one after another. Quite frequently we may need to insert one vector in the middle of another. For this operation, `c()` is not useful by itself. One could use indexing combined with `c()`, but this is not needed as R provides a function capable of directly doing this operation. Although it can be used to "insert" values, it is named `append()`, and by default, it indeed appends one vector at the end of another.

```
append(vct4, vct5)
## [1] 3 1 2 4 5 0
```

The output above is the same as for `c(a, b)`, however, `append()` accepts as an argument an index position after which to "append" its second argument. This results in an *insert* operation when the index points at any position different from the end of the vector.

```
append(vct4, values = vct5, after = 2)
## [1] 3 1 4 5 0 2
```

3.3 One can create sequences using function `seq()` or the operator `:`, or repeat values using function `rep()`. In this case, I leave it to the reader to work out the rules by running these and his/her own examples, with the help of the documentation, available through `help(seq)` and `help(rep)`.

```
-1:5
5:-1
seq(from = -1, to = 1, by = 0.1)
rep(-5, times = 4)
rep(1:2, length.out = 4)
```

How to create a vector of zeros?

```
numeric(length = 10)
##  [1] 0 0 0 0 0 0 0 0 0 0
```

or

```
rep(0, times = 10)
##  [1] 0 0 0 0 0 0 0 0 0 0
```

Next, something that makes R different from most other programming languages: vectorised arithmetic. Operators and functions that are vectorised accept, as arguments, vectors of arbitrary length, in which case the result returned is equivalent to having applied the same function or operator individually to each element of the vector.

```
log10(100)
## [1] 2
log10(c(10, 5, 100, 200))
## [1] 1.00000 0.69897 2.00000 2.30103
```

Function `sum()` accepts vectors of different lengths as input but is not vectorised, as it always returns a vector of length one as result.

```
sum(100)
## [1] 100
sum(c(10, 5, 100, 200))
## [1] 315
```

A vectorised sum, also called a parallel sum of vectors, to differentiate it from obtaining the sum of the members of a vector, as computed above with function `sum()`, is the usual way in which operators like + and other arithmetic operators and functions work in R.

```
c(3, 1, 2) + c(1, 2, 31)
## [1]  4  3 33
```

Vectorised functions and operators that operate on more than one vector simultaneously, in many cases accept vectors of mismatched length as arguments or operands. When two or more vectors are of different length, these functions and operators recycle the shorter vector(s) to match the length of the longest one. The two statements below are equivalent; in the first statement, the short vector `1` is first recycled into `c(1, 1, 1)`. The operation, addition in this example, is applied to the numbers stored at the same position in the two vectors, returning a new vector.

```
c(3, 1, 2) + 1
## [1] 4 2 3
c(3, 1, 2) + c(1, 1, 1)
## [1] 4 2 3
```

In the second code statement (line) below, `vct4` is of length 3, but the `numeric` constant 2 is a vector of length 1, this short constant vector is extended, by recycling (replicating) its value, into a longer vector of ones—i.e., a vector of the same length as the longest vector in the statement, `a`.

```
vct4 <- c(3, 1, 2)
(vct4 + 1) * 2
## [1] 8 4 6
vct4 * 0:1

## Warning in vct4 * 0:1: longer object length is not a multiple of shorter
object length
## [1] 0 1 0

vct4 - vct4
## [1] 0 0 0
```

Make sure you understand what calculations are taking place in the chunk above, and also the one below. Vectorisation and vector recycling are key features of the R language.

```
vct8 <- rep(1, 6)
vct8
## [1] 1 1 1 1 1 1
vct8 + 1:2
## [1] 2 3 2 3 2 3
vct8 + 1:3
## [1] 2 3 4 2 3 4
vct8 + 1:4

## Warning in vct8 + 1:4: longer object length is not a multiple of shorter
object length
## [1] 2 3 4 5 2 3
```

3.4 Create further variants of the statements in the code chunk above to work out when warnings or errors are issued. Does the length of the operands matter?

Most functions defined in base R apply recycling to vectors passed as arguments to at least some of their parameters. When recycling is supported, the conditions triggering warnings or errors are consistent with those you discovered in the playground above. However, if and how recycling is applied depends on how functions have been defined. Thus, there is variation, especially, but not only, in the case of functions and operators defined in contributed extension packages. For example, package 'tibble' and some other packages in the 'tidyverse' support recycling but some boundary cases that trigger a warning in base R functions, trigger an error in functions defined in these packages. See section 8.4.2 on page 247 about package 'tibble'.

As mentioned above, a vector can contain zero or more member values. Vectors of length zero may seem at first sight quite useless, but in practice they are very useful. They allow the handling of "no input" or "nothing to do" cases as normal cases, which in the absence of vectors of length zero would require to be treated as special cases. Constructors for R classes like `numeric()` return vectors of a length given by their first argument, which defaults to zero.

```
vct9 <- numeric(length = 0) # named argument
vct9
## numeric(0)
length(vct8)
## [1] 6
```

```
numeric() # default argument
## numeric(0)
```

Vectors of length zero, behave in most cases, as expected—e.g., they can be concatenated as shown here.

```
length(c(vct4, vct9, vct5))
## [1] 6
length(c(vct4, vct5))
## [1] 6
```

Many functions, such as R's maths functions and operators, will accept numeric vectors of length zero as valid input, returning also a vector of length zero, issuing neither a warning nor an error message. In other words, *these are valid operations* in R.

```
log(numeric(0))
## numeric(0)
5 + numeric(0)
## numeric(0)
```

Even when of length zero, vectors do have to belong to a class acceptable for the operation: `5 + character(0)` is an error (`character` values are described in section 3.4 on page 41).

Passing as an argument to parameter `length` a value larger than zero creates a longer vector filled with zeros in the case of `numeric()`.

```
numeric(length = 5)
## [1] 0 0 0 0 0
```

The length of a vector can be explicitly increased, with missing values filled automatically with `NA`, the marker for not available.

```
vct10 <- 1:5
length(vct10) <- 10
vct10
##  [1]  1  2  3  4  5 NA NA NA NA NA
```

If the length is decreased, the values in the *tail* of the vector are discarded.

```
vct11 <- 1:10
vct11
## [1] 1 2 3 4 5 6 7 8 9 10
length(vct11) <- 5
vct11
## [1] 1 2 3 4 5
```

There are some special values available for numbers. NA meaning "not available" is used for missing values. (NA) values play a very important role in the analysis of data, as frequently some observations are missing from an otherwise complete data set due to "accidents" during the course of an experiment or survey. It is important to understand how to interpret NA values: They are placeholders for something that is unavailable, in other words, whose value is *unknown*. NA values propagate when used, so that numerical computations yield NA when one or more input of the values is unknown.

```
vct12 <- c(NA, 5)
vct12
## [1] NA  5
vct12 + 1
## [1] NA  6
```

Calculations can also yield the following values NaN "not a number", `Inf` and `-Inf` for ∞ and $-\infty$. As you will see below, calculations yielding these values do **not** trigger errors or warnings, as they are arithmetically valid. `Inf` and `-Inf` are also valid numerical values for input and constants.

```
vct12 + Inf
## [1]  NA Inf
Inf / vct12
## [1]  NA Inf
-1 / 0
## [1] -Inf
1 / 0
## [1] Inf
Inf / Inf
## [1] NaN
Inf + 4
## [1] Inf
-Inf * -1
## [1] Inf
```

3.5 **When to use vectors of length zero, and when NAs?** Make sure you understand the logic behind the different behaviour of functions and operators with respect to NA and `numeric()` or its equivalent `numeric(0)`. What do they represent? Why NA s are not ignored, while vectors of length zero are?

```
123 + numeric()
123 + NA
```

Model answer: NA values are used to signal a value that "was lost" or "was expected" but is unavailable because of some accident. A vector of length zero, represents no values, but within the normal expectations. In particular, if vectors are expected to have a certain length, or if index positions along a vector are meaningful, then using `NA` is a must.

Any operation, even tests of equality, involving one or more `NA`'s return an `NA`. In other words, when one input to a calculation is unknown, the result of the calculation is unknown. This means that a special function is needed for testing for the presence of `NA` values.

```
is.na(c(NA, 1))
## [1]  TRUE FALSE
```

In the example above, we can also see that `is.na()` is vectorised, and that it applies the test to each of the elements of the vector individually, returning the result as `TRUE` or `FALSE`.

One needs to be aware of the consequences of numbers in computers being almost always stored with finite precision and/or range: the expectations derived from the mathematical definition of Real numbers are not always fulfilled. See the box on page 35 for an in-depth explanation.

```
1 - 1e-20
## [1] 1
```

When using `integer` values these problems do not exist, as integer arithmetic is not affected by loss of precision in calculations restricted to integers. Because of the way integers are stored in the memory of computers, within the representable range, they are stored exactly. One can think of computer integers as a subset of whole numbers restricted to a certain range of values.

```
1L + 3L
## [1] 4
1L * 3L
## [1] 3
```

Using the "usual" division operator yields a floating-point `double` result, while the integer division operator `%/%` yields an `integer` result, and the modulo operator `%%` returns the remainder from the integer division.

```
1L / 3L
## [1] 0.3333333
1L %/% 3L
## [1] 0
1L %% 3L
## [1] 1
```

If an operation would create an `integer` value that falls outside the range representable in R, the value returned is `NA` (not available).

```
1000000L * 1000000L

## Warning in 1000000L * 1000000L: NAs produced by integer overflow
## [1] NA
```

Both doubles and integers are considered numeric. In most situations, conversion is automatic and we do not need to worry about the differences between these two types of numeric values. The functions in the next chunk return `TRUE` or `FALSE`, i.e., `logical` values (see section 3.5 on page 49).

```
is.numeric(1L)
## [1] TRUE
is.integer(1L)
## [1] TRUE
is.double(1L)
## [1] FALSE
is.double(1L / 3L)
## [1] TRUE
is.numeric(1L / 3L)
## [1] TRUE
```

3.6 Study the variations of the previous example shown below, and explain why the two statements return different values. Hint: 1 is a `double` constant. You can use `is.integer()` and `is.double()` in your explorations.

```
1 * 1000000L * 1000000L
1000000L * 1000000L * 1
```

The usual way to store numerical values in computers is to reserve a fixed amount of space in memory for each value, which imposes limits on which numbers can be represented or not, and the maximum precision that can be achieved. The difference between `integer` and `double` is explained on page 27. Integers, or "whole numbers", like R `integer` values are stored always with the same resolution such that the smallest difference between two integer values is 1. The amount of memory available to store an individual value creates a limit for the size of the largest and smallest values that can be represented. Thus integers in R behave like Integers or whole numbers as defined in mathematics, but constrained to a restricted finite range of values. In the computing language C, different types of integer numbers are available `short` and `long`, these differ in the size of the space reserved for them in memory. R `integer` type is equivalent to `long` in C, thus the use of `L` for integer constant values like `5L`.

Floating point numbers like R `double` values are stored in two parts: an integer *significand* and an integer *exponent*, each part using a fixed amount of space in memory. The relative resolution is constrained by the number of digits that can be stored in the significand while the absolute size of the largest and smallest numbers that can be represented is limited by the largest and smallest values that fit in the memory reserved for the exponent. In many computing languages, different types of floating point numbers are available, these differ in the size of the space reserved for them in memory. The properties of Real numbers as defined in mathematics differ from floating point numbers in assuming unlimited resolution and an unlimited range of representable values.

In R, numbers that are not integers are stored as *double-precision floats*. Precision of numerical values in computers is usually symbolised by "epsilon" (ϵ), commonly abbreviated *eps*, defined as the largest value of ϵ for which $1 + \epsilon = 1$. The finite resolution of floats can lead to unexpected results when testing for equality or inequality. Test for equality is done with operator `==`. The use of this and other comparison operators is explained in section 3.6 on page 52.

```
1e20 == 1 + 1e20
## [1] TRUE
1 == 1 + 1e-20
## [1] TRUE
0 == 1e-20
## [1] FALSE
```

Another way of revealing the limited precision is during conversion to `character`.

```
format(5.123, digits = 16) # near maximun resolution
## [1] "5.123"
format(5.123, digits = 22) # more digits than in resolution
## [1] "5.123000000000000220268"
```

The accumulation of successive small losses of precision from multiple operations on R `double` values can be a problem. Thus when computations involve both very large and very small numbers, the returned value can depend on the order of the operations. In practice ordinary users rarely need to be concerned about losses in precision except when testing for equality and inequality. On the other hand, finite resolution of `double` numerical values can explain why sometimes returned values for equivalent computations differ, and why some computation algorithms may be preferable, and others even fail, in specific cases.

As the R program can be used on different types of computer hardware, the actual machine limits for storing numbers in memory may vary depending on the type of processor and even the compiler used to build the R program executable. However, it is possible to obtain these values at run time, i.e., while the R is being used, from the variable `.Machine`, which is part of the R language. Please see the help page for `.Machine` for a detailed and up-to-date description of the available constants. *Beware that when you run the examples below, the values returned by R in your own computer can differ from those returned in the computer I have used to typeset the book as you are reading it here.*

```
.Machine$double.eps
## [1] 2.220446e-16
.Machine$double.neg.eps
## [1] 1.110223e-16
.Machine$double.max
## [1] 1024
.Machine$double.min
## [1] -1022
.Machine$double.base
## [1] 2
```

The last two values refer to the exponents of a base number or *radix*, 2, rather than the maximum and minimum size of numbers that can be handled as objects of class `double`. The maximum size of normalised `double` values, given by `.Machine$double.xmax`, is much larger than the maximum value of `integer` values, given by `.Machine$integer.max`.

```
.Machine$double.xmax
## [1] 1.797693e+308
.Machine$integer.max
## [1] 2147483647
```

As `integer` values are stored in machine memory without loss of precision, epsilon is not defined for `integer` values. In R not all out-of-range `numeric` values behave in the same way: while off-range `double` values are stored as `-Inf` or `Inf` and enter arithmetic as infinite values according to the mathematical rules, off-range `integer` values become `NA` with a warning.

```
1e1026
## [1] Inf
1e-1026
## [1] 0

2147483699L
## [1] 2147483699
```

In those statements in the chunk below where at least one operand is `double` the `integer` operands are *promoted* to `double` before computation. A similar promotion does not take place when operations are among `integer` values, resulting in *overflow*, meaning numbers that are too big to be represented as `integer` values.

```
2147483600L + 99L

## Warning in 2147483600L + 99L: NAs produced by integer overflow
## [1] NA

2147483600L + 99
## [1] 2147483699
2147483600L * 2147483600L

## Warning in 2147483600L * 2147483600L: NAs produced by integer overflow
## [1] NA

2147483600L * 2147483600
## [1] 4.611686e+18
```

The exponentiation operator `^` forces the promotion of its arguments to `double`, resulting in no overflow. In contrast, as seen above, the multiplication operator `*` operates on `integer` values resulting in overflow.

```
2147483600L * 2147483600L

## Warning in 2147483600L * 2147483600L: NAs produced by integer overflow
## [1] NA

2147483600L^2L
## [1] 4.611686e+18
```

Both for display or as part of computations, we may want to decrease the number of significant digits or the number of digits after the decimal marker. Be aware that in the examples below, even if printing is being done by default, these functions return `numeric` values that are different from their input and can be stored and used in computations. Function `round()` is used to round numbers to a certain number of decimal places after or before the decimal marker, with a positive or negative value for `digits`, respectively. In contrast, function `signif()` rounds to the requested number of significant digits, i.e., ignoring the position of the decimal marker.

```
round(0.0124567, digits = 3)
## [1] 0.012
signif(0.0124567, digits = 3)
## [1] 0.0125
round(1789.1234, digits = -1)
## [1] 1790
round(1789.1234, digits = 3)
## [1] 1789.123
signif(1789.1234, digits = 3)
## [1] 1790
```

```
vct13 <- 0.12345
vct14 <- round(vct13, digits = 2)
vct13 == vct14
## [1] FALSE
vct13 - vct14
## [1] 0.00345
vct14
## [1] 0.12
```

Functions are described in detail in section 6.2 on page 169. Here I describe them briefly in relation to their use. Functions are objects containing R code that can be used to perform an operation on values passed as arguments to its parameters. They return the result of the operation as a single R object, or less frequently, as a side effect. Functions have a name like any other R object. If the name of a function is followed by parentheses `()` and included in a code statement, it becomes a function *call* or a "request" for the code stored in the function object to be run. Many functions, accept R objects and/or constant values as *arguments* to their *formal parameters*. Formal parameters are placeholder names in the code stored in the function object, or the *definition* of the function. In a function call, the code in its definition is evaluated (or run) with formal parameter names taking the values passed as arguments to them.

In a function definition, formal parameters can be assigned default values, which are used if no explicit argument is passed in the call. Arguments can be passed to formal parameters by name or by position. In most cases, passing arguments by name makes the code easier to understand and more robust against coding mistakes. In the examples presented in the book, I most frequently pass arguments by name, except for the first parameter.

Being `digits`, the second parameter, its argument can also be passed by position.

```
round(0.0124567, digits = 3)
## [1] 0.012
round(0.0124567, 3)
## [1] 0.012
```

When passing arguments by name, in most cases unambiguous partial matching is acceptable, but can make code difficult to read.

```
round(0.0124567, di = 3)
## [1] 0.012
```

Functions `trunc()` and `ceiling()` return the non-fractional part of a numeric value as a new numeric value. They differ in how they handle negative values, and neither of them rounds the returned value to the nearest whole number. Hint: you can use `help(trunc)` or `trunc?` at the R console, or the help tab of RStudio to find out the answer.

3.7 What does value truncation mean? Function `trunc()` truncates a numeric value, but it does not return an `integer`.

- Explore how `trunc()` and `ceiling()` differ. Test them both with positive and negative values.
- **Advanced** Use function `abs()` and operators + and - to reproduce the output of `trunc()` and `ceiling()` for the different inputs.
- Can `trunc()` and `ceiling()` be considered type conversion functions in R?

R supports complex numbers and arithmetic operations with class `complex`. As complex numbers rarely appear in user-written scripts, I give only one example of their use. Complex numbers, as defined in mathematics, have two parts, a real component and an imaginary one. Complex numbers can be used, for example, to describe the result of $\sqrt{-1} = 1i$.

```
cmp1 <- complex(real = c(-1, 1), imaginary = c(0, 0))
cmp1
## [1] -1+0i  1+0i
cmp2 <- sqrt(cmp1)
cmp2
## [1] 0+1i 1+0i
cmp2^2
## [1] -1+0i  1+0i
```

Instants in time and periods of time in computers are usually encoded as classes derived from `integer`, and thus considered in R as atomic classes and the objects vectors. Some of these encodings are standardised and supported by R classes `POSIXlt` and `POSIXct`. The computations based on times and dates are difficult because the relationship between local time at a given location and Universal Time Coordinates (UTC) has changed with time, as well as with changes in national borders. Packages 'lubridate' and 'anytime' support operations among time-related data and conversions between character strings and time and date classes, making them easier and less error prone than when using base R functions. Thus I describe classes and operations related to dates and times in section 8.8 on page 267.

It is good to *remove* from the workspace objects that are no longer needed. We use function `remove()` to delete objects stored in the current workspace.

Arguments passed to `remove()` can be bare object names as shown here.

```
an.object <- 1:4
remove(an.object) # using a bare name
```

Function `remove()` also accepts the names of the objects to remove as a `character` vector passed to its parameter `list`. In spite of its name, the argument must be a `vector` rather than a `list` (see section 3.4 on `character` and section 4.3 on `list` on pages 41 and 86).

```
an.object <- 5:2
remove(list = "an.object") # using a character vector
```

Function `objects()` returns a `character` vector containing the names of all objects visible in the current environment, or by passing an argument to parameter `pattern`, only the objects with names matching it.

```
an.object <- 1:4
another.object <- 2
objects(pattern = "*.object")
## [1] "an.object"      "another.object"
remove(an.object)
objects(pattern = "*.object")
## [1] "another.object"
```

In RStudio, all objects are listed in the **Environment** tab and the search box of this tab can be used to find a given object.

Function `remove()` accepts both bare names of objects as in the chunk above and `character` strings corresponding to object names like in `remove("any.object")`. However, While `objects()` accept patterns to be matched to object names, `remove()` does not. Because of this, these two functions have to be used together for removing all objects with names that match a pattern. The pattern can be given as a regular expression (see section 3.4 on page 46).

Both functions are available under short names matching those used in Linux and Unix for managing files: `ls()` is a synonym of `objects()` and `rm()` of `remove()`.

Using a simple search pattern we obtain the names of all objects with names `"vct1"`, `"vct2"`, and so on. When using a pattern to remove objects, it is good to first use `objects()` on its own to get a list of the objects that would be deleted by calling `remove()` when passing the names returned by `objects()` as the argument for parameter `list`.

```
objects(pattern = "^vec.*")
## character(0)
```

The code below removes all objects with names `"vct1"`, `"vct2"`, and so on. We do this at the end of the section before reusing the same names in the code examples of the next section.

```
remove(list = objects(pattern = "^vct[[:digit:]]?"))
```

Similar code chunks are included at the end of each section throughout the book to ensure that code examples are self-contained by section. The chunk about is shown above as an example, but kept hidden in later sections.

3.4 Character Values

In spite of the name `character`, values of this mode, are vectors of *character strings"*. Character constants are written by enclosing characters strings in quotation marks, i.e., `"this is a character string"`. There are three types of quotation marks in the ASCII character set, double quotes ", single quotes ', and back ticks `. The first two types of quotes can be used as delimiters of `character` constants.

```
vct1 <- "A"
vct1
## [1] "A"
vct2 <- 'A'
vct2
## [1] "A"
vct1 == vct2 # two variables holding character values, or named objects
## [1] TRUE
"A" == 'A' # two constant character values, or anonymous objects
## [1] TRUE
```

In many computer languages, vectors of characters are distinct from vectors of character strings. In these languages, character vectors store at each index position a single character, while vectors of character strings store at each index position strings of characters of various lengths, such as words or sentences. If you are familiar with C or C++, you need to keep in mind that C's `char` and R's `character` are not equivalent and that in R. In contrast to these other languages, in R there is no predefined class for vectors of individual characters and character constants enclosed in double or single quotes are not different.

Concatenating character vectors of length one does not yield a longer character string, it yields instead a longer vector of character strings.

```
vct3 <- 'ABC'
vct4 <- "bcdefg"
vct5 <- c("123", "xyz")
c(vct3, vct4, vct5)
## [1] "ABC"    "bcdefg" "123"    "xyz"
```

Having two different delimiters available makes it possible to choose the type of quotation marks used as delimiters so that other quotation marks can be easily included in a string.

```
"He said 'hello' when he came in"
## [1] "He said 'hello' when he came in"
'He said "hello" when he came in'
## [1] "He said \"hello\" when he came in"
```

The outer quotes are not part of the string, they are "delimiters" used to mark the boundaries. As you can see when `b` is printed special characters can be represented using "escape codes". There are several of them, and here we will show just four, new line (`\n`) and tab (`\t`), `\"` the escape code for a quotation mark within a string and `\\` the escape code for a single backslash `\`. I also show the different behaviour of `print()` and `cat()`, with `cat()` *interpreting* the escape sequences and `print()` displaying them as entered.

```
vct6 <- "abc\ndef\tx\"yz\"\\\tm"
print(vct6)
## [1] "abc\ndef\tx\"yz\"\\\tm"
cat(vct6)
## abc
## def x"yz"\ m
```

The *escape codes* are expanded only in some contexts, such as when using `cat()` to display text output.

How to find the length of a character string?
While function `length()` returns the number of member `character` strings in a vector, function `nchar()` returns the number of characters in each string in the vector (see below for examples).

In the example below, function `nchar()` returns the number of characters in each member string.

```
nchar(x = "abracadabra")
## [1] 11
nchar(x = c("abracadabra", "workaholic", ""))
## [1] 11 10  0
```

To convert a `character` string into upper case or lower case we use functions `toupper()` and `tolower()`, respectively.

```
toupper(x = "aBcD")
## [1] "ABCD"
tolower(x = "aBcD")
## [1] "abcd"
```

Function `strtrim()` trims a string to a maximum number of characters or width.

```
strtrim(x = "abracadabra", width = 6)
## [1] "abraca"
strtrim(x = "abra", width = 6)
## [1] "abra"
strtrim(x = c("abracadabra", "workaholic"), 6)
## [1] "abraca" "workah"
strtrim(x = c("abracadabra", "workaholic"), c(6, 3))
## [1] "abraca" "wor"
```

How to wrap long character strings?
Use R function `strwrap()` (see below for examples).

Function `strwrap()` edits a string to a maximum number of characters or width, by splitting it into a vector of shorter character strings. It can additionally insert a character string at the start or end of each of these new shorter strings.

```
strwrap(x = "This is a long sentence used to show how line wrap-
ping works.", width = 20)
## [1] "This is a long"   "sentence used to" "show how line"    "wrapping works."
```

3.8 Function `cat()` prints a character vector respecting the embedded special characters such as new line (encoded as `\n`) in `character` strings) and without issuing any additional new lines. Study the code below and the output it generates, consult the documentation of the two functions, and modify the example code until you are confident that you understand in detail how these two functions work.

```
wrapped_sentence <-
  strwrap(x = "This is a very long sentence used to show how line wrapping works.",
          width = 10,
          prefix = "\n")
print(wrapped_sentence)
cat(wrapped_sentence, "\n")
```

How to create a single character string from multiple shorter strings? While function `c()` is used to concatenate `character` vectors into longer vectors, function `paste()` is used to concatenate character strings into a single longer string (see below for examples).

Pasting together `character` strings has many uses, e.g., assembling informative messages to be printed, programmatically creating file names or file paths, etc. If we pass numbers, they are converted to `character` before pasting. The default separator is a space character, but this can be changed by passing a `character` string as an argument for parameter `sep`.

```
paste("n =", 3)
## [1] "n = 3"
paste("n", 3, sep = " = ")
## [1] "n = 3"
```

Pasting constants, as shown above, is of little practical use. In contrast, combining values stored in different variables is a very frequent operation when working with data. A simple use example follows. Assuming vector `friends` contains the names of friends and vector `fruits` the fruits they like to eat we can paste these values together into short sentences.

```
friends <- c("John ", "Yan ", "Juana ", "Mary ")
fruits <- c("apples", "lichees", "oranges", "strawberries")
paste(friends, "eats ", fruits, ".", sep = "")
## [1] "John eats apples."       "Yan eats lichees."
## [3] "Juana eats oranges."     "Mary eats strawberries."
```

3.9 Why was necessary to pass `sep = ""` in the call to `paste()` in the example above? First try to predict what will happen and then remove `, sep = ""` from the statement above and run it to learn the answer. Try your own variations of the code until you understand the role of the separator string.

We can pass an additional argument to tell that the vector resulting from the paste operation is to be collapsed into a single `character` string. The argument passed to collapse is used as the separator. I use here `cat()` so that the newline character is obeyed in the display of the single character string.

```
cat(paste(friends, "eats ", fruits, collapse = ".\n", sep = ""))
## John eats apples.
## Yan eats lichees.
## Juana eats oranges.
```

```
## Mary eats strawberries
```

When the vectors are of different length, as in the last example above, the shorter one is recycled as many times as needed, which is not always what we want. To void the recycling, we need to first collapse the members of the long vector `fruits` into a vector of length one. We can achieve this by nesting two calls to `paste()`, and passing an argument to `collapse` in the inner function call.

```
collapsed_fruits <- paste(fruits, collapse = ", ")
paste("My friends eat", collapsed_fruits, "and other fruits.")
## [1] "My friends eat apples, lichees, oranges, strawberries and other fruits."
```

The nesting of function calls is explained in section 5.5 on page 134. However, as the two statements above would in most cases be written as nested function calls, I add this example for reference.

```
paste("My friends eat", paste(fruits, collapse = ", "), "and other fruits.")
## [1] "My friends eat apples, lichees, oranges, strawberries and other fruits."
```

Function `strrep()` repeats and pastes `character` strings into a new longer `character string`, while function `rep()` repeats character strings without pasting them together, returning a longer vector with each repeat of the string as a separate member.

```
rep(x = "ABC", times = 3)
## [1] "ABC" "ABC" "ABC"
strrep(x = "ABC", times = 3)
## [1] "ABCABCABC"
strrep(x = "ABC", times = c(2, 4))
## [1] "ABCABC"       "ABCABCABCABC"
strrep(x = c("ABC", "X"), times = 2)
## [1] "ABCABC" "XX"
strrep(x = c("ABC", "X"), times = c(2, 5))
## [1] "ABCABC" "XXXXX"
```

How to trim leading and/or trailing whitespace in character strings? Use function `trimws()` (see below for examples).

Trimming leading and trailing whitespace is a frequent operation. R function `trimws()` implements this operation as shown below.

```
trimws(x = " two words ")
## [1] "two words"
trimws(x = c("  eight words and a newline at the end\n", " two words "))
## [1] "eight words and a newline at the end"
## [2] "two words"
```

3.10 Function `trimws()` has additional parameters that make it possible to select which end of the string is trimmed and which characters are considered whitespace. Use `help(trimws)` to access the help and study this documentation. Modify the example above so that only trailing whitespace is removed, and so that the newline character `\n` is not considered whitespace, and thus not trimmed away.

Within character strings, substrings can be extracted and replaced *by position* using substring() or substr().

For extraction, we can pass to x a constant as shown below or a variable.

```
substr(x = "abracadabra", start = 5, stop = 9)
## [1] "cadab"
substr(x = c("abracadabra", "workaholic"), start = 5, stop = 11)
## [1] "cadabra" "aholic"
```

Replacement is done *in place*, by having function substr() on the left-hand side (lhs) of the assignment operator <-. Thus, the argument passed to parameter x of substr() must in this case be a variable rather than a constant. This is a substitution character by character, not insertion, so the number of characters in the string passed as the argument to x remains unchanged, i.e., the value returned by nchar() does not change.

```
vct7 <- c("abracadabra", "workaholic")
substr(x = vct7, start = 5, stop = 9) <- "xxx"
vct7
## [1] "abraxxxabra" "workxxxlic"
```

If we pass values to both start and stop then only part of the value on the *rhs* of the assignment operator <- may be used.

```
vct8 <- c("abracadabra", "workaholic")
substr(x = vct8, start = 5, stop = 6) <- "xxx"
vct8
## [1] "abraxxdabra" "workxxolic"
```

3.11 Frequently, a very effective way of learning how a function behaves, is to experiment. In the example below, we set start and stop delimiting more characters than those in "xxx". In this case, is "xxx" extended, or start or stop ignored? Run this "toy example" to find out the answer.

```
VCT1 <- c("abracadabra", "workaholic")
substr(x = VCT1, start = 5, stop = 11) <- "xxx"
VCT1
remove(VCT1) # clean up
```

As in R each character value is a string comprised of zero to many characters, in addition to comparisons based on whole strings or values, partial matches among them are of interest.

To substitute part of a character string *by matching a pattern*, we can use functions sub() or gsub(). The first example uses three character constants, but values stored in variables can also be passed as arguments.

```
sub(pattern = "ab", replacement = "AB", x = "about")
## [1] "ABout"
```

The difference between sub() (substitution) and gsub() (global substitution) is that the first replaces only the first match found while the second replaces all matches.

```
sub(pattern = "ab", replacement = "x", x = "abracadabra")
## [1] "xracadabra"
gsub(pattern = "ab", replacement = "x", x = "abracadabra")
## [1] "xracadxra"
```

3.12 Functions sub() or gsub() accept character vectors as the argument for parameter x. Run the two statements below and study how the values returned differ.

```
sub(pattern = "ab", replacement = "x", x = c("abra", "cadabra"))
gsub(pattern = "ab", replacement = "x", x = c("abra", "cadabra"))
```

Function grep() returns indices to the values in a vector matching a pattern, or alternatively, the matching values themselves.

```
grep(pattern = "C", x = c("R", "C++", "C", "Perl", "Pascal"))
## [1] 2 3
grep(pattern = "C", x = c("R", "C++", "C", "Perl", "Pascal"), value = TRUE)
## [1] "C++" "C"
grep(pattern = "C", x = c("R", "C++", "C", "Perl", "Pascal"), ignore.case = TRUE)
## [1] 2 3 5
```

Function grepl() is a variation of grep() that returns a vector of logical values instead of numeric indices to the matching values in x.

```
grepl(pattern = "C", x = c("R", "C++", "C", "Perl", "Pascal"))
## [1] FALSE  TRUE  TRUE FALSE FALSE
grepl(pattern = "C", x = c("R", "C++", "C", "Perl", "Pascal"), ignore.case = TRUE)
## [1] FALSE  TRUE  TRUE FALSE  TRUE
```

In the examples above, the arguments for pattern strings matched exactly their targets. In R and other languages, *regular expressions* are used to concisely describe more elaborate and conditional patterns. Regular expressions themselves are encoded as character strings, where some characters and character sequences have special meaning. This means that when a pattern should be interpreted literally rather than specially, fixed = TRUE should be passed in the call. This, in addition, ensures faster computation. In the examples above, the patterns used contained no characters with special meaning, thus, the returned value is not affect by passing fixed = TRUE as done here.

```
sub(pattern = "ab", replacement = "AB", x = "about", fixed = TRUE)
## [1] "ABout"
```

Regular expressions are used in Unix and Linux shell scripts and programs, and are part of Perl, C++ and other languages in addition to R. This means that variations exist on the same idea, with R supporting two variations of the syntax. A description of R regular expressions can be accessed with help(regex). We here describe R's default syntax.

Regular expressions are concise, terse, and extremely powerful. They are a language in themselves. However, the effort needed to learn their use more than pays back. I will show examples of the use, rather than systematically describe them. I will use gsub() for these examples, but several other R functions including grep() and grepl() accept regular expressions as patterns.

In a regular expression, | separates alternative matching patterns.

```
gsub(pattern = "ab|t", replacement = "123", x = "about")
## [1] "123ou123"
```

Within a regular expression, we can group characters within `[ ]` as alternative, e.g., `[0123456789]`, or `[0-9]` matches any digit.

```
gsub(pattern = "a[0123456789]",
     replacement = "ab",
     x = c("a1out", "a9out", "a3out"))
## [1] "about" "about" "about"
```

Character `^` indicates that the match must be at the "head" of the string, and `$` that the match should be at its "tail".

```
gsub(pattern = "^a[0123456789]",
     replacement = "ab",
     x = c("a1out", "a9out", " a3out"))
## [1] "about"  "about"  " a3out"
```

The replacement can be an empty string.

```
gsub(pattern = "out$",
     replacement = "",
     x = c("about", "a9out", "a3outx"))
## [1] "ab"     "a9"     "a3outx"
```

A dot (`.`) matches any character. In this example, we replace the last character with "".

```
gsub(pattern = ".$",
     replacement = "",
     x = c("about", "a9out", "a3outx"))
## [1] "abou"  "a9ou"  "a3out"
```

3.13 How would you modify the last code example above to edit `c("about", "axout", "a3outx")` into `c("about", "axout", "a3out")`? Think of different ways of doing this using regular expressions.

The number of matching characters can be indicated with `+` (match 1 or more times), `?` (match 0 or 1 times), `*` (match 0 or more times) or even numerically. Matching is in most cases "greedy".

```
gsub(pattern = "^.[0-9][a-z]*$",
     replacement = "gone",
     x = c("about", "a9out", "a3outx"))
## [1] "about" "gone"  "gone"
```

Several named classes of characters are predefined, for example `[:lower:]` for lower case alphabetic characters according to the current locale (see page 49). In the regular expression in the example below, `[:lower:]` replaces only `a-z`, thus we need to keep the outer square brackets. While `a-z` includes only the unaccented letters, `[:lower:]` does include additional characters such as ä, ö, or é if they are in use in the current locale. In the case of `[:digit:]` and `0-9`, they are equivalent.

```
gsub(pattern = "^.([[:digit:]])[[:lower:]]*$",
     replacement = "gone with \\1",
     x = c("about", "a9out", "a3outx"))
## [1] "about"       "gone with 9" "gone with 3"
```

With parentheses we can isolate part of the matched string and reuse it in the replacement with a numeric back-reference. Up to a maximum of nine pairs of parentheses can be used.

```
gsub(pattern = "^.([0-9])[a-z]*$",
     replacement = "gone with \\1",
     x = c("about", "a9out", "a3outx"))
## [1] "about"        "gone with 9" "gone with 3"
```

3.14 Run the two statements below, study the returned values by creating variations of the patterns and explain why the returned values differ.

```
gsub(pattern = "^.+$",
     replacement = "",
     x = c("about", "a9out", "a3outx"))
gsub(pattern = "^.?$",
     replacement = "",
     x = c("about", "a9out", "a3outx"))
```

Splitting of character strings based on pattern matching is a frequently used operation, e..g., treatment labels containing information about two different treatment factors need to be split into their components before data analysis. Function `strsplit()` has an interface consistent with `grep()`. In the examples we will split strings containing date and time of day information in different ways.

```
strsplit(x = "2023-07-29 10:30", split = " ")
## [[1]]
## [1] "2023-07-29" "10:30"
```

Using a simple regular expression we can extract individual strings representing the numbers.

```
strsplit(x = "2023-07-29 10:30", split = " |-|:")
## [[1]]
## [1] "2023" "07"   "29"   "10"   "30"
```

The argument to `split` is by default interpreted as a regular expression, but as discussed above we can pass `fixed = TRUE` to prevent this.

One needs to be aware that the part of the string matched by the regular expression is not included in the returned vectors. If the regular expression matches more than what we consider a separator, the returned values may be surprising.

```
strsplit(x = "2023-07-29", split = "-[0-9]+$")
## [[1]]
## [1] "2023-07"
```

When the argument passed to `x` is a vector with multiple member strings, the returned value is a list of `character` vectors. This list contains as many character vectors as members had the vector passed as argument to `x`, each vector the result of splitting one character string in the input. (Lists are described in section 4.3 on page 86.)

```
strsplit(x = c("2023-07-29 10:30", "2023-07-29 19:17"), split = " ")
## [[1]]
## [1] "2023-07-29" "10:30"
##
## [[2]]
## [1] "2023-07-29" "19:17"
```

The ASCII character set is the oldest and simplest in use. In contains only 128 characters including non-printable characters. These characters support the English language. Several different extended versions with 256 characters provided support for other languages, mostly by adding accented letters and some symbols. The 128 ASCII characters were for a long time the only consistently available across computers set up for different languages and countries (or *locales*). Recently the use of much larger character sets like UTF8 has become common. Since R version 4.2.0 support for UTF8 is available under Windows 10. This makes it possible the processing of text data for many more languages than in the past. Even though now it is possible to use non-ASCII characters as part of object names, it is anyway safer to use only ASCII characters as this support is recent.

The extended character sets include additional characters, that are distinct but may produce glyphs that look very similar to those in the ASCII set. One case are em-dash (—), en-dash (-), minus sign (−) and regular dash (-), which are all different characters, with only the last one recognised by R as the minus operator. For those copying and pasting text from a word-processor into R or RStudio, a frequent difficulty is that even if one types in an ASCII quote character ("), the opening and closing quotes in many languages are automatically replaced with non-ASCII ones (“and”), which R does not accept as character string delimiters. The best solution is to use a plain text editor instead of a word processor when writing scripts or editing text files containing data to be read as code statements or numerical data.

A locale definition determines not only the language, and character set, but also date, time, and currency formats.

3.5 Logical Values and Boolean Algebra

What in Mathematics are usually called Boolean values, are called `logical` values in R. They can have only two values `TRUE` and `FALSE`, in addition to `NA` (not available). Logical values `TRUE` and `FALSE` should not be confused with text strings, they are names for the two conditions that can be stored. Logical values are always vectors as all other atomic types in R (by *atomic* we mean that each value is not composed of “parts”).

Logical values are rarely used to store data from experiments or surveys. They are used mostly to keep track of binary conditions, like results from comparisons in a script and to operate on them. Most frequent uses of `logical` values do not involve their storage in user-created variables. Most comparisons or tests return a `logical` value and Boolean algebra makes it possible to combine the results from multiple tests or conditions into a single combined outcome or binary decision, i.e., TRUE or False, Yes or No. (See section 3.6 on page 52 for examples.)

In mathematics, Boolean algebra provides the rules of the logic used to combine multiple logical values. Boolean operators like AND and OR take as operands logical values and return a logical value as a result. In R there are two “families” of Boolean operators, vectorised and not vectorised. Vectorised operators accept logical vectors of any length as operands, while non-vectorised ones accept only logical vectors of length one as operands. In the chunk below we use non-vectorised operators with two `logical` vectors of length one, `a` and `b`, as operands.

```
vct1 <- TRUE
mode(vct1)
## [1] "logical"
vct1
## [1] TRUE
!TRUE # negation
## [1] FALSE
TRUE && FALSE # logical AND
## [1] FALSE
TRUE || FALSE # logical OR
## [1] TRUE
xor(TRUE, FALSE) # exclusive OR
## [1] TRUE
```

The availability of two kinds of logical operators can be troublesome for those new to R. Pairs of "equivalent" logical operators behave differently, use similar syntax and use similar symbols! The vectorised operators have single-character names, `&` and `|` (like the vectorised arithmetic operators, such as `+`), while the non-vectorised ones have double-character names, `&&` and `||`. There is only one version of the negation operator `!` that is vectorised. In recent versions of R, an error is triggered when a non-vectorised operator is used with a vector with length > 1, which helps prevent mistakes. In some situations, vectorised `logical` operators can replace non-vectorised ones, but it is important to use the ones that match the intention of the code, as this enables relevant checks for mistakes. Once the distinction is learnt, using the most appropriate operators also contributes to make code easier to read.

```
c(TRUE, FALSE) & c(TRUE,TRUE) # vectorised AND
## [1]  TRUE FALSE
c(TRUE, FALSE) | c(TRUE,TRUE) # vectorised OR
## [1] TRUE TRUE
```

Functions `any()` and `all()` take zero or more logical vectors as their arguments, and return a single logical value "summarising" the logical values in the vectors. Function `all()` returns `TRUE` only if all values in the vectors passed as arguments are `TRUE`, and `any()` returns `TRUE` unless all values in the vectors are `FALSE`.

```
vct2 <- c(TRUE, FALSE, FALSE)
any(vct2)
## [1] TRUE
all(vct2)
## [1] FALSE
any(c(TRUE, FALSE) & c(TRUE,TRUE))
## [1] TRUE
all(c(TRUE, FALSE) & c(TRUE,TRUE))
## [1] FALSE
any(c(TRUE, FALSE) | c(TRUE,TRUE))
## [1] TRUE
all(c(TRUE, FALSE) | c(TRUE,TRUE))
## [1] TRUE
```

Another important thing to know about logical operators is that they "short-cut" evaluation. If the result is known from the first part of the statement, the rest of the statement is not evaluated. Try to understand what happens when you enter the following commands. Short-cut evaluation is useful, as the first condition can be used as a guard protecting a later condition from being evaluated when it would trigger an error.

```
TRUE || NA
## [1] TRUE
FALSE || NA
## [1] NA
TRUE && NA
## [1] NA
FALSE && NA
## [1] FALSE
TRUE && FALSE && NA
## [1] FALSE
TRUE && TRUE && NA
## [1] NA
```

3.15 Investigate how swapping the order of the operands in the code chunk above affects the values returned, e.g.., the first statement becomes `NA || TRUE`.

When using the vectorised operators on vectors of length greater than one, 'short-cut' evaluation still applies for the result obtained at each index position.

```
c(TRUE, FALSE) & c(TRUE,TRUE) & NA
## [1]    NA FALSE
c(TRUE, FALSE) & c(TRUE,TRUE) & c(NA, NA)
## [1]    NA FALSE
c(TRUE, FALSE) | c(TRUE,TRUE) | c(NA, NA)
## [1] TRUE TRUE
```

3.16 Based on the description of "recycling" presented on page 31 for `numeric` operators, explore how "recycling" works with vectorised logical operators. Create logical vectors of different lengths (including length one) and *play* by writing several code statements with operations on them. To get you started, one example is given below. Execute this example, and then create and run your own, making sure that you understand why the values returned are what they are. Sometimes, you will need to devise several examples or test cases to tease out of R an understanding of how a certain feature of the language works, so do not give up early, and make use of your imagination!

```
c(TRUE, FALSE, TRUE, NA) & FALSE
c(TRUE, FALSE, TRUE, NA) | c(TRUE, FALSE)
```

How to test if a vector contains no values other than `NA` (or `NaN`) values? A call to `is.na()` returns a `logical` vector that we can pass to `all()`. We can save the intermediate vector `temp` and pass it as argument to `is.na()`, or alternatively nest the function calls. The name `tmp`, for *temporary*, is frequently used for variables whose value is retrieved only once.

```
vct2 <- rep(NA, 5) # toy data
tmp <- is.na(vct2) # tmp for temporary
all(tmp)
## [1] TRUE
all(is.na(vct2)) # nested call
## [1] TRUE
```

How to test if a vector contains one or more NA (or NaN) values?
See previous question. We only need to replace all() by any() to obtain the answer.

```
vct2 <- rep(NA, 5)
any(is.na(vct2))
## [1] TRUE
```

3.6 Comparison Operators and Operations

Comparison operators return vectors of logical values (see section 3.5 on page 49), with values TRUE or FALSE depending on the outcome.

Equality (==) and inequality (!=) operators are defined not only for numeric values but also for character and most other atomic and many other values. Be aware that operator = is an infrequently used synonym of the assignment operator <- rather than a comparison operator!

```
# be aware that we use two = symbols
"abc" == "ab"
## [1] FALSE
"ABC" == "abc"
## [1] FALSE
"abc" != "ab"
## [1] TRUE
"ABC" != "abc"
## [1] TRUE
```

In the case of numeric values additional comparisons are meaningful and additional operators are defined.

```
1.2 > 1.0
## [1] TRUE
1.2 >= 1.0
## [1] TRUE
1.2 == 1.0
## [1] FALSE
1.2 != 1.0
## [1] TRUE
1.2 <= 1.0
## [1] FALSE
1.2 < 1.0
## [1] FALSE
```

These operators can be used on vectors of any length, returning as a result a logical vector as long as the longest operand. In other words, they behave in the same way as the arithmetic operators described on page 30: their arguments are recycled when needed. Hint: if you do not know what value is stored in numeric vector a, use `print(a)` after the first code statement below to see its contents.

```
vct3 <- 1:10
vct3 > 5
##  [1] FALSE FALSE FALSE FALSE FALSE  TRUE  TRUE  TRUE  TRUE  TRUE
vct3 < 5
##  [1]  TRUE  TRUE  TRUE  TRUE FALSE FALSE FALSE FALSE FALSE FALSE
vct3 == 5
##  [1] FALSE FALSE FALSE FALSE  TRUE FALSE FALSE FALSE FALSE FALSE
all(vct3 > 5)
## [1] FALSE
any(vct3 > 5)
## [1] TRUE
vct4 <- vct3 > 5
vct4
##  [1] FALSE FALSE FALSE FALSE FALSE  TRUE  TRUE  TRUE  TRUE  TRUE
any(vct4)
## [1] TRUE
all(vct4)
## [1] FALSE
```

Individual comparisons can be useful, but their full role in data analysis and programming is realised when we combine multiple tests using the operations of the Boolean algebra described in section 3.5 on page 49.

For example, to test if members of a numeric vector are within a range, in our example, -1 to $+1$, we can combine the results from two comparisons using the vectorised logical *AND* operator `&`, and use parentheses to override the default order of precedence of the operations.

```
vct5 <- -2:3
vct5 >= -1 & vct5 <= 1
## [1] FALSE  TRUE  TRUE  TRUE FALSE FALSE
```

If we want to find those values outside this same range, we can negate the test.

```
!(vct5 >= -1 & vct5 <= 1)
## [1]  TRUE FALSE FALSE FALSE  TRUE  TRUE
```

Or, we can combine another two comparisons using the vectorised logical *OR* operator `|`.

```
vct5 < -1 | vct5 > 1
## [1]  TRUE FALSE FALSE FALSE  TRUE  TRUE
```

In some cases, an additional advantage is that `logical` values require less space in memory for their storage than `numeric` values.

3.17 Use the statement below as a starting point in exploring how precedence works when logical and arithmetic operators are part of the same statement. *Play* with the example by adding parentheses at different positions and based on the returned values, work out the default order of operator precedence used for the evaluation of the example given below.

```
vct6 <- 1:10
vct6 > 3 | vct6 + 2 < 3
```

It is important to be aware of the consequences of "short-cut evaluation" (described on page 51). The behaviour of many of base-R's functions when NAs are present in their input arguments can be modified. If TRUE is passed as an argument to parameter na.rm, NA values are *removed* from the input **before** the function is applied.

```
vct7 <- c(1:10, NA)
all(vct7 < 20)
## [1] NA
any(vct7 > 20)
## [1] NA
all(vct7 < 20, na.rm=TRUE)
## [1] TRUE
any(vct7 > 20, na.rm=TRUE)
## [1] FALSE
```

In many situations, when writing programs one should avoid testing for equality of floating point numbers ('floats'). This is because of how numbers are stored in computers (see the box on page 35 for an in-depth explanation). Here I show how to gracefully handle rounding errors when using comparison operators. As rounding errors may accumulate, in practice .Machine$double.eps is frequently too small a value to safely use in tests for "zero.". Whenever possible according to the logic of the calculations, it is best to test for inequalities, for example using x <= 1.0 instead of x == 1.0. If this is not possible, then equality tests should be done by replacing tests like x == 1.0 with abs(x - 1.0) < k, where k is a number larger than eps. Function abs() returns the absolute value, in simpler words, makes all values positive or zero, by changing the sign of negative values, or in mathematical notation $|x| = |-x|$.

```
sin(pi) == 0 # angle in radians, not degrees!
## [1] FALSE
sin(2 * pi) == 0
## [1] FALSE
abs(sin(pi)) < 1e-15
## [1] TRUE
abs(sin(2 * pi)) < 1e-15
## [1] TRUE
sin(pi)
## [1] 1.224606e-16
sin(2 * pi)
## [1] -2.449213e-16
```

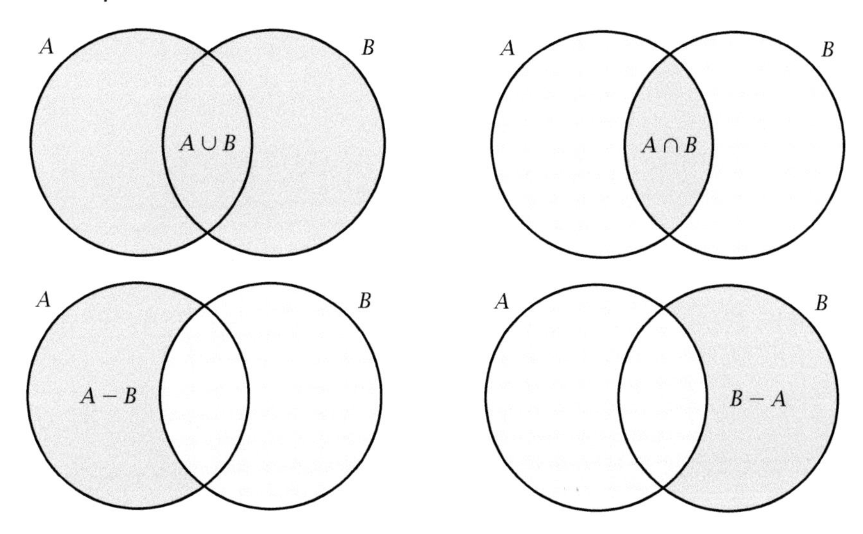

Figure 3.1
Boolean algebra. Venn diagrams for algebra of sets operations: *union*, $\cup$, `union()`; *intersection*, $\cap$, `intersect()`; *difference (asymmetrical)*, $-$, `setdiff()`; *equality test* `setequal()`; *membership*, `is.element()` and operator `%in%`

3.7 Sets and Set Operations

The R language supports set operations on vectors. They can be useful in many different contexts when manipulating and comparing vectors of values. In Bioinformatics, it is usual, for example, to make use of character vectors of gene tags. Algebra sets is implemented with functions `union()`, `intersect()`, `setdiff()`, `setequal()`, `is.element()` and operator `%in%` (Figure 3.1). The first three operations return a vector of the same mode as their inputs, and the last three a `logical` vector. The action of the first three operations is most easily illustrated with Venn diagrams, where the returned value (or result of the operation) is depicted in darker grey.

Set operations applied to vectors with values representing a mundane example, grocery shopping, demonstrate them.

```
fruits <- c("apple", "pear", "orange", "lemon", "tangerine")
bakery <- c("bread", "buns", "cake", "cookies")
dairy <- c("milk", "butter", "cheese")
shopping <- c("bread", "butter", "apple", "cheese", "orange")
intersect(fruits, shopping)
## [1] "apple"  "orange"
intersect(bakery, shopping)
## [1] "bread"
intersect(dairy, shopping)
## [1] "butter" "cheese"
"lemon" %in% dairy
## [1] FALSE
"lemon" %in% fruits
## [1] TRUE
```

```
dairy %in% shopping
## [1] FALSE  TRUE  TRUE
union(bakery, dairy)
## [1] "bread"    "buns"     "cake"     "cookies" "milk"     "butter"   "cheese"
setdiff(union(bakery, dairy), shopping) # nested call
## [1] "buns"    "cake"    "cookies" "milk"
```

Sets describe membership as a binary property, thus when vectors are interpreted as sets, duplicate members are redundant. Duplicate members although accepted as input are always simplified in the returned values.

```
union(c("a", "a", "b"), c("b", "a", "b")) # set operation
## [1] "a" "b"

setequal(c("a", "a", "b"), c("b", "a", "b")) # sets compared
## [1] TRUE
all.equal(c("a", "a", "b"), c("b", "a", "b")) # vectors compared
## [1] "1 string mismatch"
identical(c("a", "a", "b"), c("b", "a", "b")) # vectors compared
## [1] FALSE
```

We construct and save a character vector to use in the next examples.

```
vct1 <- c("a", "b", "c", "b")
```

To test if a given value belongs to a set, we use operator `%in%` or its function equivalent `is.element()`. In the algebra of sets notation, this is written $a \in A$, where A is a set and a a member. The second statement shows that the `%in%` operator is vectorised on its left-hand-side (lhs) operand, returning a logical vector.

```
is.element("a", vct1)
## [1] TRUE
"a" %in% vct1
## [1] TRUE
c("a", "a", "z") %in% vct1
## [1]  TRUE  TRUE FALSE
```

Keep in mind that inclusion, implemented in operator `%in%`, is an asymmetrical (not reflective) operation among a vector and a set. The right-hand-side (rhs) argument is interpreted as a set, while the left-hand-side (lhs) argument is interpreted as a vector of values to test for membership in the set. In other words, any duplicate member in the lhs operand is retained and tested while the rhs operand is interpreted as a set of unique values. The returned logical vector has the same length as the lhs operand.

```
vct1 %in% "a"
## [1]  TRUE FALSE FALSE FALSE
```

The negation of inclusion is $a \notin A$, and coded in R by applying the negation operator `!` to the result of the test done with `%in%` or function `is.element()`.

```
!is.element("a", vct1)
## [1] FALSE
!"a" %in% vct1
## [1] FALSE
!c("a", "a", "z") %in% vct1
## [1] FALSE FALSE  TRUE
```

Although inclusion is a set operation, it is also very useful for the simplification of `if () … else` statements by replacing multiple tests for alternative constant values of the same `mode` chained by multiple `|` operators. A useful property of `%in%` and `is.element()` is that they never return `NA`.

Operator `%in%` is equivalent to function `match()`, although the additional parameters of `match()` provide additional flexibility.

In some cases, such as when accepting partial character strings as input, the aim is not an exact match, but a partial match to target character strings. In this case, either `charmatch()` or `pmatch()` is the correct tool to use depending on the desired handling of partial, ambiguous and exact matches. Use `help()` to find the details if you need to use one of them.

3.18 Use operator `%in%` to write more concisely the following comparisons. Hint: see section 3.5 on page 49 for the difference between `|` and `||` operators.

```
vct2 <- c("a", "a", "z")
vct2 == "a" | vct2 == "b" | vct2 == "c" | xvct2 == "d"
```

Convert the `logical` vectors of length 3 into a vector of length one. Hint: see help for functions `all()` and `any()`.

With `unique()` we convert a vector of possibly repeated values into a set of unique values. In the algebra of sets, a certain object belongs or not to a set. Consequently, in a set, multiple copies of the same object or value are meaningless.

```
unique(vct1)
## [1] "a" "b" "c"
```

Function `unique()` is frequently useful, for example when we want determine the number of distinct values in a vector.

```
length(unique(vct1))
## [1] 3
```

3.19 Do the values returned by these two statements differ?

```
c("a", "a", "z") %in% vct1
c("a", "a", "z") %in% unique(vct1)
```

Function `duplicated()` is the counterpart of `unique()`, returning a logical vector, indicating which values in a vector are duplicates of values already present at positions with a lower index.

```
duplicated(vct1)
## [1] FALSE FALSE FALSE  TRUE
anyDuplicated(vct1)
## [1] 4
```

The `R` language includes many functions that simplify tasks related to data analysis. Some are well known like `unique()`, but others may need to be searched for in the documentation.

3.20 What do you expect to be the difference between the values returned by the three statements in the code chunk below? Before running them, write down your expectations about the value each one will return. Only then run the code.

Independently of whether your predictions were correct or not, write down an explanation of what each statement's operation is.

```
union(c("a", "a", "z"), vct1)
c(c("a", "a", "z"), vct1)
c("a", "a", "z", vct1)
```

Are set union and concatenation of vectors equivalent operations? why or why not?

All set algebra examples above use character vectors and character constants. This is just the most frequent use case. Sets operations are valid on vectors of any atomic class, including `integer`, and computed values can be part of statements. In the second and third statements in the next chunk, we need to use additional parentheses to alter the default order of precedence between arithmetic and set operators.

```
9 %in% 2:4
## [1] FALSE
9 %in% ((2:4) * (2:4))
## [1] TRUE
c(1, 16) %in% ((2:4) * (2:4))
## [1] FALSE  TRUE
```

Empty sets are an important component of the algebra of sets, in R they are represented as vectors of zero length. These vectors do belong to a class such as `numeric` or `character` and must be compatible with other operands in an expression.

```
c("ab", "xy") %in% character()
## [1] FALSE FALSE
character() %in% c("a", "b", "c")
## logical(0)
union("ab", character())
## [1] "ab"
```

Although set operators are defined for `numeric` vectors, rounding errors in 'floats' can result in unexpected results (see section 3.3 on page 35).

```
c(cos(pi), sin(pi)) %in% c(0, -1)
## [1]  TRUE FALSE
c(cos(pi), sin(pi))
## [1] -1.000000e+00  1.224606e-16
```

3.21 In the algebra of sets notation $A \subseteq B$, where A and B are sets, indicates that A is a subset or equal to B. For a true subset, the notation is $A \subset B$. The operators with the reverse direction are $\supseteq$ and $\supset$. Implement these four operations in four R statements, and test them on sets (represented by R vectors) with different "overlap" among set members.

3.8 The Mode and Class of Objects

Classes are abstractions, they determine the "meaning" and behaviour of objects belonging to them. New classes can be defined in user code as well as new methods, i.e., functions or operators tailored to fit them. The *class* is like a "tag" that tells how the value in an object should be interpreted and operated upon.

Variables (names given to objects) have a *class* that depends on the object stored in them. In contrast to some other languages in R assignment to a variable already in use to store an object belonging to a different class is allowed. There is a restriction that all elements in a vector, array or matrix, must be of the same mode (these are called atomic, as they contain homogeneous members). Lists and data frames can be heterogenous (to be described in chapter 4). In practice, this means that we can assign an object, such as a vector, with a different `class` to a name already in use, but we cannot use indexing to assign an object of a different mode to individual members of a vector, matrix or array.

Function `class()` is used to query the class of an object, and function `inherits()` is used to test if an object belongs to a specific class or not (including "parent" classes, to be later described).

```
vct1 <- 1:5
class(vct1)
## [1] "integer"
inherits(vct1, "character")
## [1] FALSE
inherits(vct1, "numeric")
## [1] FALSE
```

Functions with names starting with `is.` are tests returning a logical value, `TRUE`, `FALSE` or `NA`.

```
is.numeric(vct1) # no distinction of integer or double
## [1] TRUE
is.double(vct1)
## [1] FALSE
is.integer(vct1)
## [1] TRUE
is.logical(vct1)
## [1] FALSE
is.character(vct1)
## [1] FALSE
```

Functions starting with `is.` have to be individually defined and are available only for some classes. Function `inherits()` takes as its second argument a character vector containing strings to be tested against the `class` attribute of the object passed as its first argument.

```
inherits(vct1, c("numeric", "character", "logical"), which = TRUE)
## [1] 0 0 0
```

The *mode* of an object is a fundamental property, and limited to those modes defined as part of the R language. In particular, different R objects of a given mode, such as `numeric`, can belong to different `classes`. Classes and the dispatch of methods are discussed in section 6.3 on page 176, together with object-oriented programming.

```
mode(c(1, 2, 3)) # no distinction of integer or double
## [1] "numeric"
typeof(c(1, 2, 3))
## [1] "double"
class(c(1, 2, 3))
## [1] "numeric"
mode(c(1L, 2L, 3L)) # no distinction of integer or double
## [1] "numeric"
typeof(c(1L, 2L, 3L))
## [1] "integer"
class(c(1L, 2L, 3L))
## [1] "integer"

mode(factor(c("a", "b", "c"))) # no distinction of integer or double
## [1] "numeric"
typeof(factor(c("a", "b", "c")))
## [1] "integer"
class(factor(c("a", "b", "c")))
## [1] "factor"

mode(c("a", "b", "c"))
## [1] "character"
typeof(c("a", "b", "c"))
## [1] "character"
class(c("a", "b", "c"))
## [1] "character"

mode(c(TRUE, FALSE))
## [1] "logical"
typeof(c(TRUE, FALSE))
## [1] "logical"
class(c(TRUE, FALSE))
## [1] "logical"
```

3.9 Type Conversions

By type conversion we mean converting a value from one class into a value expressed in a different class. usually the meaning can be retained, at least in part. We can, for example, convert character strings into numeric values, but this conversion is possible only for character strings conformed by digits, like `"100"`. Most conversions, such as the conversion of `character` value `"100"` into `numeric` value

100 are obvious. Type conversions involving logical values are less intuitive. By convention, functions used to convert objects from one mode or class to a different one have names starting with `as.`[1].

```
as.character(102)
## [1] "102"
as.character(TRUE)
## [1] "TRUE"
as.character(3.0e10)
## [1] "3e+10"
as.numeric("203")
## [1] 203
as.logical("TRUE")
## [1] TRUE
as.logical(100)
## [1] TRUE
as.logical(0)
## [1] FALSE
as.logical(-1)
## [1] TRUE
```

Some conversions takes place automatically in expressions involving both `numeric` and `logical` values.

```
TRUE + 10
## [1] 11
1 || 0
## [1] TRUE
FALSE | -2:2
## [1]  TRUE  TRUE FALSE  TRUE  TRUE
```

3.22 There is flexibility in the conversion from character strings into `numeric` and `logical` values. Use the examples below plus your own variations to get an idea of what strings are acceptable and correctly converted and which are not. Do also pay attention at the conversion between `numeric` and `logical` values.

```
as.numeric("5E+5")
as.numeric("50e+4")
as.numeric(".12")
as.numeric("0.12")
as.numeric("A")
as.logical("TRUE")
as.logical("FALSE")
as.logical("T")
as.logical("t")
as.logical("true")
as.logical("NA")
```

3.23 Conversion of fractional numbers into whole numbers can be achieved in different ways, by truncation of the fractional part or rounding it up or down. If we consider both negative and positive numbers, how each of them is handled creates additional possibilities. All these approaches, as defined in mathematics,

[1]Except for some packages in the 'tidyverse' that use names starting with `as_` instead of `as.`.

are available through different R functions. These functions, are not conversion functions as they return a numeric value of class double. See page 37. In contrast, as.integer() is a conversion function for type double into type integer, both with mode numeric.

Compare the values returned by trunc() and as.integer() when applied to a floating point number, such as 12.34. Check for the equality of values, and for the *class* and *type* of the returned objects.

Using conversions, the difference between the length of a character vector and the number of characters composing each member "string" within a vector becomes clear.

```
vct1 <- c("1", "2", "3")
length(vct1)
## [1] 3

vct2 <- "123.1"
length(vct2)
## [1] 1

as.numeric(vct1)
## [1] 1 2 3
as.numeric(vct2)
## [1] 123.1
as.integer(vct1)
## [1] 1 2 3
as.integer(vct2)
## [1] 123
```

Other functions relevant to the "conversion" of numbers and other values are format(), and sprintf(). This is sometimes informally called "pretty printing". These two functions return character strings, instead of numeric or other values, and are useful for printed output. One could think of these functions as advanced conversion functions returning formatted, and possibly combined and annotated, character strings. However, they are usually not considered normal conversion functions, as they are very rarely used in a way that preserves the original precision of the input values. We show here the use of format() and sprintf() with numeric values, but they can also be used with values of other classes like character, logical, etc.

When using format(), the format used to display numbers is set by passing arguments to several different parameters. As print() calls format() to convert numeric values into character strings, it accepts the same options.

```
vct2 = c(123.4567890, 1.0)
format(vct2) # using defaults
## [1] "123.4568" "  1.0000"
format(123.4567890) # using defaults
## [1] "123.4568"
format(1.0) # using defaults
## [1] "1"
format(vct2, digits = 3, nsmall = 1)
## [1] "123.5" "  1.0"
format(vct2, digits = 3, scientific = TRUE)
## [1] "1.23e+02" "1.00e+00"
```

Function sprintf() is similar to C's function of the same name. The user interface is rather unusual, but very powerful, once one learns the syntax. All the formatting is specified using a character string as template. In this template, placeholders for data and the formatting instructions are embedded using special codes. These codes start with a percent character. We show in the example below the use of some of these: f is used for numeric values to be formatted according to a "fixed point", while g is used when we set the number of significant digits and e for exponential or *scientific* notation.

```
x = c(123.4567890, 1.0)
sprintf("The numbers are: %4.2f and %.0f", x[1], x[2])
## [1] "The numbers are: 123.46 and 1"
sprintf("The numbers are: %.4g and %.2g", x[1], x[2])
## [1] "The numbers are: 123.5 and 1"
sprintf("The numbers are: %4.2e and %.0e", x[1], x[2])
## [1] "The numbers are: 1.23e+02 and 1e+00"
```

In the template "The numbers are: %4.2f and %.0f", there are two placeholders for numeric values, %4.2f and %.0f; so, in addition to the template, we pass two values extracted from the first two positions of vector x. These could have been two different vectors of length one, or even numeric constants. The template itself does not need to be a character constant as in these examples, as a variable can be also passed as argument.

3.24 Function format() may be easier to use, in some cases, but sprintf() is more flexible and powerful. Those with experience in the use of the C language will already know about sprintf() and its use of templates for formatting output. Even if you are familiar with C, look up the help pages for both functions, and practice, by trying to create the same formatted output by means of the two functions. Do also play with these functions with other types of data like integer and character.

I have described above NA as a single value ignoring modes, but in reality NA values come in various flavours: NA_real_, NA_character_, etc. and NA defaults to an NA of class logical. NA is normally converted on the fly to other modes when needed, so in general NA is all we need to use. The examples below use the extraction operator to demonstrate automatic conversion on assignment. This operator is described in section 3.10 below.

```
vct3 <- c(1, NA)
is.numeric(vct3[2])
## [1] TRUE
is.numeric(NA)
## [1] FALSE

vct4 <- c("abc", NA)
is.character(vct4[2])
## [1] TRUE
class(NA_character_)
## [1] "character"
```

```
is.character(NA)
## [1] FALSE
class(NA)
## [1] "logical"
```

```
vct5 <- NA
c(vct5, 2:3)
## [1] NA  2  3
```

However, even the statement below works transparently.

```
vct3[3] <- vct4[2]
```

3.10 Vector Manipulation

If you have read earlier sections of this chapter, you already know how to create a vector. If not, see pages 28–33 before continuing.

In this section, we are going to see how to extract or retrieve, replace, and move elements such as a_2 from a vector $a_{i=1...n}$. Elements are extracted using an index enclosed in single square brackets. The index indicates the position in the vector, starting from one, following the usual mathematical tradition. While in maths notation a_1 represents the first, or leftmost, member of vector $a_{i=1...n}$, in R the equivalent notation is `a[1]` for the member and `a` for the vector.

We extract the first 10 elements of the vector `letters`, by passing an `integer` vector as argument to operator `[ ]`.

```
vct1 <- letters[1:10]
vct1
##  [1] "a" "b" "c" "d" "e" "f" "g" "h" "i" "j"
```

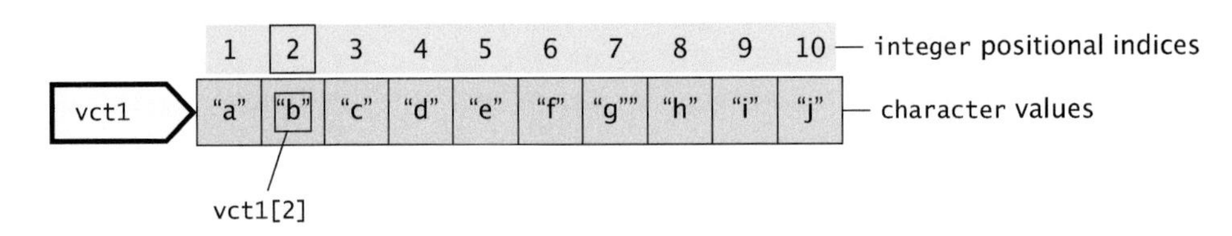


```
vct1[2]
## [1] "b"
```

Four constant vectors are available in base R: `letters`, `LETTERS`, `month.name` and `month.abb`, of which I used `letters` in the example above. These vectors are always for English, irrespective of the locale.

```
month.name
##  [1] "January"   "February"  "March"     "April"     "May"       "June"
##  [7] "July"      "August"    "September" "October"   "November"  "December"
month.name[6]
## [1] "June"
```

In R, indexes always start from one, while in some other programming languages such as C and C++, indexes start from zero. It is important to be aware of this difference, as many computation algorithms are valid only under a given indexing convention.

How to access the last value in a vector?

```
month.name[length(month.name)]
## [1] "December"
```

It is possible to extract a subset of the elements of a vector in a single operation, using a vector of indexes. The positions of the extracted elements in the result ("returned value") are determined by the ordering of the members of the vector of indexes—easier to demonstrate than to explain.

```
vct1[c(3, 2)]
## [1] "c" "b"
vct1[10:1]
##  [1] "j" "i" "h" "g" "f" "e" "d" "c" "b" "a"
```

3.25 The length of the indexing vector is *not* restricted by the length of the indexed vector. However, only numerical indexes that match positions present in the indexed vector can extract values. Those values in the indexing vector pointing to positions that are not present in the indexed vector, result in NA values. This is easier to learn by *playing* with R, than from explanations. Play with R, using the following examples as a starting point.

```
length(vct1)
vct1[c(3, 3, 3, 3)]
vct1[c(10:1, 1:10)]
vct1[c(1, 11)]
vct1[11]
```

Have you tried some of your own examples? If not yet, do *play* with additional variations of your own before continuing.

Negative indexes have a special meaning; they indicate the positions at which values should be excluded. Be aware that it is *illegal* to mix positive and negative values in the same indexing operation.

```
vct1[-2]
## [1] "a" "c" "d" "e" "f" "g" "h" "i" "j"
vct1[-c(3,2)]
## [1] "a" "d" "e" "f" "g" "h" "i" "j"
vct1[-3:-2]
## [1] "a" "d" "e" "f" "g" "h" "i" "j"
```

3.26 Results from indexing with special values and zero may be surprising. Try to build a rule from the examples below, a rule that will help you remember what to expect next time you are confronted with similar statements using special values as "subscripts" instead of integers larger or equal to one—this is likely to happen sooner or later as these special values can be returned by different R expressions depending on the value of operands or function arguments, some of them described earlier in this chapter.

```
vct1[ ]
vct1[0]
vct1[numeric(0)]
vct1[NA]
vct1[c(1, NA)]
vct1[NULL]
vct1[c(1, NULL)]
```

Another way of indexing, which is very handy, but not available in most other programming languages, is indexing with a vector of `logical` values. The `logical` vector used for indexing is usually of the same length as the vector from which elements are going to be selected. However, this is not a requirement, because if the `logical` vector of indexes is shorter than the indexed vector, it is "recycled" as discussed on page 31 in relation to other operators.

```
vct1[TRUE]
##  [1] "a" "b" "c" "d" "e" "f" "g" "h" "i" "j"
vct1[FALSE]
## character(0)
vct1[c(TRUE, FALSE)]
## [1] "a" "c" "e" "g" "i"
vct1[c(FALSE, TRUE)]
## [1] "b" "d" "f" "h" "j"
vct1 > "c"
##  [1] FALSE FALSE FALSE  TRUE  TRUE  TRUE  TRUE  TRUE  TRUE  TRUE
vct1[vct1 > "c"]
## [1] "d" "e" "f" "g" "h" "i" "j"
```

Indexing with logical vectors is very frequently used in R because comparison operators are vectorised. Comparison operators, when applied to a vector, return a `logical` vector, a vector that can be used to extract the elements for which the result of the comparison test was `TRUE`.

3.27 The examples in this text box demonstrate additional uses of logical vectors: 1) the logical vector returned by a vectorised comparison can be stored in a variable, and the variable used as a "selector" for extracting a subset of values from the same vector, or from a different vector.

```
vct1 <- letters[1:10]
vct2 <- 1:10
selector <- vct1 > "c"
selector
vct1[selector]
vct2[selector]
```

Positional indexes can be obtained from a `logical` vector by means of function `which()` as it returns a `numeric` vector with the positions of the `TRUE` values in the `logical` vector.

```
indexes <- which(vct1 > "c")
indexes
vct1[indexes]
```

Make sure to understand the examples above. These constructs are very widely used in R because they allow for concise code that is easy to understand once one is familiar with the indexing rules.

Above, `integer` or `logical` vectors were used as indices for extraction of anonymous elements, or members, from `character` vectors. In R, elements can be assigned names, and these names used in place of `numeric` indices to access them. One situation where this is very useful is the mapping of values between two representations. Let's assume we have a long vector encoding treatments using single letter codes and that we want to replace these codes with self-explanatory names.

```
treat <- c("H", "C", "H", "W", "C", "H", "H", "W", "W")
```

We can create a named vector to *map* the single letter codes onto full words. Above, we used function `c()` to concatenate several `character` strings, without assigning any names to them, thus they have to be extracted from the vector using `numeric` values, indexing by position. Below, we assign a name to each string. Using operator = we assign the name on the left-hand side (*lhs*) to the member of the vector on the right-hand-side (*rhs*).

```
treat.map <- c(H = "hot", C = "cold", W = "warm")
treat.map
##      H      C      W
##  "hot" "cold" "warm"
names(treat.map)
## [1] "H" "C" "W"
```

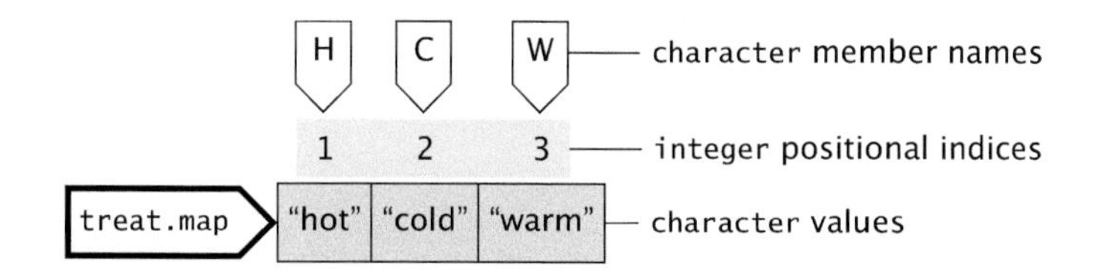

As `treat.map` is a named vector, we can use the element names, in addition to `numeric` values, as indices for element extraction.

```
treat.map["H"]
##     H
## "hot"
```

The indexing vector can be of a different length than the indexed vector, and the returned value is a new vector of the same length as the indexing vector.

```
treat.new <- treat.map[treat]
treat.new
##      H      C      H      W      C      H      H      W      W
##  "hot" "cold"  "hot" "warm" "cold"  "hot"  "hot" "warm" "warm"
```

where `treat.new` is a named vector, from which we will frequently want to remove the members' names.

```
treat.new <- unname(treat.new)
treat.new
## [1] "hot"  "cold" "hot"  "warm" "cold" "hot"  "hot"  "warm" "warm"
```

It is more common to use named members with lists than with vectors, but in R, in both cases it is possible to use both numeric positional indices and names.

Indexing can be used on either side of an assignment expression. In the code chunk below, we use the extraction operator on the left-hand side of the assignments to replace values only at selected positions in the vector. This may look rather esoteric at first sight, but it is just a simple extension of the logic of indexing described above. It works, because the low precedence of the <- operator results in both the left- and the right-hand side being fully evaluated before the assignment takes place. To make the changes to the vectors easier to compare, identical vectors are used in each of the examples below.

```
vct2 <- 1:10
vct2
##  [1]  1  2  3  4  5  6  7  8  9 10
vct2[1] <- 99
vct2
##  [1] 99  2  3  4  5  6  7  8  9 10
vct2 <- 1:10
vct2[c(2,4)] <- -99 # recycling
vct2
##  [1]   1 -99   3 -99   5   6   7   8   9  10
vct2 <- 1:10
vct2[c(2,4)] <- c(-99, 99)
vct2
##  [1]   1 -99   3  99   5   6   7   8   9  10
vct2 <- 1:10
vct2[TRUE] <- 1 # recycling
vct2
##  [1] 1 1 1 1 1 1 1 1 1 1
vct2 <- 1:10
vct2 <- 1  # no recycling
vct2
## [1] 1
```

Indexing can be used simultaneously on both sides of the assignment operator, for example, to swap two elements.

```
vct3 <- letters[1:10]
vct3[1:2] <- vct3[2:1]
vct3
##  [1] "b" "a" "c" "d" "e" "f" "g" "h" "i" "j"
```

3.28 Do play with subscripts to your heart's content, really grasping how they work and how they can be used, will be very useful in anything you do in the future with R. Even the contrived example below follows the same simple rules, just study it bit by bit. Hint: the second statement in the chunk below, modifies VCT1, so, when studying variations of this example, you will need to recreate VCT1 by executing the first statement each time you run a variation of the second statement.

```
VCT1 <- letters[1:10]
VCT1[5:1] <- VCT1[c(TRUE,FALSE)]
VCT1
```

In R, indexing with positional indexes can be done with `integer` or `numeric` values. Numeric values can be floats, but for indexing, only integer values are meaningful. Consequently, `double` values are converted into `integer` values when used as indexes. The conversion is done invisibly, but it does slow down computations slightly. When working on big data sets, explicitly using `integer` values can improve performance.

```
vct4 <- LETTERS[1:10]
vct4
##  [1] "A" "B" "C" "D" "E" "F" "G" "H" "I" "J"
vct4[1]
## [1] "A"
vct4[1.1]
## [1] "A"
vct4[1.9999] # surprise!!
## [1] "A"
vct4[2]
## [1] "B"
```

From this experiment, we can learn that if positive indexes are not whole numbers, they are truncated to the next smaller integer.

```
vct4 <- LETTERS[1:10]
vct4
##  [1] "A" "B" "C" "D" "E" "F" "G" "H" "I" "J"
vct4[-1]
## [1] "B" "C" "D" "E" "F" "G" "H" "I" "J"
vct4[-1.1]
## [1] "B" "C" "D" "E" "F" "G" "H" "I" "J"
vct4[-1.9999]
## [1] "B" "C" "D" "E" "F" "G" "H" "I" "J"
vct4[-2]
## [1] "A" "C" "D" "E" "F" "G" "H" "I" "J"
```

From this experiment, we can learn that if negative indexes are not whole numbers, they are truncated to the next larger (less negative) integer. In conclusion, `double` index values behave as if they where sanitised using function `trunc()`.

This example also shows how one can tease out of R its rules through experimentation.

A frequent operation on vectors is sorting them into an increasing or decreasing order. The most direct approach is to use `sort()`.

```
vct5 <- c(10, 4, 22, 1, 4)
sort(vct5)
## [1]  1  4  4 10 22
sort(vct5, decreasing = TRUE)
## [1] 22 10  4  4  1
```

An indirect way of sorting a vector, possibly based on a different vector, is to generate with `order()` a vector of numerical indexes that can be used to achieve the ordering.

```
order(vct5)
## [1] 4 2 5 1 3
vct5[order(vct5)]
## [1]  1  4  4 10 22
vct6 <- c("ab", "aa", "c", "zy", "e")
vct6[order(vct5)]
## [1] "zy" "aa" "e"  "ab" "c"
```

A problem linked to sorting that we may face is counting how many copies of each value are present in a vector. We need to use two functions `sort()` and `rle()` . The second of these functions computes *run length* as used in *run length encoding* for which *rle* is an abbreviation. A *run* is a series of consecutive identical values. As the objective is to count the number of copies of each value present, we need first to sort the vector.

```
vct7 <- letters[c(1, 5, 10, 3, 1, 4, 21, 1, 10)]
vct7
## [1] "a" "e" "j" "c" "a" "d" "u" "a" "j"
sort(vct7)
## [1] "a" "a" "a" "c" "d" "e" "j" "j" "u"
rle(sort(vct7))
## Run Length Encoding
##   lengths: int [1:6] 3 1 1 1 2 1
##   values : chr [1:6] "a" "c" "d" "e" "j" "u"
```

The second and third statements are only to demonstrate the effect of each step. The last statement uses nested function calls to compute the number of copies of each value in the vector.

3.11 Matrices and Multidimensional Arrays

Matrices have two dimensions, rows and columns, and like vectors all their members share the same mode, and are atomic, i.e., they are homogeneous (Figure 3.2). Most commonly, matrices are used to store `numeric`, `integer` or `logical` values. The number of rows and columns can differ, so matrices can be either square or rectangular in shape, but never ragged.

In R, the first index always denotes rows and the second index always denotes columns. The diagram below depicts a matrix, A, with m rows and n columns and size equal to $m \times n$ "cells", with individual values denoted by $a_{i,j}$. Here we use a simpler representation than that used for vectors on page 28 above, but the same concepts apply.

In R documentation, the individual dimensions of matrices and arrays are frequently called *margins*, numbered in the same order as the indices are given. Thus, in a matrix the first margin corresponds to rows and the second one to columns.

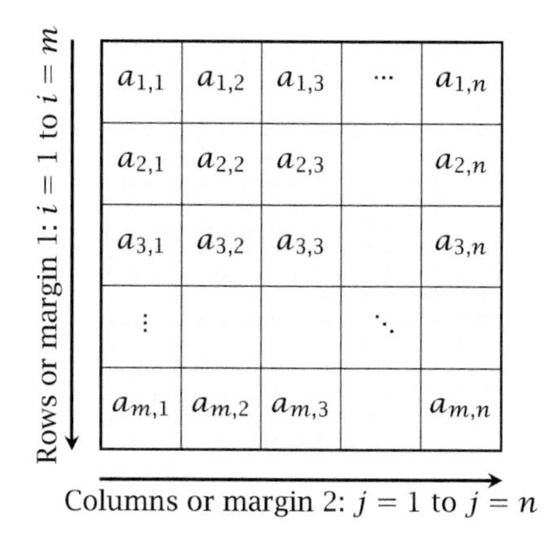

Figure 3.2
Diagram of an R matrix showing indexing of members.

In mathematical notation the same generic matrix is represented as

$$A_{m \times n} = \begin{bmatrix} a_{1,1} & a_{1,2} & \cdots & a_{1,j} & \cdots & a_{1,n} \\ a_{2,1} & a_{2,2} & \cdots & a_{2,j} & \cdots & a_{2,n} \\ \vdots & \vdots & \ddots & \vdots & & \vdots \\ a_{i,1} & a_{i,2} & \cdots & a_{i,j} & \cdots & a_{i,n} \\ \vdots & \vdots & & \vdots & \ddots & \vdots \\ a_{m,1} & a_{m,2} & \cdots & a_{m,j} & \cdots & a_{m,n} \end{bmatrix}$$

where A represents the whole matrix, $m \times n$ its dimensions, and $a_{i,j}$ its elements, with i indexing rows and j indexing columns. The lengths of the two dimensions of the matrix are given by m and n, for rows and columns.

Vectors have a single dimension, and, as described on page 28, we can query this dimension, their length, with function `length()`. Matrices have two dimensions, which can be queried individually with `ncol()` and `nrow()`, and jointly with `dim()`. As expected, `is.matrix()` can be used to query the class.

We can create a matrix using the `matrix()` or `as.matrix()` constructors. The first argument of `matrix()` must be a vector. Function `as.matrix()` is a conversion constructor, with specialisations accepting as argument objects belonging to a few other classes. The shape of the `matrix` is controlled by passing an argument to either `ncol` or `nrow`.

```
matrix(1:15, ncol = 3)
##      [,1] [,2] [,3]
## [1,]    1    6   11
## [2,]    2    7   12
## [3,]    3    8   13
## [4,]    4    9   14
## [5,]    5   10   15
matrix(1:15, nrow = 3)
##      [,1] [,2] [,3] [,4] [,5]
## [1,]    1    4    7   10   13
## [2,]    2    5    8   11   14
## [3,]    3    6    9   12   15
```

When a matrix is printed at the R console, the row and column indexes are indicated on the left and top margins, in the same way as they would be used to extract whole rows and columns.

Matrices are most useful for the storage of numeric values as matrix algebra plays an important role in statistical computations. This notwithstanding, it is possible to create matrices (and arrays) from atomic vectors of other classes such as logical or character. The only difference is the scarcity of meaningful operations other than retrieval of members using two indices.

```
matrix(letters[1:15], nrow = 3)
##      [,1] [,2] [,3] [,4] [,5]
## [1,] "a"  "d"  "g"  "j"  "m"
## [2,] "b"  "e"  "h"  "k"  "n"
## [3,] "c"  "f"  "i"  "l"  "o"
```

When a vector is converted to a matrix, R's default is to allocate the values in the vector to the matrix starting from the leftmost column, and within the column, down from the top. Once the first column is filled, the process continues from the top of the next column, as can be seen above. This order can be changed as you will discover in the playground below.

3.29 Check in the help page for the matrix constructor how to use the byrow parameter to alter the default order in which the elements of the vector are allocated to columns and rows of the new matrix.

```
help(matrix)
```

While you are looking at the help page, also consider the default number of columns and rows.

```
matrix(1:15)
```

And to start getting a sense of how to interpret error and warning messages, run the code below and make sure you understand which problem is being reported. Before executing the statement, analyse it and predict what the returned value will be. Afterwards, compare your prediction with the value actually returned.

```
matrix(1:15, ncol = 2)
```

Subscripting of matrices and arrays is consistent with that used for vectors; we only need to supply an indexing vector, or leave a blank space, for each dimension. A matrix has two dimensions, so to access an element or group of elements, we use two indices. The first index value selects rows, and the second one, columns.

```
mat1 <- matrix(1:20, ncol = 4)
mat1
##      [,1] [,2] [,3] [,4]
## [1,]    1    6   11   16
## [2,]    2    7   12   17
## [3,]    3    8   13   18
## [4,]    4    9   14   19
## [5,]    5   10   15   20
mat1[1, 2]
## [1] 6
mat1[2, 1]
## [1] 2
```

Remind yourself of how indexing of vectors works in R (see section 3.10 on page 64). We will now apply the same rules in two dimensions to extract and replace values. The first or leftmost indexing vector corresponds to rows and the second one to columns, so R uses a rows-first convention for indexing. Missing indexing vectors are interpreted as meaning *extract all rows* and *extract all columns*, respectively.

```
mat1[1, ]
## [1]  1  6 11 16
mat1[ , 1]
## [1] 1 2 3 4 5
mat1[2:3, c(1,3)]
##      [,1] [,2]
## [1,]    2   12
## [2,]    3   13
mat1[3, 4] <- 99
mat1
##      [,1] [,2] [,3] [,4]
## [1,]    1    6   11   16
## [2,]    2    7   12   17
## [3,]    3    8   13   99
## [4,]    4    9   14   19
## [5,]    5   10   15   20
mat1[4:3, 2:1] <- mat1[3:4, 1:2]
mat1
##      [,1] [,2] [,3] [,4]
## [1,]    1    6   11   16
## [2,]    2    7   12   17
## [3,]    9    4   13   99
## [4,]    8    3   14   19
## [5,]    5   10   15   20
```

Vectors are simpler than matrices, and by default when possible the "slice" extracted from a matrix is simplified into a vector by dropping one dimension. By passing `drop = FALSE`, we can prevent this.

```
is.matrix(mat1[1, ])
## [1] FALSE
is.matrix(mat1[1:2, 1:2])
## [1] TRUE

is.vector(mat1[1, ])
## [1] TRUE
is.vector(mat1[1:2, 1:2])
## [1] FALSE

is.matrix(mat1[1, , drop = FALSE])
## [1] TRUE
is.matrix(mat1[1:2, 1:2, drop = FALSE])
## [1] TRUE
```

Matrices, like vectors, can be assigned names that function as "nicknames" for indices for assignment and extraction. Matrices can have row names and/or column names.

```
colnames(mat1)
## NULL
rownames(mat1)
## NULL
colnames(mat1) <- c("a", "b", "c", "d")
mat1
##      a  b  c  d
## [1,] 1  6 11 16
## [2,] 2  7 12 17
## [3,] 9  4 13 99
## [4,] 8  3 14 19
## [5,] 5 10 15 20
rownames(mat1) <- c("A", "B", "C", "D", "E")
mat1
##   a  b  c  d
## A 1  6 11 16
## B 2  7 12 17
## C 9  4 13 99
## D 8  3 14 19
## E 5 10 15 20
mat1[c("E", "A", "D"), c("b", "a")]
##    b a
## E 10 5
## A  6 1
## D  3 8
colnames(mat1) <- NULL
mat1
##   [,1] [,2] [,3] [,4]
## A    1    6   11   16
## B    2    7   12   17
## C    9    4   13   99
## D    8    3   14   19
## E    5   10   15   20
```

Matrices can be indexed as vectors, without triggering an error or warning.

```
mat1 <- matrix(1:20, ncol = 4)
mat1
##      [,1] [,2] [,3] [,4]
## [1,]    1    6   11   16
## [2,]    2    7   12   17
## [3,]    3    8   13   18
## [4,]    4    9   14   19
## [5,]    5   10   15   20
dim(mat1)
## [1] 5 4
mat1[10]
## [1] 10
mat1[5, 2]
## [1] 10
```

The next code example demonstrates that indexing as a vector with a single index, always works column-wise even if matrix `B` was created by assigning vector elements by row.

```
mat2 <- matrix(1:20, ncol = 4, byrow = TRUE)
mat2
##      [,1] [,2] [,3] [,4]
## [1,]    1    2    3    4
## [2,]    5    6    7    8
## [3,]    9   10   11   12
## [4,]   13   14   15   16
## [5,]   17   18   19   20
dim(mat2)
## [1] 5 4
mat2[10]
## [1] 18
mat2[5, 2]
## [1] 18
```

In R, a `matrix` can have a single row, a single column, a single element, or no elements. However, in all cases, a `matrix` will have as *dimensions* attribute an `integer` vector of length two.

```
vct1 <- 1:6
dim(vct1)
## NULL

one.col.matrix <- matrix(1:6, ncol = 1)
dim(one.col.matrix)
## [1] 6 1

two.col.matrix <- matrix(1:6, ncol = 2)
dim(two.col.matrix)
## [1] 3 2

one.elem.matrix <- matrix(1, ncol = 1)
dim(one.elem.matrix)
## [1] 1 1

no.elem.matrix <- matrix(numeric(), ncol = 0)
dim(no.elem.matrix)
## [1] 0 0
```

Arrays are similar to matrices, but can have one or more dimensions (Figure 3.3). The dimensions of an array can be queried with `dim()`, similarly as with matrices. Whether an R object is an array can be found out with `is.array()`. The diagram below depicts an array, A with three dimensions giving a size equal to $l \times m \times n$, and individual values denoted by $a_{i,j,k}$.

When calling the constructor `array()`, dimensions are specified with the argument passed to parameter `dim`.

```
ary1 <- array(1:27, dim = c(3, 3, 3))
ary1
## , , 1
##
##      [,1] [,2] [,3]
## [1,]    1    4    7
## [2,]    2    5    8
## [3,]    3    6    9
##
```

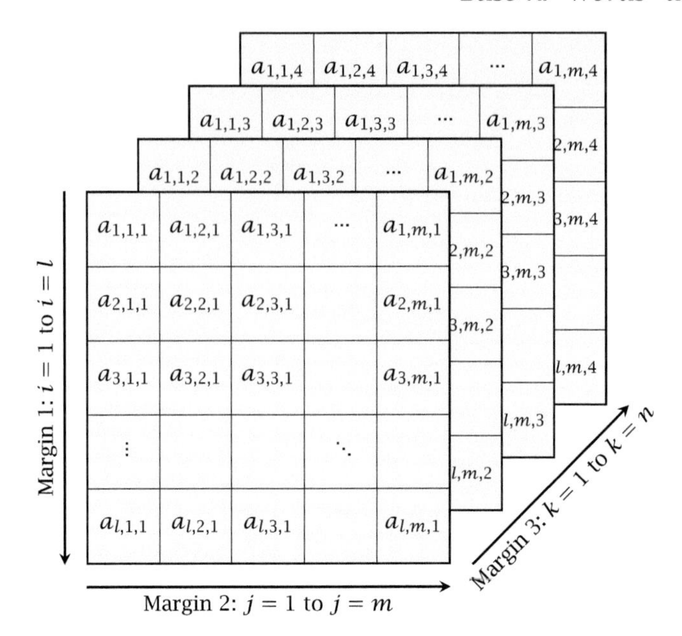

Figure 3.3
Diagram of an R array with three dimensions showing indexing of members.

```
## , , 2
##
##      [,1] [,2] [,3]
## [1,]   10   13   16
## [2,]   11   14   17
## [3,]   12   15   18
##
## , , 3
##
##      [,1] [,2] [,3]
## [1,]   19   22   25
## [2,]   20   23   26
## [3,]   21   24   27
ary1[2, 2, 2]
## [1] 14
```

In the chunk above, the length of the supplied vector is the product of the dimensions, $27 = 3 \times 3 \times 3 = 3^3$. Arrays are printed in slices, where slices across 3rd and higher dimensions are shown separately, with their corresponding indexes above each slice and the first two dimensions on the margins of the individual slices, similarly to how matrices are displayed.

3.30 How do you use indexes to extract the second element of the original vector, in each of the following matrices and arrays?

```
VCT2 <- 1:10
MAT1 <- matrix(VCT2, ncol = 2)
MAT2 <- matrix(VCT2, ncol = 2, byrow = TRUE)
MAT3 <- matrix(VCT2, nrow = 2)
MAT4 <- matrix(VCT2, nrow = 2, byrow = TRUE)
```

```
ARY1 <- array(VCT2, dim = c(5, 2))
ARY2 <- array(VCT2, dim = c(5, 2), dimnames = list(NULL, c("c1", "c2")))
ARY3 <- array(VCT2, dim = c(2, 5))
```

Be aware that vectors and one-dimensional arrays are not the same thing, while two-dimensional arrays are matrices.

1. Use the different constructors and query functions to explore this, and its consequences.
2. Convert a matrix into a vector using `as.vector()` and compare the returned values to those in the matrix. Are values extracted by columns or by rows first?

Operators and functions for matrix algebra are available in R as matrices are used in statistical algorithms. I describe below only some of these matrix-specific functions and operators. I also give examples of the use of some of the usual arithmetic operators together with objects of class `matrix`.

Recycling applies to the usual arithmetic operators when applied to matrices. This is similar to their behaviour when all operands are vectors (see page 31).

```
mat3 <- matrix(1:20, ncol = 4)
mat3 + 2
##      [,1] [,2] [,3] [,4]
## [1,]    3    8   13   18
## [2,]    4    9   14   19
## [3,]    5   10   15   20
## [4,]    6   11   16   21
## [5,]    7   12   17   22
mat3 * 0:1
##      [,1] [,2] [,3] [,4]
## [1,]    0    6    0   16
## [2,]    2    0   12    0
## [3,]    0    8    0   18
## [4,]    4    0   14    0
## [5,]    0   10    0   20
mat3 * 1:0
##      [,1] [,2] [,3] [,4]
## [1,]    1    0   11    0
## [2,]    0    7    0   17
## [3,]    3    0   13    0
## [4,]    0    9    0   19
## [5,]    5    0   15    0
```

3.31 When a `matrix` and a `vector` are operands in an arithmetic operation, how the positions of the `vector` are mapped to positions in the `matrix` affects the result of the operation. Run the code below to find out. What is the logic behind?

```
matrix(rep(1, 6)) * 1:6
```

Function `t()` transposes a matrix, by swapping columns and rows.

```
mat3
##      [,1] [,2] [,3] [,4]
## [1,]    1    6   11   16
## [2,]    2    7   12   17
## [3,]    3    8   13   18
```

```
## [4,]    4    9   14   19
## [5,]    5   10   15   20
t(mat3)
##      [,1] [,2] [,3] [,4] [,5]
## [1,]    1    2    3    4    5
## [2,]    6    7    8    9   10
## [3,]   11   12   13   14   15
## [4,]   16   17   18   19   20
```

In the examples above with the usual multiplication operator `*`, the operation described is not a matrix product, but instead, the products between individual elements of the matrix and vectors. Operators and functions implementing the operations of matrix algebra are distinct. Matrix algebra gives the rules for operations where both operands are matrices. For example, matrix multiplication is indicated by the operator `%*%`.

```
mat4 <- matrix(1:16, ncol = 4)
mat4 * mat4
##      [,1] [,2] [,3] [,4]
## [1,]    1   25   81  169
## [2,]    4   36  100  196
## [3,]    9   49  121  225
## [4,]   16   64  144  256
mat4 %*% mat4
##      [,1] [,2] [,3] [,4]
## [1,]   90  202  314  426
## [2,]  100  228  356  484
## [3,]  110  254  398  542
## [4,]  120  280  440  600
```

Function `diag()` makes it possible to easily create a diagonal matrix.

```
mat5 <- diag(4)
mat5
##      [,1] [,2] [,3] [,4]
## [1,]    1    0    0    0
## [2,]    0    1    0    0
## [3,]    0    0    1    0
## [4,]    0    0    0    1
mat4 %*% mat5
##      [,1] [,2] [,3] [,4]
## [1,]    1    5    9   13
## [2,]    2    6   10   14
## [3,]    3    7   11   15
## [4,]    4    8   12   16
```

The inverse of a matrix can be found by means of function `solve()`.

```
mat6 <- matrix(c(3, 2, 0, 1, 3, 2, 7, 2, 4), ncol = 3)
solve(mat6)
##             [,1]       [,2]       [,3]
## [1,]  0.18181818  0.2272727 -0.4318182
## [2,] -0.18181818  0.2727273  0.1818182
## [3,]  0.09090909 -0.1363636  0.1590909
```

Additional operators and functions for matrix algebra like cross-product (`crossprod()`) and Cholesky root (`chol()`) are available in base R. Packages, including 'matrixStats', provide additional functions and operators for matrices.

3.12 Factors

In data analysis and Statistics, the distinction between values measured on continuous vs. discrete *scales* is crucial. In a continuous scale, any values are in theory possible. In a discrete scale, the observations are values from a few categories.

In contrast to other statistical software in which a variable is set as continuous or discrete when defining a model to be fitted or when setting up a test, in R this distinction is based on whether the explanatory variable is `numeric` (continuous) or a `factor` (discrete). This approach makes sense because in most cases considering an explanatory variable as categorical or not, depends on the quantity stored and/or the design of the experiment or survey. In other words, being categorical is a property of the data. The order of the levels in an unordered `factor` does not affect simple calculations or the values plotted, but as we will see in chapters 7 and 9, it can affect the contrasts used by some tests of significance, and the arrangement or positions of the levels along axes and keys in plots.

In an R `factor`, values indicate discrete unordered categories, most frequently the treatments in an experiment, or categories in a survey. Factor can be created either from numerical or character vectors. The different possible values are called *levels*. Factors created with `factor()` are always unordered or categorical. R also supports `ordered` factors, created with function `ordered()` with identical user interface. The distinction, however, only affects how they are interpreted in statistical tests as discussed in chapter 7.

When using `factor()` or `ordered()` we create a factor from a vector, but this vector can be created on-the-fly and anonymous as shown in this example. When the vector is `numeric` and no labels are supplied, level labels are character strings matching the numbers. The default ordering of the levels is alphanumerical.

```
factor(x = c(1, 2, 2, 1, 2, 1, 1))
## [1] 1 2 2 1 2 1 1
## Levels: 1 2
ordered(x = c(1, 2, 2, 1, 2, 1, 1))
## [1] 1 2 2 1 2 1 1
## Levels: 1 < 2
factor(x = c(1, 2, 2, 1, 2, 1, 1), ordered = TRUE)
## [1] 1 2 2 1 2 1 1
## Levels: 1 < 2
```

When the pattern of levels is regular, it is possible to use function `gl()`, *generate levels*, to construct a factor. Nowadays, it is usual to read data into R from files in which the treatment codes are already available as character strings or numeric values, however, when we need to create a factor within R, `gl()` can save some typing. In this case, instead of passing a vector as argument, we pass a *recipe* to create it: `n` is the number of levels, `k` the number of contiguous repeats (called "replicates" in R documentation), and `length` the length of the factor to be created.

```
gl(n = 2, k = 5, labels = c("A", "B"))
##  [1] A A A A A B B B B B
## Levels: A B
```

```
gl(n = 2, k = 1, length = 10, labels = c("A", "B"))
##  [1] A B A B A B A B A B
## Levels: A B
```

It is always preferable to use meaningful labels for levels, even if R does not require it. Here the vector is stored in a variable named `my.vector`. In a real data analysis situation, in most cases, the vector would have been read from a file on disk and would be longer.

```
vct1 <- c("treated", "treated", "control", "control", "control", "treated")
factor(vct1)
## [1] treated treated control control control treated
## Levels: control treated
```

The ordering of levels is established at the time a factor is created and by default it is alphabetical. This default ordering of levels is frequently not the one needed. We can pass an argument to parameter `levels` of function `factor()` to set a different ordering of the levels.

```
factor(x = vct1, levels = c("treated", "control"))
## [1] treated treated control control control treated
## Levels: treated control
```

The labels ("names") of the levels can be set when calling `factor()`. Two vectors are passed as arguments to parameters `levels` and `labels` with levels and matching labels in the same position. The argument passed to `levels` determines the order of the levels based on their old names or values, and the argument passed to `labels` gives new names to the levels.

```
factor(x = c("a", "a", "b", "b", "b", "a"), levels = c("a", "b"), la-
bels = c("treated", "control"))
## [1] treated treated control control control treated
## Levels: treated control
```

The argument passed to `labels` can be a named vector that *maps* new labels onto the values stored in the vector passed as the argument to `x` (see named vectors and mapping on page 67).

```
factor(x = c("a", "a", "b", "b", "b", "a"), labels = c(a = "treated", b = "con-
trol"))
## [1] treated treated control control control treated
## Levels: treated control
```

In the examples above, we passed a numeric vector or a character vector as an argument for parameter `x` of function `factor()`. It is also possible to pass a `factor` as an argument to parameter `x`. This makes it possible to modify the ordering of levels or replace the labels in a factor.

```
fct1 <- factor(x = vct1)
fct1
## [1] treated treated control control control treated
## Levels: control treated
factor(x = fct1, levels = c("treated", "control"))
## [1] treated treated control control control treated
## Levels: treated control
factor(x = fct1, labels = c(control = "cooled", treated = "heated"))
## [1] heated heated cooled cooled cooled heated
## Levels: cooled heated
```

```
factor(x = fct1,
       levels = c("treated", "control"),
       labels = c("heated", "cooled"))
## [1] heated heated cooled cooled cooled heated
## Levels: heated cooled
```

Merging factor levels. We use `factor()` as shown below, setting the same label for the levels we want to merge.

```
fct2 <- gl(4, 3, labels = c("A", "F", "B", "Z"))
fct2
##  [1] A A A F F F B B B Z Z Z
## Levels: A F B Z
factor(fct2,
       levels = c("A", "B", "F", "Z"),
       labels = c("A", "B", "C", "C"))
##  [1] A A A C C C B B B C C C
## Levels: A B C
```

3.32 Edit the code in the chunk above to use only a named vector for `labels` instead of separate vectors passed to `levels` and `labels`.

We can use indexing on factors in the same way as with vectors. In the next example, we use a test returning a logical vector to extract all “controls”. We use function `levels()` to look at the levels of the factors, as with vectors, `lengtgh()` to query the number of values stored.

```
fct1
## [1] treated treated control control control treated
## Levels: control treated
levels(fct1)
## [1] "control" "treated"
length(fct1)
## [1] 6
fct1.control <- fct1[fct1 == "control"]
fct1.control
## [1] control control control
## Levels: control treated
levels(fct1.control) # same as in my.factor
## [1] "control" "treated"
length(fct1.control) # shorter than my.factor
## [1] 3
```

How to drop unused levels in a factor?
It can be seen above that subsetting does not drop unused factor levels. Constructor function `factor()` can be used to explicitly drop the unused factor levels.

```
fct1.control <- factor(fct1.control)
levels(fct1.control) # the unused level was dropped
## [1] "control"
```

How to convert a factor into a vector with matching values?
This operation is not obvious, specially when the factor was created from a `numeric` vector.

```
vct3 <- rep(3:5, 4)
vct3
## [1] 3 4 5 3 4 5 3 4 5 3 4 5
fct3 <- factor(vct3)
fct3
## [1] 3 4 5 3 4 5 3 4 5 3 4 5
## Levels: 3 4 5
as.numeric(fct3)
## [1] 1 2 3 1 2 3 1 2 3 1 2 3
as.numeric(as.character(fct3))
## [1] 3 4 5 3 4 5 3 4 5 3 4 5
```

Why is a double conversion needed? Internally, factor values are stored as running integers starting from one, each distinct integer value corresponding to a level. These underlying integer values are returned by `as.numeric()` when applied to a factor. The labels of the factor levels are always stored as character strings, even when these characters are digits. In contrast to `as.numeric()`, `as.character()` returns the character labels of the levels for each of the values stored in the factor. If these character strings represent numbers, they can be converted, in a second step, using `as.numeric()` into the original numeric values. Use of `class` and `mode` is described on section 3.8 on page 59, and `str()` on page 91.

```
class(fct3)
## [1] "factor"
mode(fct3)
## [1] "numeric"
str(fct3)
## Factor w/ 3 levels "3","4","5": 1 2 3 1 2 3 1 2 3 1 ...
```

3.33 Create a factor with levels labelled with words. Create another factor with the levels labelled with the same words, but ordered differently. After this convert both factors to numeric vectors using `as.numeric()`. Explain why the two numeric vectors differ or not from each other.

Safely reordering and renaming factor levels. The simplest approach is to use `factor()` and its `levels` parameter as shown on page 80. In these more advanced examples, we use `levels()` to retrieve the names of the levels from the factor itself to protect from possible bugs due to typing mistakes, or for changes in the naming conventions used.

Reverse previous order using `rev()`.

```
fct4 <- factor(c("treated", "treated", "control", "control", "control", "treated"))
levels(fct4)
## [1] "control" "treated"
fct4 <- factor(fct4, levels = rev(levels(fct4)))
levels(fct4)
## [1] "treated" "control"
```

Sort in decreasing order, i.e., opposite to default.

```
fct5 <- factor(fct4,
               levels = sort(levels(fct4), decreasing = TRUE))
levels(fct5)
## [1] "treated" "control"
```

Alter ordering using subscripting; especially useful with three or more levels.

```
fct6 <- factor(fct4, levels = levels(fct4)[c(2, 1)])
levels(fct6)
## [1] "control" "treated"
```

Reordering the levels of a factor based on summary quantities from data stored in a numeric vector is very useful, especially when plotting. Function `reorder()` can be used in this case. It defaults to using `mean()` for summaries, but other suitable summary functions, such as `median()` can be supplied in its place.

```
fct7 <- gl(2, 5, labels = c("A", "B"))
vct4 <- c(5.6, 7.3, 3.1, 8.7, 6.9, 2.4, 4.5, 2.1, 1.4, 2.0)
fct7
##  [1] A A A A A B B B B B
## Levels: A B
fct7ord <- reorder(fct7, vct4)
levels(fct7ord)
## [1] "B" "A"
fct7rev <- reorder(fct7, -vct4) # a simple trick: change sign
levels(fct7rev)
## [1] "A" "B"
```

In the last statement, using the unary negation operator, which is vectorised, allows us to easily reverse the ordering of the levels, while still using the default function, `mean()`, to summarise the data.

3.34 **Reordering factor values.** It is possible to arrange the values stored in a factor either alphabetically according to the labels of the levels or according to the order of the levels. (The use of `rep()` is explained on page 30.)

```
# gl() keeps order of levels
FCT1 <- gl(4, 3, labels = c("A", "F", "B", "Z"))
FCT1
as.integer(FCT1)

# factor() orders levels alphabetically
FCT2 <- factor(rep(c("A", "F", "B", "Z"), times = rep(3, times = 4))) # nes-
ted calls
FCT2
as.integer(FCT2)
levels(FCT2)[as.integer(FCT2)]
```

We see above that the integer values by which levels in a factor are stored, are equivalent to indices or "subscripts" referencing the vector of labels. Function `sort()` operates on the values' underlying integers and sorts according to the order of the levels while `order()` operates on the values' labels and returns a vector of indices that arrange the values alphabetically.

```
sort(FCT2)
FCT2[order(FCT2)]
FCT2[order(as.integer(FCT2))]
```

Run the examples in the chunk above and work out why the results differ.

Factors encode levels as `integer` values in a vector. In many cases, statistical computations, require the same information to be encoded as binary values using multiple *dummy variables*. Factors are much friendlier for the user to manage. They are converted into the equivalent dummy variables when a model formula is translated into a *model matrix*. This is handled transparently by most functions implementing fitting of statistical models to data (see sections 7.8 and 7.13 on pages 199 and 226).

3.13 Further Reading

For further reading on the aspects of R discussed in the current chapter, I suggest the book *The Art of R Programming: A Tour of Statistical Software Design* (Matloff 2011).

4

Base R: "Collective Nouns"

> The information that is available to the computer consists of a selected set of *data* about the real world, namely, that set which is considered relevant to the problem at hand, that set from which it is believed that the desired results can be derived. The data represent an abstraction of reality...
>
> Niklaus Wirth
> *Algorithms + Data Structures = Programs*, 1976

4.1 Aims of This Chapter

Data set organisation and storage is one of the keys to efficient data analysis. How to keep together all the information that belongs together, say all measurements from an experiment and corresponding metadata such as treatments applied and/or dates. The title "collective nouns" is based on the idea that a data set is a collection of data objects.

In this chapter, you will familiarise with how data sets are usually managed in R. I use both abstract examples to emphasise the general properties of data sets and the R classes available for their storage and a few more specific examples to exemplify their use in a more concrete way. While in chapter 3 the focus was on atomic data types and objects, like vectors, useful for the storage of collections of values of a given type, like numbers, in the present chapter the focus is on the storage within a single object of heterogeneous data, such as a combination of factors, and character and numeric vectors. Broadly speaking, heterogeneous *data containers*.

To describe the structure of R objects I use diagrams similar to those in the previous chapter.

DOI: 10.1201/9781003404187-4

4.2 Data from Surveys and Experiments

The data we plot, summarise, and analyse in R, in most cases, originate from measurements done as part of experiments or surveys. Data collected mechanically from user interactions with websites or by crawling through internet content originate from a statistical perspective from surveys. The value of any data comes from knowing their origin, say treatments applied to plants, or the country from where website users connect; sometimes several properties are of interest to describe the origin of the data and in other cases observations consist in the measurement of multiple properties on each subject under study. Consequently, all software designed for data analysis implements ways of dealing with data sets as a whole both during storage and when passing them as arguments to functions. A data set is a usually heterogeneous collection of data with related information.

In R, lists are the most flexible type of objects useful for storing whole data sets. In most cases, we do not need this much flexibility, so rectangular collections of observations are most frequently stored in a variation upon lists called data frames. These objects can have as their members the vectors and factors described in chapter 3.

Any R object can have attributes, allowing objects to carry along additional bits of information. Some like comments are part of R and aimed at storage of ancillary information or metadata by users. Other attributes are used internally by R and finally users can store arbitrary ancillary data using attributes created *ad hoc*.

4.3 Lists

In R, `list` objects are in several respects similar the vectors described in chapter 3 but differently to vectors, the members they contain can be heterogeneous, i.e., different members of the same list can belong to different classes. In addition, while the member elements of a vector must be *atomic* values like numbers or character strings, any R object can be a list member including other lists.

In R, the members of a list can be considered as following a sequence, and accessible through numerical indexes, the same as the members of vectors. Members of a list as well as members of a vector can be named, and retrieved (indexed) through their names. In practice, named lists are more frequently used than named vectors. R lists are created, or constructed, with function `list()` similarly as vectors are constructed with function `c()`.

R lists can have as members not only objects storing data on observations and categories, but also function definitions, model formulas, unevaluated expressions, matrices, arrays, and objects of user-defined classes.

List and list-like objects are widely used in R because they make it possible to keep, for example, the data, instructions for operations, and results from operations together in a single R object that can be saved, copied, etc. as a unit. This avoids the proliferation of multiple disconnected objects with their interrelations being encoded only by their names, or even worse in separate notes or even in a person's memory—all approaches that are error-prone. Model fit functions described in chapter 7 are good examples of this approach. Objects used to store the instructions to build plots with multiple layers as described in chapter 9 are also good examples.

Our first list has as its members three different vectors, each one belonging to a different class: `numeric`, `character` and `logical`. The three vectors also differ in their length: 6, 1, and 2, respectively.

```
lst1 <- list(x = 1:3, y = "ab", z = c(TRUE, FALSE))
```

```
str(lst1)
## List of 3
##  $ x: int [1:3] 1 2 3
##  $ y: chr "ab"
##  $ z: logi [1:2] TRUE FALSE
names(lst1)
## [1] "x" "y" "z"
```

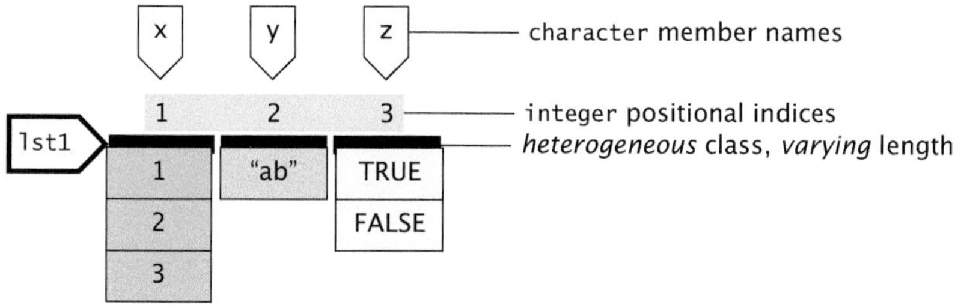

It is best to use informative names for accessing `list` members, as their members are heterogenous, usually containing loosely related/connected data. Names make code easier to understand and mistakes more visible. Using names also makes code more robust to future changes in the position of list members in lists created upstream of our own R code. Below, we use both positional indices and names to highlight the similarities between lists and vectors.

Lists can behave as vectors with heterogeneous elements as members, as we will describe next. Lists can also be nested, so tree-like structures are also possible (see section 4.3.2 on page 91).

How to create an empty list?

In the same way as `numeric()` by default creates a `numeric` vector of length zero, `list()` by default creates a `list` object with no members.

```
list()
## list()
```

4.3.1 Member extraction, deletion and insertion

In section 3.10 on page 64, we saw that the extraction operator `[ ]` applied to a vector, returns a vector, longer or shorter, possibly of length one, or even length zero. Similarly, applying operator `[ ]` to a list returns a list, possibly of different length: `lst1["x"]` or `lst[1]` return a list containing only one member, the numeric vector stored at the first position of `lst1`. In the last statement in the chunk below, `lst1[c(1, 3)]` returns a list of length two as expected.

```
lst1["x"]
## $x
## [1] 1 2 3
lst1[1]
## $x
## [1] 1 2 3
lst1[c(1, 3)]
## $x
## [1] 1 2 3
##
## $z
## [1]  TRUE FALSE
```

As with vectors negative positional indices remove members instead of extracting them. See page 90 for a safer approach to the deletion of list members.

```
lst1[-1]
## $y
## [1] "ab"
##
## $z
## [1]  TRUE FALSE
lst1[c(-1, -3)]
## $y
## [1] "ab"
```

Using operator `[[ ]]` (double square brackets) for indexing a list extracts the element stored in the list, in its original mode. In the example below, `lst1[["x"]]` and `lst1[[1]]` return a numeric vector. We might say that extraction operator `[[ ]]` reaches "deeper" into the list than operator `[ ]`. Operator `$`, used in the second statement below, provides a shorthand notation, equivalent to calling `[[ ]]` with a single constant `character` value as argument.

```
lst1$x
## [1] 1 2 3
lst1[["x"]]
## [1] 1 2 3
lst1[[1]]
## [1] 1 2 3
```

We mentioned above that indexing by name can be done either with double square brackets, `[[ ]]`, or with `$`. Operators `[ ]` and `[[ ]]` work like normal R functions, accepting as arguments passed to them both constant values and variables for indexing. In contrast, `$` mainly intended for use when typing at the console, accepts only bare member names on its *rhs*. With `[[ ]]`, the name of the variable or column is given as a character string, enclosed in quotation marks, or as a variable with mode `character`. A number as a positional index is also accepted.

```
lst1a <- list(abcd = 123, xyzw = 789)
lst1a[[1]]
## [1] 123
lst1a[["abcd"]]
## [1] 123
vct1 <- "abcd"
lst1a[[vct1]]
## [1] 123
```

When using `$`, the name is entered as a constant, without quotation marks, and cannot be a variable or a number.

```
lst1a$abcd
## [1] 123
lst1a$ab
## [1] 123
lst1a$a
## [1] 123
```

Both in the case of lists and data frames (see section 4.4 on page 94), when using double square brackets, by default an exact match is required between the name in the object and the name used for indexing. In contrast, with `$`, an unambiguous partial match is silently accepted. For interactive use, partial matching decreases the extent of the text typed at the console. However, in scripts, and especially R code in packages, it is best to avoid the use of `$` as partial matching to a wrong variable present at a later time, e.g., when someone else revises the script, misdirected partial matching can lead to difficult-to-diagnose errors.

In addition, as `$` is implemented by first attempting a match to the name and then calling `[[ ]]`, using `$` for indexing can result in slightly slower performance compared to using `[[ ]]`. It is possible to set R option `warnPartialMatchDollar` so that partial matching triggers a warning when using `$` to extract a member, which can be very useful when debugging.

```
is.vector(lst1[1])
## [1] TRUE
is.list(lst1[1])
## [1] TRUE
is.vector(lst1[[1]])
## [1] TRUE
is.list(lst1[[1]])
## [1] FALSE
```

The two extraction operators can be used together as shown below, with `lst1[[1]]` extracting the vector from `lst1` and `[3]` extracting the member at position 3 of the vector.

```
lst1[[1]][3]
## [1] 3
```

Extraction operators can be used on the *lhs* as well as on the *rhs* of an assignment, and lists can be empty, i.e., be of length zero. The example below makes use of this to build a list step by step.

```
lst2 <- list()
lst2[["x"]] <- 1:3
lst2[["y"]] <- "ab"
lst2[["z"]] <- c(TRUE, FALSE)
```

4.1 Compare `lst2` to `lst1`, used for the examples above. Then run the code below and compare them again. Try to understand why `lst2` has changed as it did. Pay also attention to possible changes to the members' names.

```
lst2[["y"]] <- lst2[["x"]]
```

Lists, as usually defined in languages like C, are based on pointers to memory locations, with pointers stored at each node. These pointers chain or link the different member nodes (this allows, for example, sorting of lists in place by modifying the pointers). In such implementations, indexing by position is not possible, or at least requires "walking" down the list, node by node. R does not implement pointers to "addresses", or locations, in memory. In R, `list` members can be accessed through positional indexes or member names, similarly to vector members. Of course, as with vectors, insertions and deletions in the middle of a list, shift the position of members, and change which member is pointed at by indexes for positions past the modified location. The names, in contrast, remain valid.

```
list(a = 1, b = 2, c = 3)[-2]
## $a
## [1] 1
##
## $c
## [1] 3
```

Three frequent operations on lists are concatenation, insertions, and deletions. The same functions as with vectors are used: `c()`, to concatenate, and `append()`, to append and insert. Lists can be combined only with other lists, otherwise, these operations work as with vectors (see pages 28–30).

```
lst3 <- append(lst1, list(yy = 1:10, zz = letters[5:1]), after = 2)
lst3
## $x
## [1] 1 2 3
##
## $y
## [1] "ab"
##
## $yy
##  [1]  1  2  3  4  5  6  7  8  9 10
##
## $zz
## [1] "e" "d" "c" "b" "a"
##
## $z
## [1]  TRUE FALSE
```

To delete a member from a list, we assign `NULL` to it.

```
lst1$y <- NULL
lst1
## $x
## [1] 1 2 3
```

```
##
## $z
## [1]  TRUE FALSE
```

To investigate the members contained in a list, function `str()` (*structure*), used above, is convenient, especially when lists have many members. Structure formats lists more compactly than `print()` applied directly to a list.

```
print(lst1)
## $x
## [1] 1 2 3
##
## $z
## [1]  TRUE FALSE
str(lst1)
## List of 2
##  $ x: int [1:3] 1 2 3
##  $ z: logi [1:2] TRUE FALSE
```

4.3.2 Nested lists

Lists can be nested, i.e., lists of lists can be constructed to an arbitrary depth. In the example below, `lst4` and `lst5` are members of `lst6`, i.e., `lst4` and `lst5` are nested within `lst6`.

```
lst4 <- list("a", "aa", 10)
lst5 <- list("b", TRUE)
lst6 <- list(A = lst4, B = lst5) # nested
str(lst6)
## List of 2
##  $ A:List of 3
##   ..$ : chr "a"
##   ..$ : chr "aa"
##   ..$ : num 10
##  $ B:List of 2
##   ..$ : chr "b"
##   ..$ : logi TRUE
```

A nested list can alternatively be constructed within a single statement in which several member lists are created. Here we combine the first three statements in the earlier chunk into a single one.

```
lst7 <- list(A = list("a", "aa", 10), B = list("b", TRUE))
str(lst7)
## List of 2
##  $ A:List of 3
##   ..$ : chr "a"
##   ..$ : chr "aa"
##   ..$ : num 10
##  $ B:List of 2
##   ..$ : chr "b"
##   ..$ : logi TRUE
```

A list can contain a combination of `list` and `vector` members.

```
lst8 <- list(A = list("a", "aa", 10),
             B = list("b", TRUE),
```

```
             C = c(1, 3, 9),
             D = 4321)
str(lst8)
## List of 4
##  $ A:List of 3
##   ..$ : chr "a"
##   ..$ : chr "aa"
##   ..$ : num 10
##  $ B:List of 2
##   ..$ : chr "b"
##   ..$ : logi TRUE
##  $ C: num [1:3] 1 3 9
##  $ D: num 4321
```

The logic behind the extraction of members of nested lists using indexing is the same as for simple lists, but applied recursively—e.g., `lst7[[2]]` extracts the second member of the outermost list, which is another list. As, this is a list, its members can be extracted using again the extraction operator: `lst7[[2]][[1]]`. It is important to remember that these concatenated extraction operations are written so that the leftmost operator is applied to the outermost list.

The example above uses the `[[ ]]` operator, but the left-to-right precedence also applies to concatenated calls to `[ ]` and to calls combining both operators.

4.2 What do you expect each of the statements below to return? *Before running the code*, predict what value and of which mode each statement will return. You may use implicit or explicit calls to `print()`, or calls to `str()` to visualise the structure of the different objects.

```
LST9 <- list(A = list("a", "aa", "aaa"), B = list("b", "bb"))
# str(LST9)
LST9[2:1]
LST9[1]
LST9[[1]][2]
LST9[[1]][[2]]
LST9[2]
LST9[2][[1]]
```

When dealing with deep lists, it is sometimes useful to limit the number of levels of nesting returned by `str()` by passing a `numeric` argument to parameter `max.levels`.

```
str(lst8, max.level = 1)
## List of 4
##  $ A:List of 3
##  $ B:List of 2
##  $ C: num [1:3] 1 3 9
##  $ D: num 4321
```

Sometimes we need to flatten a list, or a nested structure of lists within lists. Function `unlist()` is what should be normally used in such cases.

The list `lst10` is a nested system of lists, but all the "terminal" members are character strings. In other words, terminal nodes are all of the same `mode`, allowing the list to be "flattened" into a character vector.

```
lst10 <- list(A = list("a", "aa", "aaa"), B = list("b", "bb"))
vct1 <- unlist(lst10)
vct1
##    A1    A2    A3    B1    B2
##   "a"  "aa" "aaa"   "b"  "bb"
is.list(lst10)
## [1] TRUE
is.list(vct1)
## [1] FALSE
mode(lst10)
## [1] "list"
mode(vct1)
## [1] "character"
names(lst10)
## [1] "A" "B"
names(vct1)
## [1] "A1" "A2" "A3" "B1" "B2"
```

The returned value is a vector with named member elements. We use function `str()` to figure out how this vector relates to the original list. The names, always of mode character, are based on the names of list elements when available, while characters depicting positions as numbers are used for anonymous nodes. We can access the members of the vector either through numeric indexes or names.

```
str(vct1)
##  Named chr [1:5] "a" "aa" "aaa" "b" "bb"
##  - attr(*, "names")= chr [1:5] "A1" "A2" "A3" "B1" ...
vct1[2]
##   A2
## "aa"
vct1["A2"]
##   A2
## "aa"
```

4.3 Function `unlist()` has two additional parameters, with default argument values, which we did not modify in the example above. These parameters are `recursive` and `use.names`, both of them expecting a `logical` value as an argument. Modify the statement `c.vec <- unlist(c.list)`, by passing `FALSE` as an argument to these two parameters, in turn, and in each case, study the value returned and how it differs with respect to the one obtained above.

Function `unname()` can be used to remove names safely—i.e., without risk of altering the mode or class of the object.

```
unname(vct1)
## [1] "a"   "aa"  "aaa" "b"   "bb"
unname(lst10)
## [[1]]
## [[1]][[1]]
## [1] "a"
##
## [[1]][[2]]
## [1] "aa"
##
```

```
## [[1]][[3]]
## [1] "aaa"
##
##
## [[2]]
## [[2]][[1]]
## [1] "b"
##
## [[2]][[2]]
## [1] "bb"
```

4.4 Data Frames

Data frames are a special type of list, in which all members have the same length, giving origin to a matrix-like object, in which columns can belong to different classes. Most commonly the member "columns" are vectors or factors, but they can also be matrices with the same number of rows as the enclosing data frame, or lists with the same number of members as rows in the enclosing data frame.

Data frames are central to most data manipulation and analysis procedures in R. They are commonly used to store observations, with `numeric` columns holding data for continuous variables and `factor` columns data for categorical variables. Binary variables can be stored in `logical` columns. Text data can be stored in `character` columns. Date and time can be stored in columns of specific classes, such as `POSIXct`. In the diagram below, column `treatment` is a factor with two levels encoding two conditions, `hot` and `cold`. Columns `height` and `weight` are numeric vectors containing measurements.

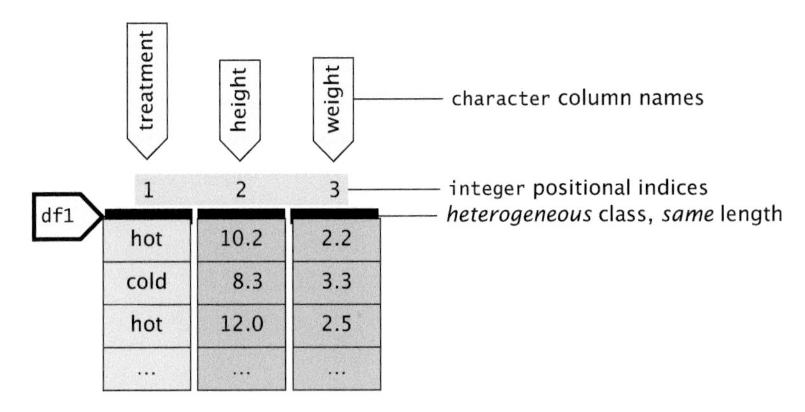

Data frames are created with constructor function `data.frame()` with a syntax similar to that used for lists.

```
df1 <- data.frame(treatment = factor(rep(c("hot", "cold"), 3)),
                  height = c(10.2, 8.3, 12.0, 9.0, 11.2, 8.7),
                  weight = c(2.2, 3.3, 2.5, 2.8, 2.4, 3.0))
df1
##   treatment height weight
```

```
## 1       hot  10.2    2.2
## 2      cold   8.3    3.3
## 3       hot  12.0    2.5
## 4      cold   9.0    2.8
## 5       hot  11.2    2.4
## 6      cold   8.7    3.0
colnames(df1)
## [1] "treatment" "height"    "weight"
rownames(df1)
## [1] "1" "2" "3" "4" "5" "6"
str(df1)
## 'data.frame': 6 obs. of  3 variables:
##  $ treatment: Factor w/ 2 levels "cold","hot": 2 1 2 1 2 1
##  $ height   : num  10.2 8.3 12 9 11.2 8.7
##  $ weight   : num  2.2 3.3 2.5 2.8 2.4 3
class(df1)
## [1] "data.frame"
mode(df1)
## [1] "list"
is.data.frame(df1)
## [1] TRUE
is.list(df1)
## [1] TRUE
```

We can see above that when printed each row of a `data.frame` is preceded by a row name. Row names are character strings, just like column names. The `data.frame()` constructor adds by default row names representing running numbers. Default row names are rarely of much use, except to track insertions and deletions of rows during debugging.

4.4 As the expectation is that all member variables (or "columns") have equal length, if vectors of different lengths are supplied as arguments, the shorter vector(s) is/are recycled, possibly several times, until the required full length is reached, as shown below for `treatment`.

```
df2 <- data.frame(treatment = factor(c("hot", "cold")),
                  height = c(10.2, 8.3, 12.0, 9.0, 11.2, 8.7),
                  weight = c(2.2, 3.3, 2.5, 2.8, 2.4, 3.0))
```

Are `df1` crated above and `df2` created here equal?

With function `class()` we can query the class of an R object (see section 3.8 on page 59). As we saw in the previous chunk, `list` and `data.frame` objects belong to two different classes. However, their `mode` is the same. Consequently, data frames inherit the methods and characteristics of lists, as long as they have not been hidden by new ones defined for data frames (for an explanation of *methods*, see section 6.3 on page 176).

Extraction of individual member variables or "columns" can be done like in a list with operators `[[ ]]` and `$` (see call-out in 88).

```
df1$height
## [1] 10.2  8.3 12.0  9.0 11.2  8.7
df1[["height"]]
## [1] 10.2  8.3 12.0  9.0 11.2  8.7
```

```
df1[[2]]
## [1] 10.2  8.3 12.0  9.0 11.2  8.7
class(df1[["height"]])
## [1] "numeric"
```

In the same way as with lists, we can add member variables to data frames. Recycling takes place if needed.

```
df1$x2 <- 6:1
df1[["x3"]] <- "b"
str(df1)
## 'data.frame':	6 obs. of  5 variables:
##  $ treatment: Factor w/ 2 levels "cold","hot": 2 1 2 1 2 1
##  $ height   : num  10.2 8.3 12 9 11.2 8.7
##  $ weight   : num  2.2 3.3 2.5 2.8 2.4 3
##  $ x2       : int  6 5 4 3 2 1
##  $ x3       : chr  "b" "b" "b" "b" ...
```

4.5 We have added two columns to the data frame, and in the case of column `x3` recycling took place. This is where lists and data frames differ substantially in their behaviour. In a data frame, although class and mode can be different for different member variables (columns), they are required to be vectors or factors of the same length (or a matrix with the same number of rows, or a list with the same number of members). In the case of lists, there is no such requirement, and recycling never takes place when adding a member. Compare the values returned below for `LST1`, to those in the example above for `df1`.

```
LST1 <- list(x = 1:6, y = "a", z = c(TRUE, FALSE))
str(LST1)
## List of 3
##  $ x: int [1:6] 1 2 3 4 5 6
##  $ y: chr "a"
##  $ z: logi [1:2] TRUE FALSE
LST1$x2 <- 6:1
LST1$x3 <- "b"
str(LST1)
## List of 5
##  $ x : int [1:6] 1 2 3 4 5 6
##  $ y : chr "a"
##  $ z : logi [1:2] TRUE FALSE
##  $ x2: int [1:6] 6 5 4 3 2 1
##  $ x3: chr "b"
```

How to create an empty data frame?

In the same way as `numeric()` creates a `numeric` vector of length zero, `data.frame()` by default creates a `data.frame` with zero rows and no columns.

```
data.frame()
## data frame with 0 columns and 0 rows
```

How to make a list of data frames?

We create a list of data frames in the same way as we create a nested list of lists, or in fact of a list of any other R objects. See section 4.3.2 on page 91.

```
list(df1, df2)
## [[1]]
##   treatment height weight x2 x3
## 1       hot   10.2    2.2  6  b
## 2      cold    8.3    3.3  5  b
## 3       hot   12.0    2.5  4  b
## 4      cold    9.0    2.8  3  b
## 5       hot   11.2    2.4  2  b
## 6      cold    8.7    3.0  1  b
##
## [[2]]
##   treatment height weight
## 1       hot   10.2    2.2
## 2      cold    8.3    3.3
## 3       hot   12.0    2.5
## 4      cold    9.0    2.8
## 5       hot   11.2    2.4
## 6      cold    8.7    3.0
```

? How to add a new column to a data frame (to the front and end)?
In the same way as we can assign a new member to a list using the extraction operator `[[ ]]`, we can add a new column to a data frame (see page 89). In this case, if the column name does not already exist, the assigned vector or factor is appended as the last column (no recycling applied to short vectors or factors unless of length one).

```
DF1 <- data.frame(A = 1:5, B = factor(5:1))
DF1[["C"]] <- 11:15
DF1
##   A B  C
## 1 1 5 11
## 2 2 4 12
## 3 3 3 13
## 4 4 2 14
## 5 5 1 15
```

To add a column at the front, we can use function `cbind()` (column bind).

```
DF2 <- data.frame(A = 1:5, B = factor(5:1))
cbind(C = 11:15, DF2)
##    C A B
## 1 11 1 5
## 2 12 2 4
## 3 13 3 3
## 4 14 4 2
## 5 15 5 1
```

Being two-dimensional and rectangular in shape, data frames, in relation to indexing and dimensions, behave similarly to a matrix. They have two margins, rows, and columns, and, thus, two indices are used to indicate the location of a member "cell". We provide some examples here, but please consult section 3.10 on page 64 and section 3.11 on page 70 for additional details.

Matrix-like notation allows simultaneous extraction from multiple columns, which is not possible with lists. The value returned is in most cases a "smaller" data frame as in this example.

```
df1[2:3, 1:2]
##   treatment height
## 2      cold    8.3
## 3       hot   12.0
# first column, df1[[1]] preferred
df1[ , 1]
## [1] hot  cold hot  cold hot  cold
## Levels: cold hot
# first column, df1[["x"]] or df1$x preferred
df1[ , "treatment"]
## [1] hot  cold hot  cold hot  cold
## Levels: cold hot
# first row
df1[1, ]
##   treatment height weight x2 x3
## 1       hot   10.2    2.2  6  b
# first two rows of the third and fourth columns
df1[1:2, c(FALSE, FALSE, TRUE, TRUE, FALSE)]
##   weight x2
## 1    2.2  6
## 2    3.3  5
# the rows for which comparison is true
df1[df1$treatment == "hot" , ]
##   treatment height weight x2 x3
## 1       hot   10.2    2.2  6  b
## 3       hot   12.0    2.5  4  b
## 5       hot   11.2    2.4  2  b
# the heights > 8
df1[df1$height > 8, "height"]
## [1] 10.2  8.3 12.0  9.0 11.2  8.7
```

As explained earlier for vectors (see section 3.10 on page 64), indexing can be present both on the right- and left-hand sides of an assignment, allowing the replacement of both individual values and rectangular regions.

The next few examples do assignments to "cells" of `df1`, either to one whole column, or individual values. The last statement in the chunk below copies a number from one location to another by using indexing of the same data frame both on the right side and left side of the assignment.

```
df1[1, 2] <- 99
df1
##   treatment height weight x2 x3
## 1       hot   99.0    2.2  6  b
## 2      cold    8.3    3.3  5  b
## 3       hot   12.0    2.5  4  b
## 4      cold    9.0    2.8  3  b
## 5       hot   11.2    2.4  2  b
## 6      cold    8.7    3.0  1  b
df1[ , 2] <- -99
df1
##   treatment height weight x2 x3
## 1       hot    -99    2.2  6  b
## 2      cold    -99    3.3  5  b
## 3       hot    -99    2.5  4  b
```

```
## 4      cold    -99    2.8  3  b
## 5       hot    -99    2.4  2  b
## 6      cold    -99    3.0  1  b
df1[["height"]] <- c(10, 12)
df1
##   treatment height weight x2 x3
## 1       hot     10    2.2  6  b
## 2      cold     12    3.3  5  b
## 3       hot     10    2.5  4  b
## 4      cold     12    2.8  3  b
## 5       hot     10    2.4  2  b
## 6      cold     12    3.0  1  b
df1[1, 2] <- df1[6, 3]
df1
##   treatment height weight x2 x3
## 1       hot      3    2.2  6  b
## 2      cold     12    3.3  5  b
## 3       hot     10    2.5  4  b
## 4      cold     12    2.8  3  b
## 5       hot     10    2.4  2  b
## 6      cold     12    3.0  1  b
df1[3:6, 2] <- df1[6, 3]
df1
##   treatment height weight x2 x3
## 1       hot      3    2.2  6  b
## 2      cold     12    3.3  5  b
## 3       hot      3    2.5  4  b
## 4      cold      3    2.8  3  b
## 5       hot      3    2.4  2  b
## 6      cold      3    3.0  1  b
```

Similarly as with matrices, if we extract a single column from a data frame using matrix-like indexing, it is by default simplified into a vector or factor, i.e., the column-dimension is dropped. By passing `drop = FALSE`, we can prevent this. Contrary to matrices, rows are not simplified in the case of data frames.

```
is.data.frame(df1[1, ])
## [1] TRUE
is.data.frame(df1[ , 2])
## [1] FALSE
is.data.frame(df1[ , "treatment"])
## [1] FALSE
is.data.frame(df1[1:2, 2:3])
## [1] TRUE
is.vector(df1[1, ])
## [1] FALSE
is.vector(df1[ , 2])
## [1] TRUE
is.factor(df1[ , "treatment"])
## [1] TRUE
is.vector(df1[1:2, 2:3])
## [1] FALSE
```

```
is.data.frame(df1[ , 1, drop = FALSE])
## [1] TRUE
is.data.frame(df1[ , "treatment", drop = FALSE])
## [1] TRUE
```

In contrast to matrices and data frames, the extraction operator `[ ]` of tibbles—defined in package 'tibble'—never simplifies returned one-column tibbles into vectors (see section 8.4.2 on page 247 for details on the differences between data frames and tibbles).

Usually data frames are created from lists or by passing individual vectors and factors to the constructors. It is also possible to construct data frames starting from matrices, other data frames and named vectors, in which case, the identity function `I()` can be used to protect them from interpretation by the `data.frame()` constructor. In these cases, additional nuances become important. The details are well described in `help(data.frame)`.

With a named numeric vector, and a factor as arguments, the names are moved from the vector to the rows of the data frame!

```
vct1 <- c(one = 1, two = 2, three = 3, four = 4)
fct1 <- as.factor(c(1, 2, 3, 2))
df1 <- data.frame(fct1, vct1)
df1
##       fct1 vct1
## one      1    1
## two      2    2
## three    3    3
## four     2    4
df1$vct1
## [1] 1 2 3 4
```

If the vector is protected with R's identity function `I()` the names are not moved as can be seen by extracting the column `vct1` from data frame `df2`.

```
df2 <- data.frame(fct1, I(vct1))
df2
##       fct1 vct1
## one      1    1
## two      2    2
## three    3    3
## four     2    4
df2$vct1
##   one   two three  four
##     1     2     3     4
```

With a matrix instead of a vector, the matrix is split into separate columns in the data frame. If the matrix has no column names, new ones are created.

```
mat1 <- matrix(1:12, ncol = 3)
df4 <- data.frame(fct1, mat1)
```

```
df4
##   fct1 X1 X2 X3
## 1    1  1  5  9
## 2    2  2  6 10
## 3    3  3  7 11
## 4    2  4  8 12
```

If the matrix is protected with function `I()`, it is not split, and the whole matrix becomes a column in the data frame.

```
df5 <- data.frame(fct1, I(mat1))
df5
##   fct1 mat1.1 mat1.2 mat1.3
## 1    1      1      5      9
## 2    2      2      6     10
## 3    3      3      7     11
## 4    2      4      8     12
df5$mat1
##      [,1] [,2] [,3]
## [1,]    1    5    9
## [2,]    2    6   10
## [3,]    3    7   11
## [4,]    4    8   12
```

With a list, whose member are vectors, each member of the list becomes a column in the data frame. In the case of too short members, recycling is applied.

```
lst1 <- list(a = 4:1, b = letters[4:1], c = "n", d = "z")
df6<- data.frame(fct1, lst1)
df6
##   fct1 a b c d
## 1    1 4 d n z
## 2    2 3 c n z
## 3    3 2 b n z
## 4    2 1 a n z
```

If the list is protected with `I()`, the list is added in whole as a variable or column in the data frame. In this case, the length of the list must match the number of rows in the data frame, while the length and class of the individual members of the list can vary. The names of the list members are used to set the `rownames` of the data frame. This is similar to the default behaviour of tibbles, while R data frames require explicit use of `I()` for lists not to be split (see chapter 8 on page 243 for details about package 'tibble').

```
df7<- data.frame(fct1, I(lst1))
df7
##   fct1       lst1
## a    1 4, 3, 2, 1
## b    2 d, c, b, a
## c    3          n
## d    2          z
```

```
df7$lst1
## $a
## [1] 4 3 2 1
##
## $b
## [1] "d" "c" "b" "a"
##
## $c
## [1] "n"
##
## $d
## [1] "z"
```

4.6 What do we gain using `I()`? Check the documentation carefully and think of uses where the flexibility gained by the option to protect or not the arguments passed to the `data.frame()` constructor can be useful. In addition, write R statements to extract individual members of embedded matrices or lists using indexing. Finally, test if the behaviour of `I()` is the same when assigning new member variables (or "columns") to an existing data frame.

4.4.1 Sub-setting data frames

When the names of data frames are long, complex conditions become awkward to write using indexing—i.e., subscripts. In such cases, `subset()` is handy because it evaluates the condition with the data frame as the "environment", i.e., the names of the columns are recognised if entered directly when writing the condition. Function `subset()` "filters" rows, usually corresponding to observations or experimental units. The condition is computed for each row, and if it returns `TRUE`, the row is included in the returned data frame, and excluded if `FALSE`.

We create a data frame with six rows and three columns. For column `y`, we rely on R automatically extending `"a"` by repeating it six times, while for column `z`, we rely on R automatically extending `c(TRUE, FALSE)` by repeating it three times.

```
df8 <- data.frame(x = 1:6, y = "a", z = c(TRUE, FALSE))
subset(df8, x > 3)
##   x y     z
## 4 4 a FALSE
## 5 5 a  TRUE
## 6 6 a FALSE
```

4.7 What is the behaviour of `subset()` when the condition is `NA`? Find the answer by writing code to test this, for a case where tests for different rows return `NA`, `TRUE` and `FALSE`.

When calling functions that return a vector, data frame, or other structure, the extraction operators `[ ]`, `[[ ]]`, or `$` can be appended to the rightmost parenthesis of the function call, in the same way as to the name of a variable holding the same data.

```
subset(df8, x > 3)[ , -3]
##   x y
## 4 4 a
## 5 5 a
```

```
## 6 6 a
subset(df8, x > 3)[ , "x", drop = FALSE]
##   x
## 4 4
## 5 5
## 6 6
subset(df8, x > 3)[ , "x"]
## [1] 4 5 6
```

4.8 When do extraction operators applied to data frames return a vector or factor, and when do they return a data frame? Please, experiment with your own code examples to work out the answer.

In the case of `subset()`, we can select columns directly as shown below, while for most other functions, extraction using operators `[ ]`, `[[ ]]`, or `$` is needed.

```
subset(df8, x > 3, select = 2)
##   y
## 4 a
## 5 a
## 6 a
```

```
subset(df8, x > 3, select = x)
##   x
## 4 4
## 5 5
## 6 6
```

```
subset(df8, x > 3, select = "x")
##   x
## 4 4
## 5 5
## 6 6
```

None of the examples in the last four code chunks alters the original data frame `df8`. We can store the returned value using a new name if we want to preserve `df8` unchanged, or we can assign the result to `df8`, deleting in the process, the previously stored value.

In the examples above, the names in the expression passed as the second argument to `subset()` were searched within `df8` and found. However, if not found in the data frame, objects with matching names are searched for in the global environment (outside the data frame, and visible in the user's workspace or enclosing environment). With no variable `A` present in data frame `df8`, vector `A` from the environment is silently used in the chunk below resulting in a returned data frame with no rows as `A > 3` returns `FALSE`.

```
A <- 1
subset(df8, A > 3)
## [1] x y z
## <0 rows> (or 0-length row.names)
```

This also applies to the expression passed as argument to parameter `select`, here shown as a way of selecting columns based on names stored in a character vector.

```
columns <- c("x", "z")
subset(df8, select = columns)
##   x     z
## 1 1  TRUE
## 2 2 FALSE
## 3 3  TRUE
## 4 4 FALSE
## 5 5  TRUE
## 6 6 FALSE
```

The use of `subset()` is convenient, but more prone to bugs compared to directly using the extraction operator `[ ]`. This same "cost" to achieving convenience applies to functions like `attach()` and `with()` described below. The longer time that a script is expected to be used, adapted, and reused, the more careful we should be when using any of these functions. An alternative way of avoiding excessive verbosity is to keep the names of data frames short.

A frequently used way of deleting a column by name from a data frame is to assign `NULL` to it—i.e., in the same way as members are usually deleted from `lists`. This approach modifies `df9` in place, rather than returning a modified copy of `df9`.

```
df9 <- df8
head(df9)
##   x y     z
## 1 1 a  TRUE
## 2 2 a FALSE
## 3 3 a  TRUE
## 4 4 a FALSE
## 5 5 a  TRUE
## 6 6 a FALSE
df9[["y"]] <- NULL
head(df9)
##   x     z
## 1 1  TRUE
## 2 2 FALSE
## 3 3  TRUE
## 4 4 FALSE
## 5 5  TRUE
## 6 6 FALSE
```

Alternatively, negative indexing can be used to remove columns from a copy of a data frame. In this example, a single column is removed. As base R does not support negative indexing by name with the extraction operator, the numerical index of the column to delete needs to be obtained first. (See the examples above using `subset()` with bare names to delete columns.)

```
df8[ , -which(colnames(df8) == "y")]
##   x     z
## 1 1  TRUE
## 2 2 FALSE
## 3 3  TRUE
## 4 4 FALSE
## 5 5  TRUE
## 6 6 FALSE
```

Instead of using the equality test, we can use the operator `%in%` or function `grepl()` to create a `logical` vector useful for deleting or selecting multiple columns in a single statement.

4.9 In the previous code chunk, we deleted the last column of the data frame `df8`, but using the extraction operator, we modified only the returned copy of `df8`, leaving `df8` unchanged. Thus we reuse it here for a surprising trick. You should first untangle how it changes the positions of columns and rows, and afterwards think how and why indexing with the extraction operator `[ ]` on both sides of the assignment operator `<-` can be useful when working with data.

```
df8[1:6, c(1,3)] <- df8[6:1, c(3,1)]
df8
```

Although in this last example we used numeric indexes to make it more interesting, in practice, especially in scripts or other code that will be reused, do use column or member names instead of positional indexes whenever possible. This makes code much more reliable, as changes elsewhere in the script could alter the order of columns and *invalidate* numerical indexes. In addition, using meaningful names makes programmers' intentions easier to understand.

4.4.2 Summarising and splitting data frames

Function `summary()` can be used to obtain a summary from objects of most R classes, including data frames. It is also possible to use `sapply()`, `lapply()` or `vapply()` to apply any suitable function to data by columns (see section 5.8 on page 154 for a description of these functions and their use).

```
summary(df8)
##        x              y                  z
##  Min.   :1.00   Length:6           Mode :logical
##  1st Qu.:2.25   Class :character   FALSE:3
##  Median :3.50   Mode  :character   TRUE :3
##  Mean   :3.50
##  3rd Qu.:4.75
##  Max.   :6.00
```

R function `split()` makes it possible to split a data frame into a list of data frames, based on the levels of a factor, even if the rows are not ordered according to factor levels.

We create a data frame with six rows and three columns. In the case of column z, we rely on R to automatically extend `c("a", "b")` by repeating it three times so as to fill the six rows.

```
df10 <- data.frame(x1 = 1:6, x2 = c(1, 5, 4, 2, 6, 3), z = c("a", "b"))

split(df10, df10$z)
## $a
##   x1 x2 z
## 1  1  1 a
## 3  3  4 a
## 5  5  6 a
##
## $b
```

```
##   x1 x2 z
## 2  2  5 b
## 4  4  2 b
## 6  6  3 b
```

The same operation can be specified using a one-sided formula `~z` to indicate the grouping.

```
split(df10, ~ z)
## $a
##   x1 x2 z
## 1  1  1 a
## 3  3  4 a
## 5  5  6 a
##
## $b
##   x1 x2 z
## 2  2  5 b
## 4  4  2 b
## 6  6  3 b
```

Function `unsplit()` can be used to reverse splitting done by `split()`.

`split()` is sometimes used in combination with apply functions (see section 5.8 on page 154) to compute group or treatment summaries. However, in most cases it is simpler to use `aggregate()` for computing such summaries.

Related to splitting a data frame is the calculation of summaries based on a subset of cases, or more commonly summaries for all observations but after grouping them based on the values in a column or the levels of a factor.

How to summarise one variable from a data frame by group?
To summarise a single variable by group, we can use `aggregate()`.

```
aggregate(x = iris$Petal.Length,
          by = list(iris$Species), FUN = mean)
##      Group.1     x
## 1     setosa 1.462
## 2 versicolor 4.260
## 3  virginica 5.552
```

How to summarise numeric variables from a data frame by group?
To summarise variables, we can use `aggregate()` (see section 8.7.2 on page 262 for an alternative approach using package 'dplyr').

```
aggregate(x = iris[ , sapply(iris, is.numeric)],
          by = list(iris$Species), FUN = mean)
##      Group.1 Sepal.Length Sepal.Width Petal.Length Petal.Width
## 1     setosa        5.006       3.428        1.462       0.246
## 2 versicolor        5.936       2.770        4.260       1.326
## 3  virginica        6.588       2.974        5.552       2.026
```

For these data, as the only non-numeric variable is `Species`, we could have also used formula notation as shown below.

There is also a formula-based `aggregate()` method (or "variant") available (R *formulas* are described in depth in section 7.13 on page 226). In `aggregate()`, the left-hand side (*lhs*) of the formula indicates the variable to summarise and its right-hand side (*rhs*) the factor used to split or group the data before summarising them.

```
aggregate(x1 ~ z, FUN = mean, data = df10)
##   z x1
## 1 a  3
## 2 b  4
```

We can summarise more than one column at a time.

```
aggregate(cbind(x1, x2) ~ z, FUN = mean, data = df10)
##   z x1       x2
## 1 a  3 3.666667
## 2 b  4 3.333333
```

If all the columns not used for grouping are valid input to the function passed as the argument to `FUN` the formula can be simplified using a point (`.`) with meaning "all columns except those on the *rhs* of the formula".

```
aggregate(. ~ z, FUN = mean, data = df10)
##   z x1       x2
## 1 a  3 3.666667
## 2 b  4 3.333333
```

Function `aggregate()` can be also used to aggregate time series data based on time intervals (see `help(aggregate)`).

4.4.3 Re-arranging columns and rows

As with members of vectors and lists, to change the position of columns or rows in a data frame we use the extraction operator and indexing by name or position. In a matrix-like object, such as a data frame, the first index corresponds to rows and the second to columns.

The most direct way of changing the order of columns and/or rows in data frames (as for matrices and arrays) is to use subscripting. Once we know the original position and target position we can use column names or positions as indexes on the right-hand side, listing all columns to be retained, even those remaining at their original position.

```
df11 <- data.frame(A = 1:10, B = 3, C = c("A", "B"))
head(df11, 2)
##   A B C
## 1 1 3 A
## 2 2 3 B
df11 <- df11[ , c("B", "A", "C")]
head(df11, 2)
##   B A C
## 1 3 1 A
## 2 3 2 B
```

When using the extraction operator `[ ]` on both the left- and right-hand-sides, with a `numeric` vector as an argument to swap two columns, the vectors or factors are swapped, while the names of the columns are not! To retain the correspondence between column naming and column contents after swapping or rearranging the columns *using numeric indices*, we need to separately move the names of the columns. This may seem counter-intuitive, unless we think in terms of positions being named rather than the contents of the columns being linked to the names.

```
df11 <- data.frame(A = 1:10, B = 3, C = c("A", "B"))
head(df11, 2)
##   A B C
## 1 1 3 A
## 2 2 3 B
df11[ , 1:2] <- df11[ , 2:1]
head(df11, 2)
##   A B C
## 1 3 1 A
## 2 3 2 B
colnames(df11)[1:2] <- colnames(df11)[2:1]
head(df11, 2)
##   B A C
## 1 3 1 A
## 2 3 2 B
```

Taking into account that `order()` returns the indexes needed to sort a vector (see page 69), we can use `order()` to generate the indexes needed to sort the rows of a data frame. In this case, the argument to `order()` is usually a column of the data frame being arranged. However, any vector of suitable length, including the result of applying a function to one or more columns, can be passed as an argument to `order()`. Function `order()` is not useful for sorting columns of data frames *based on data from the columns* as it requires a vector across columns as input, which is possible only when all columns are of the same class. (In the case of `matrix` and `array` this approach can be applied to any of their dimensions as all their elements homogenously belong to one class.)

? How to order columns or rows in a data frame?

We use column names or numeric indexes with the extraction operator `[ ]` only on the *rhs* of the assignment. For example, to arrange the columns of data set `iris` in decreasing alphabetical order, we use `sort()` as shown, or `order()` (see page 69).

```
sorted_cols_iris <- iris[ , sort(colnames(iris), decreasing = TRUE)]
head(sorted_cols_iris, 5)
##   Species Sepal.Width Sepal.Length Petal.Width Petal.Length
## 1  setosa         3.5          5.1         0.2          1.4
## 2  setosa         3.0          4.9         0.2          1.4
## 3  setosa         3.2          4.7         0.2          1.3
## 4  setosa         3.1          4.6         0.2          1.5
## 5  setosa         3.6          5.0         0.2          1.4
```

Similarly, we can use values in a column as argument to `order()` to obtain the `numeric` indices to sort rows.

```
sorted_rows_iris <- iris[order(iris$Petal.Length), ]
head(sorted_rows_iris, 5)
##    Sepal.Length Sepal.Width Petal.Length Petal.Width Species
## 23          4.6         3.6          1.0         0.2  setosa
## 14          4.3         3.0          1.1         0.1  setosa
## 15          5.8         4.0          1.2         0.2  setosa
## 36          5.0         3.2          1.2         0.2  setosa
## 3           4.7         3.2          1.3         0.2  setosa
```

4.10 Create a new data frame containing three numeric columns with three different haphazard sequences of values and a factor with two levels. Call these columns `A`, `B`, `C` and `F`. 1) Sort the rows of the data frame so that the values in `A` are in decreasing order. 2) Sort the rows of the data frame according to increasing values of the sum of `A` and `B` without adding a new column to the data frame or storing the vector of sums in a variable. In other words, do the sorting based on sums calculated on-the-fly. 1) Sort the rows by level of factor `F`, and 2) by level of factor `F` and by values in `B` within each factor level. Hint: revisit the exercise on page 83 were the use of `order()` on factors is described.

4.4.4 Re-encoding or adding variables

It is common that some variables need to be added to an existing data frame based on existing variables, either as a computed value or based on mapping, for example, treatments to sample codes already in a data frame. In the second case, named vectors can be used to replace values in a variable or to add a variable to a data frame.

Mapping is possible because the length of the value returned by the extraction operator `[ ]` is given by the length of the indexing vector (see section 3.10 on page 64). Although we show toy-like examples, this approach is most useful with data frames containing many rows.

If the existing variable is a character vector or factor, we need to create a named vector with the new values as data and the existing values as names.

```
df12 <-
  data.frame(genotype = rep(c("WT", "mutant1", "mutant2"), 2),
             value = c(1.5, 3.2, 4.5, 8.2, 7.4, 6.2))
mutant <- c(WT = FALSE, mutant1 = TRUE, mutant2 = TRUE)
df12$mutant <- mutant[df12$genotype]
df12
##   genotype value mutant
## 1       WT   1.5  FALSE
## 2  mutant1   3.2   TRUE
## 3  mutant2   4.5   TRUE
## 4       WT   8.2  FALSE
## 5  mutant1   7.4   TRUE
## 6  mutant2   6.2   TRUE
```

If the existing variable is an `integer` vector, we can use a vector without names, being careful that the positions in the *mapping* vector match the values of the existing variable

```
df13 <- data.frame(individual = rep(1:3, 2),
                   value = c(1.5, 3.2, 4.5, 8.2, 7.4, 6.2))
genotype <- c("WT", "mutant1", "mutant2")
df13$genotype <- genotype[df13$individual]
df13
##   individual value genotype
## 1          1   1.5       WT
## 2          2   3.2  mutant1
## 3          3   4.5  mutant2
## 4          1   8.2       WT
## 5          2   7.4  mutant1
```

```
## 6          3   6.2  mutant2
```

4.11 Add a variable named `genotype` to the data frame below so that for individual `4` its value is `"WT"`, for individual `1` its value is `"mutant1"`, and for individual `2` its value is `"mutant2"`.

```
DF1 <- data.frame(individual = rep(c(2, 4, 1), 2),
                  value = c(1.5, 3.2, 4.5, 8.2, 7.4, 6.2))
```

4.4.5 Operating within data frames

In the case of computing new values from existing variables, named vectors are of limited use. Instead, variables in a data frame can be added or modified with R functions `transform()`, `with()` and `within()`. These functions can be thought as convenience functions as the same computations can be done using the extraction operators to access individual variables, in the lhs, rhs, or both lhs and rhs (see section 3.10 on page 64).

In the case of `with()`, only one, possibly compound code statement is affected and this statement is passed as an argument. As before, we need to fully specify the left-hand side of the assignment. The value returned is the one returned by the statement passed as an argument, in the case of compound statements, the value returned by the last contained simple code statement to be executed. Consequently, if the intent is to modify the container, assignment to an individual member variable (column in this case) is required.

In this example, column `A` of `df14` takes precedence, and the returned value is the expected one.

```
df14 <- data.frame(A = 1:10, B = 3)
df14$C <- with(df14, (A + B) / A) # add column
head(df14, 3)
##   A B   C
## 1 1 3 4.0
## 2 2 3 2.5
## 3 3 3 2.0
```

In the case of `within()`, assignments in the argument to its second parameter affect the object returned, which is a copy of the container (in this case, a whole data frame), which still needs to be saved through assignment. Here the intention is to modify it, so we assign it back to the same name, but it could have been assigned to a different name so as not to overwrite the original data frame.

```
df14$C <- NULL
df15 <- within(df14,  C <- (A + B) / A) # midified copy
head(df15, 3)
##   A B   C
## 1 1 3 4.0
## 2 2 3 2.5
## 3 3 3 2.0
```

In the example above, using `within()` instead of `with()` makes little difference to the amount of typing or clarity of the code, but with multiple member variables being operated upon, as shown below, using `within()` results in more concise and easier to understand code.

```
df16 <- within(df14,
               {C <- (A + B) / A
                D <- A * B
                E <- A / B + 1}
              )
head(df16, 3)
##   A B        E D   C
## 1 1 3 1.333333 3 4.0
## 2 2 3 1.666667 6 2.5
## 3 3 3 2.000000 9 2.0
```

Repeatedly pre-pending the name of a *container*, such as a list or data frame, to the name of each member variable being accessed can make R code verbose and difficult to understand. Functions `attach()` and its matching `detach()` allow us to change where R first looks for the names of objects mentioned in a code statement. When using a long name for a data frame, entering a simple calculation can easily result in a difficult-to-read statement. Here even with a very short name for the data frame, the verbosity compared to the last chunk above is clear.

```
df14$C <- (df14$A + df14$B) / df14$A
df14$D <- df14$A * df14$B
df14$D <- df14$A / df14$B + 1
head(df14, 3)
##   A B   C        D
## 1 1 3 4.0 1.333333
## 2 2 3 2.5 1.666667
## 3 3 3 2.0 2.000000
```

Using `attach()` we can alter where R looks up names and consequently simplify the statement. With `detach()` we can restore the original state. It is important to remember that here we can only simplify the right-hand side of the assignment, while the "destination" of the result of the computation still needs to be fully specified on the left-hand side of the assignment operator. We include below only one statement between `attach()` and `detach()` but multiple statements are allowed. Furthermore, if variables with the same name as the columns exist in the search path, these will take precedence, something that can result in bugs or crashes, or as seen below, a message warns that variable `A` from the global environment will be used instead of column `A` of the attached `df17`. The returned value is, of course, not the desired one.

```
df17 <- data.frame(A = 1:10, B = 3)
A
## [1] 1
attach(df17)

## The following object is masked _by_ .GlobalEnv:
##
##     A

A
## [1] 1
detach(df17)
A
## [1] 1
```

```
attach(df17)

## The following object is masked _by_ .GlobalEnv:
##
##     A

df17$C <- (A + B) / A
detach(df17)
head(df17, 2)
##   A B C
## 1 1 3 4
## 2 2 3 4
```

Use of `attach()` and `detach()`, which work as a pair of ON and OFF switches, can result in an undesired after-effect on name lookup if the script terminates after `attach()` is executed but before `detach()` is called, as the attached object is not detached. In contrast, `with()` and `within()`, being self-contained, guarantee that cleanup takes place. Consequently, the usual recommendation is to give preference to the use of `with()` and `within()` over `attach()` and `detach()`.

4.5 Reshaping and Editing Data Frames

As mentioned above, in most cases, in R data rows represent measurement events or observations possibly on multiple response variables and factors describing groupings, i.e., a "long" shape. However, when measurements are repeated in time, columns rather frequently represent observations of the same response variable at different times, i.e., a "wide" shape. Other cases exist where reshaping is needed. Function `reshape()` can convert wide data frames into long data frames and vice versa. See section 8.6 on page 256 on package 'tidyr' for an alternative approach to reshaping data with a friendlier user interface.

We start by creating a data frame of hypothetical data measured on two occasions. With these data, for example, if we wish to compute the growth of each subject by computing the difference in `weight` and `height` between the two time points, one approach is to reshape the data frame into a wider shape and subsequently subtract the columns.

```
# artifical data
df1 <- data.frame(id = rep(1:4, rep(2,4)),
                  Time = factor(rep(c("Before","After"), 4)),
                  Weight = rnorm(n = 4, mean = c(20.1, 30.8)),
                  Height = rnorm(n = 4, mean = c(9.5, 14.2)))

df1
##   id   Time   Weight   Height
## 1  1 Before 21.05859 10.57587
## 2  1  After 29.57182 14.15418
## 3  2 Before 21.52375 10.12694
## 4  2  After 30.15447 14.91599
## 5  3 Before 21.05859 10.57587
## 6  3  After 29.57182 14.15418
## 7  4 Before 21.52375 10.12694
```

```
## 8   4  After 30.15447 14.91599
# make it wider
df2 <- reshape(df1, timevar = "Time", idvar = "id", direction = "wide")
df2
##    id Weight.Before Height.Before Weight.After Height.After
## 1  1      21.05859      10.57587     29.57182     14.15418
## 3  2      21.52375      10.12694     30.15447     14.91599
## 5  3      21.05859      10.57587     29.57182     14.15418
## 7  4      21.52375      10.12694     30.15447     14.91599
# possible further calculation
within(df2,
       {
        Height.growth <- Height.After - Height.Before
        Weight.growth <- Weight.After - Weight.Before
       })
##    id Weight.Before Height.Before Weight.After Height.After Weight.growth
## 1  1      21.05859      10.57587     29.57182     14.15418      8.513234
## 3  2      21.52375      10.12694     30.15447     14.91599      8.630720
## 5  3      21.05859      10.57587     29.57182     14.15418      8.513234
## 7  4      21.52375      10.12694     30.15447     14.91599      8.630720
##    Height.growth
## 1       3.578307
## 3       4.789053
## 5       3.578307
## 7       4.789053
```

Alternatively, we may want to convert `df1` into a longer shape, with a single column with measurements, and a new column indicating whether the measured variable was `height` or `weight`. For this operation to succeed, we need to add a column with a unique value for each row in `df1`, and one easy way is to copy row names into a column. The names of the parameters of function `reshape()` are meaningful only when dealing with time series. Thus, reading the code below becomes rather difficult. It is also to be noted that the user is responsible of passing the values to `times` in the correct order.

```
df1$ID <- rownames(df1) # unique ID for each row
# make it longer
reshape(df1,
        idvar = "ID",
        timevar = "Quantity",
        times = c("Weight", "Height"),
        v.names = "Value",
        direction = "long",
        varying = c("Weight", "Height"))
##          id   Time ID Quantity    Value
## 1.Weight  1 Before  1   Weight 21.05859
## 2.Weight  1  After  2   Weight 29.57182
## 3.Weight  2 Before  3   Weight 21.52375
## 4.Weight  2  After  4   Weight 30.15447
## 5.Weight  3 Before  5   Weight 21.05859
## 6.Weight  3  After  6   Weight 29.57182
## 7.Weight  4 Before  7   Weight 21.52375
## 8.Weight  4  After  8   Weight 30.15447
## 1.Height  1 Before  1   Height 10.57587
## 2.Height  1  After  2   Height 14.15418
## 3.Height  2 Before  3   Height 10.12694
```

```
## 4.Height  2  After  4   Height 14.91599
## 5.Height  3 Before  5   Height 10.57587
## 6.Height  3  After  6   Height 14.15418
## 7.Height  4 Before  7   Height 10.12694
## 8.Height  4  After  8   Height 14.91599
```

To edit a data frame programmatically, one can use the approaches already discussed, using the extraction operators `[ ]` or `[[ ]]` on the *lhs* of <- to replace member elements. This in combination with functions like `gsub()` makes it possible to "edit" the contents of data frames.

Methods `View()`, `edit()` and `fix()` can be used interactively to display and edit R objects. When using R from within IDEs like RStudio, calling these functions with a data frame as argument opens in most cases the IDE's own worksheet-like data editors, and for other types of objects a text editor pane. Output is not included for this chunk, as the use of these functions requires user interaction. Please, run these examples in R and in an IDE like RStudio.

```
View(cars)
edit(cars)
```

These functions can be used at the R console also when R is used on its own, but the editors activated are different ones. In any case, the use of scripts has made the interactive use of R at the console less frequent and the need to edit R objects previously saved in the user's current workspace nearly disappear. `View()`, `edit()` and `fix()` are unusual in that their definitions are dependent on system variables that at least when using R on its own, can be modified by the user.

4.6 Attributes of R Objects

R objects can have attributes. Attributes are named *slots* normally used to store ancillary data such as object properties functioning as additional fields where to store additional information in any R object. There are no restrictions on the class of what is assigned to an attribute. They can be used to store metadata accompanying the data stored in an object, which is important for reproducible research and data sharing. They can be set and read by user code and they are also used internally by R among other things to store the class an object belongs to, column and row names in data frames and matrices and the labels of levels in factors. Although most R objects have attributes, they are rarely displayed explicitly when an object is printed, while the structure of objects as displayed by function `str()` includes them.

Although we rarely need to set or extract values stored in attributes explicitly, many of the features of R that we take for granted are implemented using attributes: columns names in data frames are stored in an attribute. Matrices are vectors with additional attributes.

```
df1 <- data.frame(x = 1:6, y = c("a", "b"), z = c(TRUE, FALSE, NA))
df1
```

```
##   x y     z
## 1 1 a  TRUE
## 2 2 b FALSE
## 3 3 a    NA
## 4 4 b  TRUE
## 5 5 a FALSE
## 6 6 b    NA
attributes(df1)
## $names
## [1] "x" "y" "z"
##
## $class
## [1] "data.frame"
##
## $row.names
## [1] 1 2 3 4 5 6
str(df1)
## 'data.frame': 6 obs. of  3 variables:
##  $ x: int  1 2 3 4 5 6
##  $ y: chr  "a" "b" "a" "b" ...
##  $ z: logi  TRUE FALSE NA TRUE FALSE NA
```

Attribute "comment" is meant to be set by users to store a character string—e.g., to store metadata as text together with data. As comments are frequently used, R has functions for accessing and setting comments.

```
comment(df1)
## NULL
comment(df1) <- "this is stored as a comment"
comment(df1)
## [1] "this is stored as a comment"
```

Functions like names(), dim() or levels() return values retrieved from attributes stored in R objects, whereas names()<-, dim()<- or levels()<- set (or unset with NULL) the value of the respective attributes. Dedicated query and set functions do not exist for all attributes. Functions attr(), attr()<- and attributes() can be used with any attribute. With attr() we query, and with attr()<- we set individual attributes by name. With attributes() we retrieve all attributes of an object as a named list. In addition, method str() displays the structure of an R object with all its components, including their attributes.

Continuing with the previous example, we can retrieve and set the value stored in the "comment" attribute using these functions. In the second statement, we delete the value stored in the attribute by assigning NULL to it.

```
attr(df1, "comment")
## [1] "this is stored as a comment"
attr(df1, "comment") <- NULL
attr(df1, "comment")
## NULL
comment(df1) # same as previous line
## NULL
```

The "names" attribute of df1 was set by the data.frame() constructor when it was created above. In the next example, in the first statement we retrieve the names and implicitly print them. In the second statement, read from right to left, we retrieve the names, convert them to upper case, and save them back to the same attribute.

```
names(df1)
## [1] "x" "y" "z"
colnames(df1) # same as names()
## [1] "x" "y" "z"
colnames(df1) <- toupper(colnames(df1))
colnames(df1)
## [1] "X" "Y" "Z"
attr(df1, "names") # same as previous line
## [1] "X" "Y" "Z"
```

4.12 In general, `R` objects do not have by default names assigned to members. As seen on page 67, we can give names to vector members during construction with a call to `c()` or we can assign names (set attribute `names`) with function `names()<-` to existing vectors. Lists behave almost the same as vectors, although members of nested objects can also be named. Data frames have attributes `names` and `row.names`, that can be accessed with functions `names()` or `colnames()`, and function `rownames()`, respectively. The attributes can be set with functions `names()<-` or `colnames()<-`, and `rownames()<-`. The `data.frame()` constructor sets (column) names and row names by default. The `matrix()` constructor by default does not set `dimnames` or `names` attributes. When names are assigned to a `matrix` with `names()<-`, the matrix behaves like a vector, and the names are assigned to individual members. Functions `dimnames()<-`, `colnames()<-`, and `rownames()<-` are used to assign names to columns and rows. The matching functions `dimnames()`, `colnames()` and `rownames()` are used to access these values.

When no names have been set, `names()`, `colnames()`, `rownames()`, and `dimnames()` return `NULL`. In contrast, `labels()`, intended to be used for printing, returns made-up names based on positions.

Run the examples below and write similar examples for a `list` and a `data.frame`. For `matrix`, write an additional statement that uses `dimnames()<-` to set row and column names simultaneously.

```
VCT1 <- 5:10
names(VCT1)
labels(VCT1)
names(VCT1) <- letters[5:10]
names(VCT1)
labels(VCT1)

MAT1 <- matrix(1:10, ncol = 2)
dimnames(MAT1)
labels(MAT1)
colnames(MAT1) <- c("a", "b")
colnames(MAT1)
dimnames(MAT1)
labels(MAT1)
```

We can add a new attribute, under our own control, as long as its name does not clash with those of existing attributes.

```
attr(df1, "my.attribute") <- "this is stored in my attribute"
attributes(df1)
## $names
## [1] "x" "y" "z"
##
## $class
## [1] "data.frame"
##
## $row.names
## [1] 1 2 3 4 5 6
##
## $my.attribute
## [1] "this is stored in my attribute"
```

The attributes used internally by R can be directly modified by user code. In most cases, this is unnecessary as R provides pairs of functions to query and set the relevant attributes. This is true for the attributes `dim`, `names` and `levels`. In the example below, we read the attributes from a matrix.

```
mat1 <- matrix(1:10, ncol = 2)
attributes(mat1)
## $dim
## [1] 5 2
dim(mat1)
## [1] 5 2
dimnames(mat1)
## NULL

labels(mat1)
## [[1]]
## [1] "1" "2" "3" "4" "5"
##
## [[2]]
## [1] "1" "2"
mat1
##      [,1] [,2]
## [1,]    1    6
## [2,]    2    7
## [3,]    3    8
## [4,]    4    9
## [5,]    5   10

attr(mat1, "dim")
## [1] 5 2
attr(mat1, "dim") <- c(2, 5)
mat1
##      [,1] [,2] [,3] [,4] [,5]
## [1,]    1    3    5    7    9
## [2,]    2    4    6    8   10

attr(mat1, "dim") <- NULL
is.vector(mat1 )
## [1] TRUE
mat1
##  [1]  1  2  3  4  5  6  7  8  9 10
```

In this case we can also use dim().

```
dim(mat1) <- NULL
is.vector(mat1 )
## [1] TRUE
```

There is no restriction to the creation, setting, resetting, and reading of attributes, but not all functions and operators that can be used to modify objects will preserve non-standard attributes. This can be a problem when using some R packages, such as the 'tidyverse'. So, using private attributes is a double-edged sword that usually is worthwhile considering only when designing a new class together with the corresponding methods for it. The values returned by model fitting functions like lm() are good examples of the extensive use of class-specific attributes (see section 7.9 on page 200).

4.7 Saving and Loading Data

4.7.1 Data sets in R and packages

To be able to present more meaningful examples, we need some real data. Here we use cars, one of the many data sets included in base R. Function data() is used to load data objects that are included in R or contained in packages (whether a call to data() is needed or not depends on how the package where the data objects are defined was configured). It is also possible to import data saved in files with *foreign* formats, defined by other software or commonly used for data exchange. Package 'foreign', included in the R distribution, as well as contributed packages make available functions capable of reading and decoding various foreign formats. How to read or import 'foreign' data is discussed in the R documentation, in the manual *R Data Import/Export*, and in this book, in chapter 10 on page 383. It is also good to keep in mind that in R, URLs (Uniform Resource Locators) are accepted as arguments to the file or path parameter of many functions (see section 10.12 on page 410).

In the next example, we load data available in R package 'datasets' as R objects by calling function data(). The loaded R object cars is a data frame. (Package 'datasets' is part of the R distribution and is always available).

```
data(cars)
```

4.7.2 .rda files

By default, at the end of a session, the current workspace containing the results of one's work is saved into a file called .RData. In addition to saving the whole workspace, it is possible to save one or more R objects present in the workspace to disk using the same file format (with file name tag .rda or .Rda). One or more objects, belonging to any mode or class can be saved into a single file using function save(). Reading the file restores all the saved objects into the current workspace with their original names. These files are portable across most R versions—i.e., old formats can be read and written by newer versions of R, although the newer, default format may be not readable with earlier R versions. Whether compression is used, and whether the "binary" data are encoded into ASCII characters, allowing maximum portability at the expense of increased size can be controlled by passing suitable arguments to `save()`.

We create a data frame object and then save it to a file. The file name used can be any valid one in the operating system, however to ensure compatibility with multiple operating systems, it is good to use only ASCII characters. Although not enforced, using the name tag `.rda` or `.Rda` is recommended.

```
df1 <- data.frame(x = 1:5, y = 5:1)
df1
##   x y
## 1 1 5
## 2 2 4
## 3 3 3
## 4 4 2
## 5 5 1
save(df1, file = "df1.rda")
```

We delete the data frame object and confirm that it is no longer present in the workspace (see page 39 for details about `remove()` and `objects()`).

```
remove(df1)
objects(pattern = "df1")
## character(0)
```

We read the file we earlier saved to restore the object.

```
load(file = "df1.rda")
objects(pattern = "df1")
## [1] "df1"
df1
##   x y
## 1 1 5
## 2 2 4
## 3 3 3
## 4 4 2
## 5 5 1
```

The default format used is binary and compressed, which results in smaller files.

4.13 In the example above, only one object was saved, but one can simply give the bare names of additional objects as arguments separated by commas ahead of `file`. Just try saving more than one data frame to the same file. Then the data frames plus a few vectors. After creating each file, clear the workspace and then restore from the file the objects you saved.

Sometimes it is easier to supply the names of the objects to be saved as a vector of `character` strings passed as an argument to parameter `list` (in spite of the name the argument passed must be a `vector`, not a `list`). One use case is saving a group of objects based on their names. In this case, one can use `objects()` (also available as `ls()`) to obtain a vector of `character` strings with the names of objects matching a simple `pattern` or a complex *regular expression* (see section 3.4 on page 46).

The example below uses this approach in two steps, first saving in variable `dfs` a `character vector` with the names of the objects matching a pattern, and then using this saved vector as an argument to parameter `list` in the call to `save()`.

```
dfs <- objects(pattern = "*.df")
save(list = dfs, file = "my-dfs.rda")
```

The two statements above can be combined into a single statement by nesting the function calls.

```
save(list = objects(pattern = "*.df"), file = "my-dfs.rda")
```

4.14 Practice using different patterns with `objects()`. You do not need to save the objects to a file. Just have a look at the list of object names returned.

As a coda, I show how to clean up by deleting the two files we created. Function `file.remove()` can be used to delete files stored in the operating system file system, usually on a hard disk drive or a solid state drive, as long as the user has enough rights. No confirmation is requested, so care not to delete valuable files is required. Function `unlink()`, is not an exact equivalent, as it can also delete folders and supports recursion through nested folders. The name *unlink* is borrowed from that of the equivalent function in Unix and Linux.

```
file.remove(c("my-dfs.rda", "df1.rda"))
## [1] TRUE TRUE
```

4.7.3 .rds files

The RDS format can be used to save individual objects instead of multiple objects (usually using file name tag `.rds`). They are read and saved with functions `readRDS()` and `saveRDS()`, respectively. The value returned by a call to `readRDS()` is the object read from the file on disk. When RDS files are read, different from when RDA files are loaded, assigning the object read to a name is frequently the first step. This name can be any valid R name. Of course, it is also possible to use the object returned by `readRDS()` as an argument to a function by nesting the function calls.

```
saveRDS(df1, "df1.rds")
```

If we read the file at the R console, by default the read R object will be printed at the console.

```
readRDS("df1.rds")
##   x y
## 1 1 5
## 2 2 4
## 3 3 3
## 4 4 2
## 5 5 1
```

If we assign the read object to a different name, it is possible to check if the object read is identical to the one saved.

```
df2 <- readRDS("df1.rds")
identical(df1, df2)
## [1] TRUE
```

As above, we clean up by deleting the file.

```
file.remove("df1.rds")
## [1] TRUE
```

4.7.4 dput()

In general, the use of `.rda` and .rds files is preferred. Function `dput()` is sometimes used to share data as part of a code chunk at StackOverflow, mostly as a convenient way of converting a data frame or list into plain text that can be pasted into the code chunk listing to reconstruct the object. If no argument is passed to parameter `file`, the result of deparsing an object is printed at the R console.

```
dput(df1)
## structure(list(x = 1:5, y = 5:1), class = "data.frame", row.names = c(NA,
## -5L))
```

There exists a companion function `dget()` to recreate the object.

Output to, and input from, text-based file formats as well as to and from various binary formats *foreign* to R is described in chapter 10 on page 383.

4.8 Plotting

In most cases, the most effective way of obtaining an overview of a data set is by plotting it using multiple approaches. The base-R generic method `plot()` can be used to plot different data. It is a generic method that has specialisations suitable for different kinds of objects (see section 6.3 on page 176 for a brief introduction to objects, classes and methods). In this section, I very briefly demonstrate the use of the most common base-R graphics functions. They are well described in the book *R Graphics* (Murrell 2019). I describe in detail the use of the *layered grammar of graphics* and plotting with package 'ggplot2' in chapter 9 on page 271.

4.8.1 Plotting data

It is possible to pass two vectors (here columns from a data frame) directly as arguments to the `x` and `y` parameters of function `plot()`. (The plot is shown farther down, as the three approaches create identical plots.)

```
plot(x = cars$speed, y = cars$dist)
```

It is also possible to use `with()` or `attach()` as described in section 4.4.5 on page 110.

```
with(cars, plot(x = speed, y = dist))
```

However, it is better to use a *formula* to specify the variables to be plotted on the x and y axes, passing as an argument to parameter `data` a data frame containing these variables as columns. The formula `dist ~ speed`, is read as `dist` explained by `speed`—i.e., `dist` is mapped to the y-axis as the dependent variable and `speed` to the x-axis as the independent variable. The names used in the formula, are those of columns in the data.frame. As described in section 7.8 on page 199, the same syntax is used to describe models to be fitted to observations.

```
plot(dist ~ speed, data = cars)
```

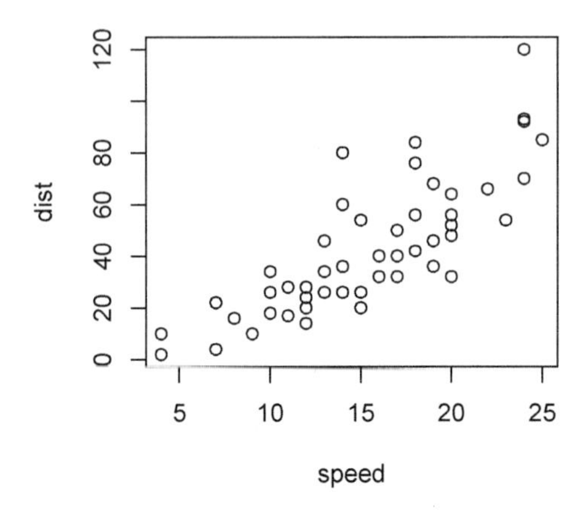

Within R there exist different specialisations, or "flavours", of method plot() that become active depending on the class of the variables passed as arguments: passing two numerical variables results in a scatter plot as seen above. In contrast, passing one factor and one numeric variable to plot() results in a box-and-whiskers plot being produced. Use help("chickwts") to learn more about this data set, also included in R.

```
plot(weight ~ feed, data = chickwts)
```

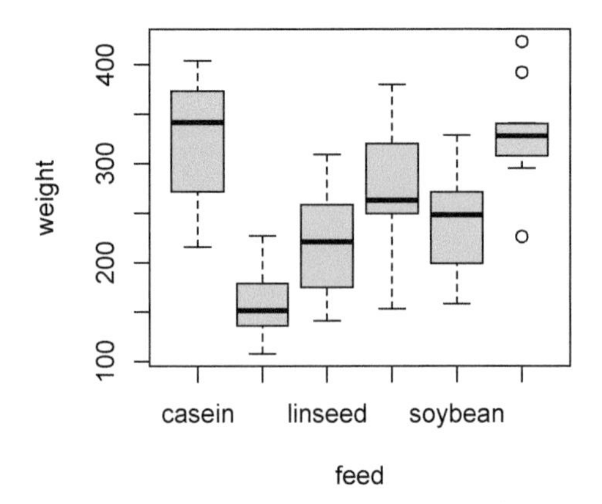

4.8.2 Graphical output

Graphical output, such as produced by plot(), is rendered by *graphical output devices*. When R is used interactively, a software device is opened automatically to output the graphical output to a physical device, usually the computer screen. The name of the R software device used may depend on the operating system (e.g., MS-Windows or Linux), or on the IDE (e.g., RStudio).

In R, software graphical devices not necessarily generate output on a physical device like a printer, as several of these devices translate the plotting commands into a file format and save it to disk. Graphical devices in R differ in the kind of output they produce: raster or bitmap files (e.g., TIFF, PNG, and JPEG formats), vector graphics files (e.g., SVG, EPS, and PDF), or output to a physical device like the screen of a computer. Additional devices are available through contributed R packages.

RStudio makes it possible to export plots into graphic files through a menu-based interface in the *Plots* viewer tab. This interface uses some of the some graphic devices that are available at the console and through scripts. For reproducibility, it is preferable to include the R commands used to export plots in the scripts used for data analysis.

Devices follow the paradigm of ON and OFF switches, opening and closing a destination for `print()`, `plot()` and related functions. Some devices producing a file as output, save their output one plot at a time to single-page graphic files, while others write the file only when the device is closed, possibly as a multi-page file.

When opening a device the user supplies additional information. For the PDF and SVG devices that produce output in a vector-graphics format, width and height of the output are specified in *inches*. A default file name is used unless we pass a `character` string as an argument to parameter `file`.

```
pdf(file = "output/my-file.pdf", width = 6, height = 5, onefile = TRUE)
plot(dist ~ speed, data = cars)
plot(weight ~ feed, data = chickwts)
dev.off()
## cairo_pdf
##         2
```

Raster devices return bitmaps and `width` and `height` are specified in most cases in *pixels*.

```
png(file = "output/my-file.png", width = 600, height = 500)
plot(weight ~ feed, data = chickwts)
dev.off()
## cairo_pdf
##         2
```

The approach of direct output to a software device is used in base R by `plot()` and its companions `text()`, `lines()`, and `points()`. `plot()` outputs a graph, and the other three functions can add elements to it. The addition of plot components, as shown below, is done directly to the output device, i.e., when output is to the computer screen the partial plot is visible at each step.

```
png(file = "output/my-file.png", width = 600, height = 500)
plot(dist ~ speed, data = cars)
text(x = 10, y = 110, labels = "some texts to be added")
dev.off()
## cairo_pdf
##         2
```

This is not the only approach available in R for building complex plots. As we will see in chapter 9 on page 271, an alternative approach is to build a *plot object* as a list of member components, that can be saved as any other R object. This object functions as a "recipe" that is later rendered as a whole on a graphical device by calling `print()` to display it.

4.9 Further Reading

For further reading on the aspects of R discussed in the current chapter, I suggest the book *The Art of R Programming: A Tour of Statistical Software Design* (Matloff 2011), with emphasis on the R language and programming. The new, open-source, book *Deep R Programming* (Gagolewski 2023) provides a free alternative. This book also covers base R plotting giving more advanced examples than *Learn R: As a Language*. An in-depth description of plotting and graphic devices in R is available in the book *R Graphics* (Murrell 2019).

5

Base R: "Paragraphs" and "Essays"

> An R script is simply a text file containing (almost) the same commands that you would enter on the command line of R.
>
> Jim Lemon
> *Kickstarting R*

5.1 Aims of This Chapter

For those who have mainly used graphical user interfaces, understanding why and when scripts can help in communicating a certain data analysis protocol can be revelatory. As soon as a data analysis stops being trivial, describing the steps followed through a system of menus and dialogue boxes becomes extremely tedious.

Moreover, graphical user interfaces tend to be difficult to extend or improve in a way that keeps step-by-step instructions valid across program versions and operating systems.

Many times, exactly the same sequence of commands needs to be applied to different data sets, and scripts make both implementation and validation of such a requirement easy.

In this chapter, I will walk you through the use of R scripts, starting from an extremely simple script.

5.2 Writing Scripts

In R language, the closest match to a natural language essay is a script. A script is built from multiple interconnected code statements needed to complete a given task. Simple statements, equivalent to sentences, can be combined into compound statements, equivalent to natural language paragraphs. Frequently, we combine simple sequences of statements into a sequence of actions necessary to complete

DOI: 10.1201/9781003404187-5

a task. The sequence is not necessarily linear, as branching and repetition are also available.

Scripts can vary from simple scripts containing only a few code statements, to complex scripts containing hundreds of code statements. In the rest of the present section I discuss how to write readable and reliable scripts and how to use them.

5.2.1 What is a script?

A *script* is a text file that contains (almost) the same commands that you would type at the R console prompt. A true script is not, for example, an MS-Word file where you have pasted or typed some R commands.

When typing commands/statements at the R console, we “feed” one line of text at a time. When we end the line by typing the enter key, the line of text is interpreted and evaluated. We then type the next line of text, which gets in turn interpreted and evaluated, and so on. In a script we write nearly the same text in an editor and save multiple lines containing commands into a text file. Interpretation takes place only later, when we *source* the file as a whole into R.

A script file has the following characteristics.

- The script is a plain text file, i.e., a file containing bytes that represent alphanumeric characters in a standardised character set like UTF8 or ASCII.
- The text in the file contains valid R statements (including comments) and nothing else.
- Comments start at a `#` and end at the end of the line.
- The R statements are in the file in the order that they must be executed, and respecting the line continuation rules of R.
- R scripts customarily have file names ending in `.r` or `.R`.

The statements in the text file, are read, interpreted, and evaluated sequentially, from the start to the end of the file, as represented in the diagram (Figure 5.1).

As we will see later in the chapter, code statements can be combined into larger statements and evaluated conditionally and/or repeatedly, which allows us to control the realised sequence of evaluated statements.

In addition to being valid, it is important that scripts are also understandable to humans. Consequently, a clear writing style and consistent adherence to it are important.

It is good practice to write scripts so that they are self-contained. To make a script self-contained, one must include code to load the packages used, load or import data from files, perform the data analysis, and display and/or save the results of the analysis. Such scripts can be used to apply the same analysis algorithm to other data by reading data from a different file and/or to reproduce the same analysis at a later time using the same data. Such scripts document all steps used for the analysis.

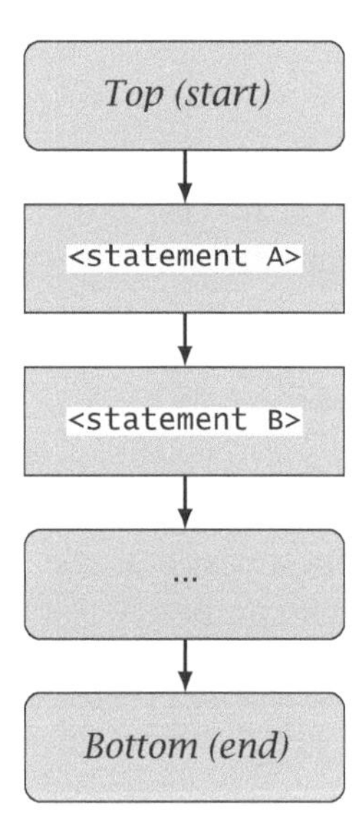

Figure 5.1
Diagram of script showing sequentially evaluated code statements; ⋯ represent additional statements in the script.

5.2.2 How do we use a script?

A script can be "sourced" using function `source()`. If a text file called `my.first.script.r` contains the text

```
# this is my first R script
print(3 + 4)
```

it can be sourced by typing at R console

```
source("my.first.script.r")
## [1] 7
```

Execution of the statements in the file makes R display `[1] 7` at the console, below the command we typed in. The commands themselves are not shown (by default the sourced file is not *echoed* to the console) and the results of computations are not printed unless one includes explicit `print()` commands in the script.

Scripts can be run both by sourcing them into an open R session, or at the operating system command prompt (see section 2.3 on page 12). In RStudio, the script in the currently active editor tab can be sourced using the "source" button. The drop-down menu of this button has three entries: "Source" , quietly to the R console; "Source with echo" showing the code as it is run; and "Source as local job", using a new instance of R in the background. In the last case, the R console remains free for other uses while the script is running.

When a script is *sourced*, the output can be saved to a text file instead of being shown in the console. It is also easy to call R with the R script file as an argument directly at the operating system shell or command-interpreter prompt—and obviously also from shell scripts. The next two chunks show commands entered at the OS shell command prompt rather than at the R command prompt.

```
RScript my.first.script.r
```

You can open an operating system's *shell* from the Tools menu in RStudio, to run this command. The output will be printed to the shell console. If you would like to save the output to a file, use output redirection using the operating system's syntax.

```
RScript my.first.script.r > my.output.txt
```

While developing or debugging a script, one usually wants to run (or *execute*) one or a few statements at a time. This can be done in RStudio using the "run" button after either positioning the cursor in the line to be executed, or selecting the text to be run (the selected text can be part of a line, a whole line, or a group of lines, as long as it is syntactically valid). The key-shortcut Ctrl-Enter is equivalent to pressing the "run" button.

5.2.3 How to write a script

As with any type of writing, different approaches may be preferred by different R users. In general, the approach used, or mix of approaches, will also depend on how confident one is that the statements will work as expected—one already knows the best approach vs. one is exploring different alternatives.

Three approaches are listed below. They all can result in equally good code, but as work in progress, they differ. In the first approach, the script as a whole is likely to contain some bugs until being thoroughly tested. In the middle approach, only the most recently added statements are likely to contain bugs. In the last one, the script contains at all times only valid R code, even if incomplete. This third approach also has the advantage that code remains in the R console *History* and can be retrieved with a delay, e.g., after comparison against an alternative statement.

If one is very familiar with similar problems, one can create a new text file and write the whole script in the editor, testing it only afterwards. Use of this approach is uncommon.

If one is moderately familiar with the problem, one can write a script as above, but testing it, step by step, while writing it, i.e., running parts of the script before continuing with the writing. This is the approach I use most frequently.

If one is mostly playing around, one can type statements at the console prompt to try them. As every statement ran at the console is saved to the "History", these previously entered statement(s) can be copied and pasted into the script. In this way one can build a script from statements already known to work correctly.

5.1 By now you should be familiar enough with R to be able to write your own script.

1. Create a new R script (in RStudio, from the File menu, leftmost "+" icon, or by typing "Ctrl + Shift + N").
2. Save the file as `my.second.script.r`.

3. Use the editor pane in RStudio to type some R commands and comments.
4. *Run* individual commands.
5. *Source* the whole file.

5.2.4 The need to be understandable to people

It is not enough for program code to be understood by a computer and that it returns the correct answer. Both large programs and small scripts have to be readable to humans, and the intention of the code understandable. In most cases, R code will be maintained, reused, and modified over time. In many cases, this code also serves to document a given computation and to make it possible to reproduce it.

When one writes a script, it is either because one wants to document what has been done or because one plans to use it again in the future. In the first case, other persons will read it, and in the second case, one rarely remembers all the details. Thus, spending time and effort on the writing style, paying special attention to the following recommendations, is important.

- Avoid the unusual. People using a certain programming language tend to use some implicit or explicit rules of style—style includes *indentation* of statements, *capitalisation* of variable and function names. As a minimum try to be consistent with yourself.
- Use meaningful names for variables, and any other object. What is meaningful depends on the context. Depending on common use, a single letter may be more meaningful than a long word. However self-explanatory names are usually better: e.g., using `n.rows` and `n.cols` is much clearer than using `n1` and `n2` when dealing with a matrix of data. Probably `number.of.rows` and `number.of.columns` would make the script verbose, and take longer to type without gaining anything in return. Sometimes, short textual explanations in comments (ignored by R) are needed to achieve readability for humans.
- How to make the words visible in names: traditionally in R one would use dots to separate the words and use only lower case. Some years ago, it became possible to use underscores. The use of underscores is common nowadays because it is "safer", as in some situations a dot may have a special meaning. Names like `NumCols`, using "camel case", are only infrequently used in R programming but are frequently used in other languages like Pascal.

The *Tidyverse style guide* for writing R code (`https://style.tidyverse.org/`) provides more detailed "rules". However, more important than strictly following a published guideline is to be consistent in the style one, a team of programmers or data analysts, or even members of an organisation use. In the current book, I have not followed this guide in all respects, instead following in some cases the style used in R documentation. However, I have attempted to be consistent.

5.2 Here is an example of bad style in a script. Edit the code in the chunk below so that it becomes easier to read.

```
a <- 2 # height
b <- 4 # length
C <-
    a *
b
C -> variable
      print(
"area: ", variable
)
```

The points discussed above already help a lot. However, one can go further in achieving the goal of human readability by interspersing explanations and code "chunks" and using all the facilities of typesetting, even of formatted maths formulas and equations, within the listing of the script. Furthermore, by including the results of the calculations and the code itself in a typeset report built automatically one ensures that they match each other. This greatly contributes to data analysis reproducibility, which is becoming a widespread requirement both in academia and in industry.

This approach is called literate programming and was first proposed by Knuth (1984) through his WEB system. In the case of R programming, the first support of literate programming was in 'Sweave', which has been superseded by 'knitr' (Xie 2013). This package supports the use of Markdown or LaTeX (Lamport 1994) as the markup language for the textual contents and also formats and applies syntax highlighting to code. R markdown is an extension to Markdown that makes it easier to include R code in documents (see `http://rmarkdown.rstudio.com/`). It is the basis of R packages that support typesetting large and complex documents ('bookdown'), web sites ('blogdown'), package vignettes ('pkgdown'), and slides for presentations (Xie 2016; Xie et al. 2018). Quarto, which provides an enhanced version of R markdown, is implemented in R package 'quarto' together with the Quarto program as a separate executable. The use of 'knitr' and 'quarto' is very well integrated into the RStudio IDE. The generation of typeset reports is outside the scope of the book, but it is an important skill to learn. It is well described in the books and web sites cited.

5.2.5 Debugging scripts

The use of the word *bug* to describe a problem in computer hardware and software started in 1946 when a real bug, more precisely a moth, got between the contacts of a relay in an electromechanical computer causing it to malfunction and Grace Hooper described the first computer *bug*. The use of the term bug in engineering predates the use in computer science, and consequently, the use of the word bug in computing caught on easily.

A suitable quotation from a letter written by Thomas Alva Edison in 1878 (as given by Hughes 2004):

> It has been just so in all of my inventions. The first step is an intuition, and comes with a burst, then difficulties arise-this thing gives out and [it is] then that "Bugs"—as such little faults and difficulties are called—show them-

selves and months of intense watching, study and labour are requisite before commercial success or failure is certainly reached.

The quoted paragraph above makes clear that only very exceptionally does any new design fully succeed. The same applies to R scripts as well as any other non-trivial piece of computer code. From this it logically follows that testing and debugging are fundamental steps in the development of R scripts and packages. Debugging, as an activity, is outside the scope of this book. However, clear programming style and good documentation are indispensable for efficient testing and reuse.

Even for scripts used for analysing a single data set, we need to be confident that the algorithms and their implementation are valid, and able to return correct results. This is true both for scientific reports, expert reports, and any data analysis related to assessment of compliance with legislation or regulations. Of course, even in cases when we are not required to demonstrate validity, say for decision making purely internal to a private organisation, we will still want to avoid costly mistakes.

The first step in producing reliable computer code is to accept that any code that we write needs to be tested and, if possible, validated. Another important step is to make sure that input is validated within the script and a suitable error produced for bad input (including valid input values falling outside the range that can be reliably handled by the script).

If during testing, or during normal use, a wrong value or no value is returned by a calculation (e.g., the script crashes or triggers a fatal error), debugging consists in finding the cause of the problem. The cause can be either a mistake in the implementation of an algorithm or in the algorithm itself. However, many apparent *bugs* are caused by bad, or missing, code for handling of special cases, such as invalid input values, rounding errors, and division by zero, making a function or script crash instead of elegantly issuing a helpful message.

Diagnosing the source of bugs is, in most cases, like detective work. One uses hunches based on common sense and experience to try to locate the lines of code causing the problem. One follows different *leads* until the case is solved. In most cases, at the very bottom, we rely on some sort of divide-and-conquer strategy. For example, we may check the value returned by intermediate calculations until we locate the earliest code statement producing a wrong value. Another common case is when some input values trigger a bug. In such cases, it is frequently best to start by testing if different "cases" of input lead to errors/crashes or not. Boundary input values are usually the telltale ones: for numbers, zero, negative and positive values, very large values, very small values, missing values (`NA`), vectors of length zero (`numeric()`), etc.

⚠ **Error messages** When debugging, keep in mind that in some cases a single bug can lead to a whole cascade of error messages. Do also keep in mind that typing mistakes, originating when code is entered through the keyboard, can wreak havock in a script: usually there is little correspondence between the number of error messages and the seriousness of the bug triggering them. When several errors are triggered, start by reading the error message printed first, as later errors can be an indirect consequence of earlier ones.

There are special tools, called debuggers, available, and they help enormously. Debuggers allow one to step through the code, executing one statement at a time, allowing inspection of the objects present in the R environment. It is even possible to execute additional statements at the R console, e.g., to modify the value of a variable, while execution is paused. An R debugger is available within RStudio and also through the R console.

When writing your first scripts, you will manage perfectly well, and learn more by running the script one line at a time, and when needed temporarily inserting `print()` statements to "look" at how the value of variables changes at each step. A debugger allows a lot more control, as one can "step in" and "step out" of function definitions, and set and unset break points where execution will stop. However, using a debugger is not as simple as using `print()`.

If you get stuck trying to find the cause of a bug, do extend your search both to the most trivial of possible causes, and later on to the least likely ones (such as a bug in a package installed from CRAN or R itself). Of course, when suspecting a bug in code you have not written, it is wise to very carefully read the documentation, as the "bug" may be just a misunderstanding of what a certain piece of code is expected to do. Also keep in mind that as discussed on page 6, you will be able to find online already-answered questions to many of your likely problems and doubts. For example, searching with Google for the text of an error message is usually well rewarded. Most important to remember is that bugs do pop up frequently in newly written code, and occasionally in old code. No coding is immune to them, thus, the code you write, packages you use or R itself can contain bugs.

5.3 Compound Statements

Individual statements can be grouped into *compound statements* by enclosing them in curly braces (Figure 5.2). Conceptually, is like putting these statements into a box that allows us to operate with them as an anonymous whole.

```
print("...")
## [1] "..."
{
  print("A")
  print("B")
}
## [1] "A"
## [1] "B"
print("...")
## [1] "..."
```

The grouping of the two middle statements above is of no consequence, as it does not alter sequential evaluation. In the example above, only side effects are of interest. In the example below, the value returned by a compound statement is that returned by the last statement evaluated within it. Individual statements

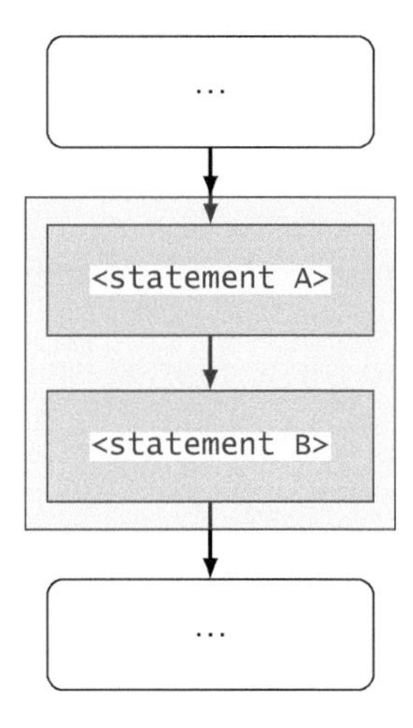

Figure 5.2
Diagram of a compound code statement is a grouping of statements that in some contexts behaves as a single statement. In the diagram, statements A and B have been grouped into a compound statement.

can be separated by an end-of-line as above, or by a semicolon (;) as below: two statements, each of them implementing an arithmetic operation.

```
{1 + 2; 3 + 4}
## [1] 7
```

The example above demonstrates that only the value returned by the compound statement as a whole is displayed automatically at the R console, i.e., the implicit call to `print()` is applied to the compound statement. Thus, even though both statements were evaluated, we only see the result returned by the second one.

5.3 Nesting is also possible. Before running the compound statement below try to predict the value it will return, and then run the code and compare your prediction to the value returned.

```
{1 + 2; {a <- 3 + 4; a + 1}}
```

Grouping is of little use by itself. It becomes useful together with control-of-execution constructs, when defining functions, and in similar cases where we need to treat a group of code statements as if they were a single statement. We will see several examples of the use of compound statements in the current chapter and in chapter 6 on page 169.

5.4 Function Calls

We will describe functions in detail and how to create new ones in chapter 6. We have already been using functions since chapter 3. Functions are structurally R statements, in most cases, compound statements, using formal parameters as placeholders. When one calls a function, one passes arguments for the different parameters (or placeholder names) and the (compound) statement conforming the *body* of the function is evaluated after "replacing" the placeholders by the values passed as arguments.

In the first example, we use two statements. In the first statement, $log(100)$ is computed by calling function `log10()` with `100` as argument and the returned value is assigned to variable `a`. In the second statement, the value 2 is displayed as a side effect of calling `print()` with variable `a` as argument.

```
a <- log10(100)
print(a)
## [1] 2
```

The two statements in example above can be rewritten as a single statement using a nested function call.

```
print(log10(100))
## [1] 2
```

The difference is that we avoid the explicit creation of a variable. Whether this is an advantage or not depends on whether we use variable `a` in later statements or not.

Statements with more levels of nesting than shown above become very difficult to read, so alternative notations can help.

5.5 Data Pipes

Pipes have been at the core of shell scripting in Unix since early stages of its design (Kernigham and Plauger 1981) as well as in Linux distributions. Within an OS, pipes are chains of small programs or "tools" that carry out a single well-defined task (e.g., `ed`, `sub`, `gsub`, `grep`, and `more`). Data such as text is described as flowing from a source into a sink through a series of steps at which a specific transformations take place. In Unix and Linux shells like `sh` or `bash`, sinks and sources are files, but in Unix and Linux files are an abstraction that includes all devices and connections for input or output, including physical ones such as terminals and printers.

```
stdin | grep("abc") | more
```

How can *pipes* exist within a single R script? When chaining functions into a pipe, data is passed between them through temporary R objects stored in memory, which are created and destroyed automatically. Conceptually, there is little difference between Unix shell pipes and pipes in R scripts, but the implementations are different.

What do pipes achieve in R scripts? They relieve us from the responsibility of creating and deleting the temporary objects. By chaining the statements they enforce their sequential execution. Pipes usually improve the readability of scripts by allowing more concise code.

Since 2021, starting from version 4.1.0, R has had a native pipe operator (`|>`) as part of the language. Subsequently, the placeholder (`_`) was implemented in version 4.2.0 and its functionality expanded in version 4.3.0. Another two implementations of pipes, that have been available as R extensions for some years in packages 'magrittr' and 'wrapr', are described in chapter 8 on page 243.

I describe R's pipe syntax based on R 4.3.0. I start by showing the same operations coded using nested function calls, using explicit saving of intermediate values in temporary objects, and using the pipe operator.

Nested function calls are concise, but difficult to read when the depth of nesting increases.

```
sum(sqrt(1:10))
## [1] 22.46828
```

Saving intermediate results explicitly results in clear but verbose code.

```
data.in <- 1:10
data.tmp <- sqrt(data.in)
sum(data.tmp)
## [1] 22.46828
rm(data.tmp) # clean up!
```

A pipe using operator `|>` makes the data flow clear and keeps the code concise.

```
1:10 |> sqrt() |> sum()
## [1] 22.46828
```

We can assign the result of the computation to a variable, most elegantly using the `->` operator on the *rhs* of the pipe.

```
1:10 |> sqrt() |> sum() -> my_rhs.var
my_rhs.var
## [1] 22.46828
```

We can also use the `<-` operator on the *lhs* of the pipe, i.e., for assignments a pipe behaves as a compound statement.

```
my_lhs.var <- 1:10 |> sqrt() |> sum()
my_lhs.var
## [1] 22.46828
```

Formally, the `|>` operator from base R takes two operands, just like operator `+` does. The value returned by the *lhs* (left-hand side) operand, which can be any R expression, is passed as argument to the function-call operand on *rhs* (right-hand side). The called function must accept at least one argument. This default syntax that implicitly passes the argument by position to the first parameter of the function would limit which functions could be used in a pipe construct. However, it is also possible to pass the piped argument explicitly by name to any parameter of the function on the *rhs* using an underscore (`_`) as a placeholder.

```
1:10 |> sqrt(x = _) |> sum(x = _)
## [1] 22.46828
```

The placeholder can be also used with extraction operators.

```
1:10 |> sqrt(x = _) |> _[2:8] |> sum(x = _)
## [1] 15.306
```

Base R functions like `subset()` have formal parameters in an order that is suitable for implicitly passing the piped value as an argument to their first parameter, while others like `assign()` do not. For example, when calling function `assign()` to save a value using a name available as a character string, we would like to pass the piped value as an argument to parameter `value` which is not the first. In such cases, we can use _ as a placeholder and pass it by name.

```
obj.name <- "data.out"
1:10 |> sqrt() |> sum() |> assign(x = obj.name, value = _)
```

Alternatively, we can define a wrapper function, with the desired order for the formal parameters. This approach can be worthwhile when the same function is called repeatedly within a script.

```
value_assign <- function(value, x, ...) {
  assign(x = x, value = value, ...)
}
obj.name <- "data.out"
1:10 |> sqrt() |> sum() |> value_assign(obj.name)
```

In general, whenever we use temporary variables to store values that are passed as arguments only once, we can nest or chain the statements making the saving of intermediate results into a temporary variable implicit instead of explicit. Examples of some useful idioms follow.

Addition of computed variables to a data frame using `within()` (see section 4.4.5 on page 110) and selecting rows with `subset()` (see section 4.4.1 on page 102) are combined in our first simple example. For clarity, we use the _ placeholder to indicate the value returned by the preceding function in the pipe.

```
data.frame(x = 1:10, y = rnorm(10)) |>
  within(data = _,
         {
           x4 <- x^4
           is.large <- x^4 > 1000
         }) |>
  subset(x = _, is.large)
##     x           y is.large    x4
## 6   6  0.24821983     TRUE  1296
## 7   7 -1.32684062     TRUE  2401
## 8   8 -0.05789719     TRUE  4096
## 9   9  0.29258553     TRUE  6561
## 10 10 -0.63177298     TRUE 10000
```

5.4 Without using the _ placeholder, but using a more compact layout, the code above becomes that shown below. Compare it to that above to work out how I simplified the code.

```
data.frame(x = 1:10, y = rnorm(10)) |>
  within({x4 <- x^4; is.large <- x^4 > 1000}) |>
  subset(is.large)
```

Subset can be also used to select variables or columns from data frames and matrices.

```
data.frame(x = 1:10, y = rnorm(10)) |>
  within(data = _,
         {
           x4 <- x^4
           is.large <- x^4 > 1000
         }) |>
```

```
  subset(x = _, is.large, select = -x)
##               y is.large    x4
## 6    0.37246559     TRUE  1296
## 7    0.94549288     TRUE  2401
## 8    1.06913611     TRUE  4096
## 9   -0.06230746     TRUE  6561
## 10  -0.47022979     TRUE 10000
```

```
data.frame(x = 1:10, y = rnorm(10)) |>
  within(data = _,
         {
           x4 <- x^4
           is.large <- x^4 > 1000
         }) |>
  subset(x = _, select = c(y, x4))
##              y    x4
## 1  -0.92142448     1
## 2   0.62913424    16
## 3   0.42764317    81
## 4   0.04825671   256
## 5   0.56229201   625
## 6  -3.73168145  1296
## 7   0.16993802  2401
## 8  -0.40038974  4096
## 9  -0.80676551  6561
## 10  0.13977301 10000
```

```
data.frame(group = factor(rep(c("T1", "T2", "Ctl"), each = 4)),
           y = rnorm(12)) |>
  subset(x = _, group %in% c("T1", "T2")) |>
  aggregate(data = _, y ~ group, mean)
##   group          y
## 1    T1 -0.2764033
## 2    T2  0.3205809
```

The extraction operators are accepted on the *rhs* of a pipe only starting from `R` 4.3.0. With these versions `_[["y"]]`, as shown below, as well as its equivalent `_$y` can be used. Function `getElement()` used as `getElement("y")`, being a normal function, can be used in situations where operators are not accepted, like on the *rhs* of `|>` in older versions of `R`.

```
data.frame(group = factor(rep(c("T1", "T2", "Ctl"), each = 4)),
           y = rnorm(12)) |>
  subset(x = _, group %in% c("T1", "T2")) |>
  aggregate(data = _, y ~ group, mean) |>
  _[["y"]]
## [1] -0.2216849 -0.7040829
```

Additional functions designed to be used in pipes are available through packages as described in chapter 8.

5.5 In the last three examples, in which function calls is the explicit use of the placeholder needed, and in which ones is it optional? Hint: edit the code, removing the parameter name, `=`, and `_`, and test whether the edited code works and returns the same value as before.

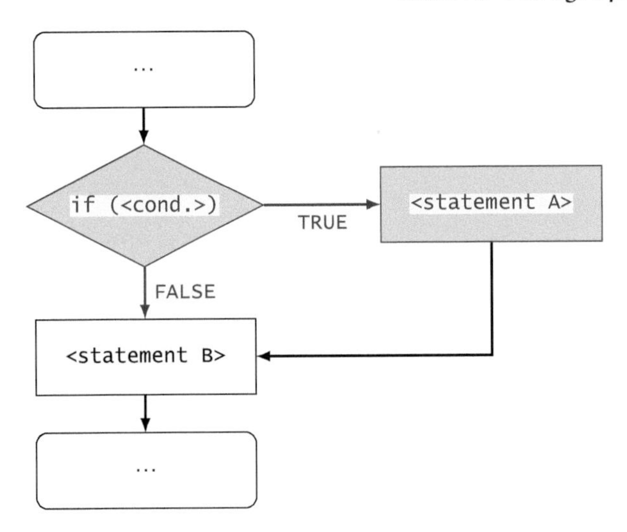

Figure 5.3
Flowchart for `if` construct.

5.6 Conditional Evaluation

By default, R statements in a script are evaluated (or executed) in the sequence they appear in the script *listing* or text. We give the name *control of execution constructs* to those special statements that allow us to alter this default sequence, by either skipping or repeatedly evaluating individual statements. The statements whose evaluation is controlled can be either simple or compound. Some of the control of execution flow statements, function like *ON-OFF switches* for program statements. Others allow statements to be executed repeatedly while or until a condition is met, or until all members of a list or a vector are processed.

These *control of execution constructs* can be also used at the R console, but it is usually awkward to do so as they can extend over several lines of text. In simple scripts, the *flow of execution* can be fixed and linear from the first to the last statement in the script. However, *control of execution constructs* are a crucial part of most useful scripts. As we will see next, a compound statement can include multiple simple or nested compound statements. R has two types of *if* statements, non-vectorised and vectorised.

5.6.1 Non-vectorised `if`, `else` and `switch`

The `if` construct "decides", depending on a `logical` value, whether the next code statement is executed (if `TRUE`) or skipped (if `FALSE`) (Figure 5.3). The flow chart shows how `if` works: `<statement A>` is either evaluated or skipped depending on the value of `<condition>`, while `<statement B>` is always evaluated.

The usefulness of *if* statements stems from the possibility of computing the `logical` value used as `<condition>` with comparison operators (see section 3.6 on page 52) and logical operators (see section 3.5 on page 49).

We start with toy examples demonstrating how *if* statements work. Later we will see examples closer to real use cases. Here `if` controls the evaluation or not of the simple statement `print("Hello!")`.

We use the name *flag* for a `logical` variable set manually, preferably near the top of the script. Real flags were used in railways to indicate to trains whether to stop or continue at stations and which route to follow at junctions. Use of `logical` flags in scripts is most useful when switching between two behaviours that depend on multiple separate statements.

```
flag <- TRUE
if (flag) print("Hello!")
## [1] "Hello!"
```

5.6 Play with the code above by changing the value assigned to variable `flag`, `FALSE`, `NA`, and `logical(0)`.

In the example above we use variable `flag` as the *condition*.

Nothing in the R language prevents this condition from being a `logical` constant. Explain why `if (FALSE)` in the syntactically correct statement below is of no practical use.

```
if (FALSE) print("Hello!")
```

Conditional execution is much more useful than what could be expected from the previous examples, because the statement whose execution is being controlled can be a compound statement of almost any length or complexity. A very simple example follows, with a compound statement containing two statements, each one, a call to function `print()` with a different argument.

```
printing <- TRUE
if (printing) {
  print("A")
  print("B")
}
## [1] "A"
## [1] "B"
```

The condition passed as an argument to `if`, enclosed in parentheses, can be anything yielding a `logical` vector of length one. As this condition is *not* vectorised, a longer vector will trigger an R warning or error depending on R's version.

The `if` ... `else` ... construct "decides", depending on a `logical` value, which of two code statements is executed (Figure 5.4). The flow chart shows how it works: either `<statement A>` or `<statement B>` is evaluated and the other skipped depending on the value of `<condition>`, while `<statement C>` is always evaluated.

```
a <- 10
if (a < 0) print("'a' is negative") else print("'a' is not negative")
## [1] "'a' is not negative"
print("This is always printed")
## [1] "This is always printed"
```

As can be seen above, the statement immediately following `if` is executed if the condition returns `TRUE` and that following `else` is executed if the condition

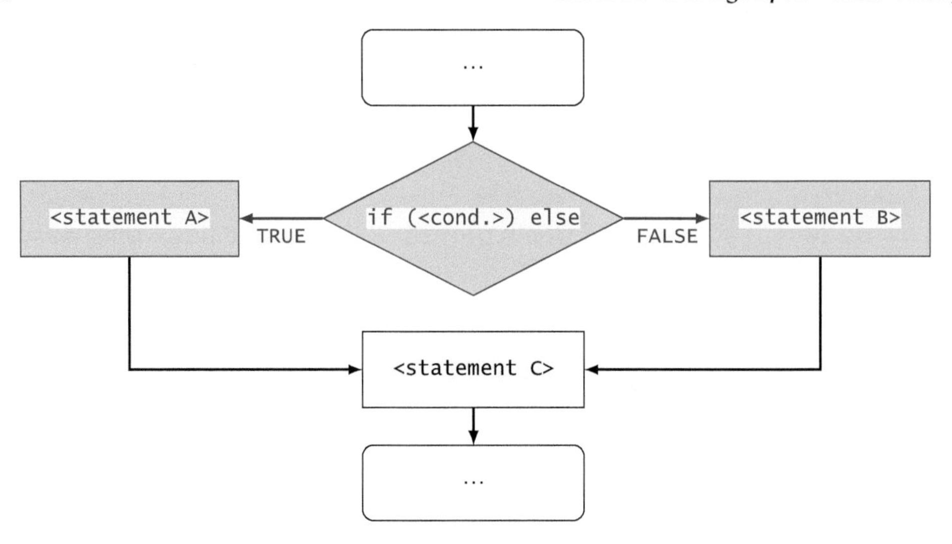

Figure 5.4
Flowchart for `if` ...`else` construct.

returns FALSE. Statements after the conditionally executed `if` and `else` statements are always executed, independently of the value returned by the condition.

5.7 Play with the code in the chunk above by assigning different numeric vectors to a.

Do you still remember the rules about continuation lines?

```
# 1
a <- 1
if (a < 0) print("'a' is negative") else print("'a' is not negative")
## [1] "'a' is not negative"
```

Why does the statement below (not evaluated here) trigger an error while the one above does not?

```
# 2 (not evaluated here)
if (a < 0) print("'a' is negative")
else print("'a' is not negative")
```

How do the continuation line rules apply when we add curly braces as shown below.

```
# 1
a <- 1
if (a < 0) {
    print("'a' is negative")
  } else {
    print("'a' is not negative")
  }
## [1] "'a' is not negative"
```

In the example above, we enclosed a single statement between each pair of curly braces, but as these braces create compound statements, multiple statements could have been enclosed between each pair.

5.8 Play with the use of conditional execution, with both simple and compound statements, and also think how to combine `if` and `else` to select among more than two options.

In R, the value returned by any compound statement is the value returned by the last simple statement executed within the compound one. This means that we can assign the value returned by an `if` and `else` statement to a variable. This style is less frequently used, but occasionally can result in easier-to-understand scripts.

```
a <- 1
my.message <-
  if (a < 0) "'a' is negative" else "'a' is not negative"
print(my.message)
## [1] "'a' is not negative"
```

If the condition statement returns a value of a class other than `logical`, R will attempt to convert it into a logical. This is sometimes used instead of a comparison to zero, as the conversion from `integer` yields `TRUE` for all integers except zero. The code below illustrates a rather frequently used idiom for checking if there is something available to display.

```
message <- "abc"
if (length(message)) print(message)
## [1] "abc"
```

5.9 Study the conversion rules between `numeric` and `logical` values, run each of the statements below, and explain the output based on how type conversions are interpreted, remembering the difference between *floating-point numbers* as implemented in computers and *real numbers* as defined in mathematics (see page 27).

```
if (0) print("hello")
if (-1) print("hello")
if (0.01) print("hello")
if (1e-300) print("hello")
if (1e-323) print("hello")
if (1e-324) print("hello")
if (1e-500) print("hello")
if (as.logical("true")) print("hello")
if (as.logical(as.numeric("1"))) print("hello")
if (as.logical("1")) print("hello")
if ("1") print("hello")
```

Hint: if you need to refresh your understanding of the type conversion rules, see section 3.9 on page 60.

In addition to `if` and `if...else`, there is in R a `switch()` statement (Figure 5.5). It can be used to select among several *cases*, or alternative statements, based on an expression that returns a `numeric` or a `character` value of length one when evaluated.

A `switch()` statement returns a value, just like `if` does. The value passed as argument to `switch()` functions as an index selecting one of the statements. The value returned by the `switch()` statement is the value returned by the selected *case* statement.

In the first example below, we use a `character` variable as the condition, named cases, and a final unlabelled case as default in case of no match. In real use, a computed value or user input would be used in place of `my.object`. As with the

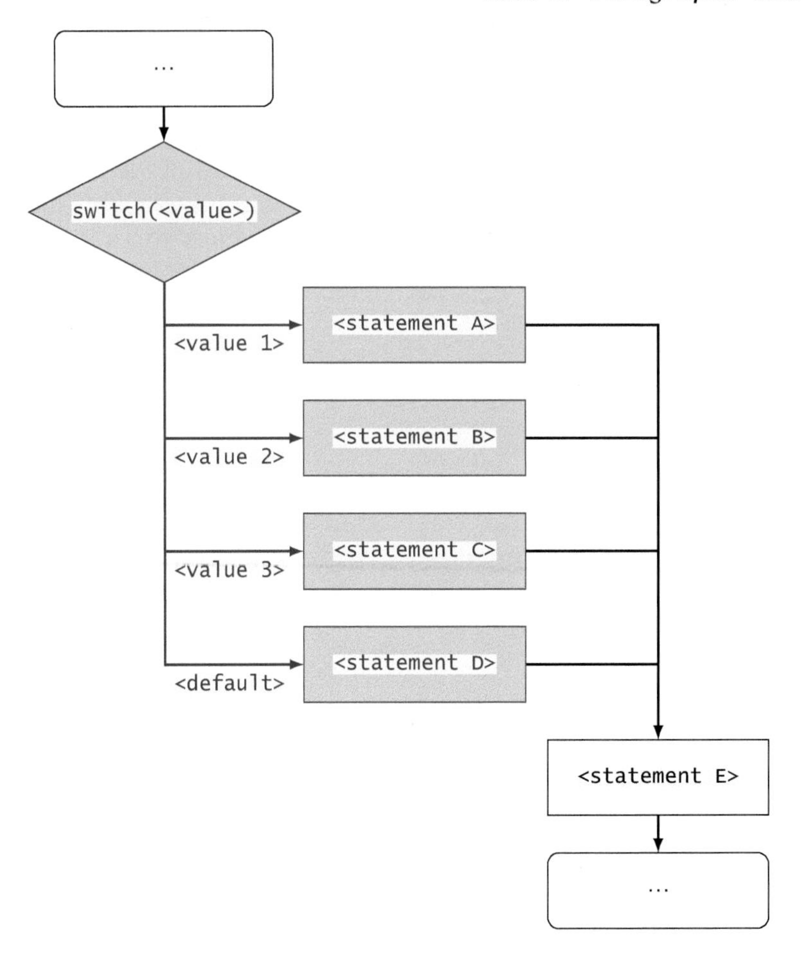

Figure 5.5
Flowchart for a `switch` construct with four cases.

`logical` argument to `if`, the `character` string value passed as argument must be a vector of length one.

```
my.object <- "two"
b <- switch(my.object,
            one = 1,
            two = 1 / 2,
            four = 1 / 4,
            0
)
b
## [1] 0.5
```

Multiple condition values can share the same statement.

```
my.object <- "two"
b <- switch(my.object,
            one =, uno = 1,
            two =, dos = 1 / 2,
            four =, cuatro = 1 / 4,
            0
```

```
)
b
## [1] 0.5
```

5.10 Do play with the use of the switch statement. Look at the documentation for `switch()` using `help(switch)` and study the examples at the end of the help page. Explore what happens if you set `my.object <- "ten"`, `my.object <- "three"`, `my.object <- NA_character_` or `my.object <- character()`. Then remove the `, 0` as default value, and repeat.

When the expression used as a condition returns a value that is not a `character`, it will be interpreted as an `integer` index. In this case, no names are used for the cases and the last case is always interpreted as the default.

```
my.number <- 2
b <- switch(my.number,
            1,
            1 / 2,
            1 / 4,
            0
)
b
## [1] 0.5
```

5.11 Continue playing with the use of the switch statement. Explore what happens if you set `my.number <- 10`, `my.number <- 3`, `my.number <- NA`, or `my.object <- numeric()`. Afterwards, remove the `, 0` as default value, and repeat.

The statements for the cases in a `switch()` statement can be compound statements as in the case of `if`, and they can even be used for a side effect. The code example above can edited to print a message when the default value is returned.

```
my.object <- "ten"
b <- switch(my.object,
            one = 1,
            two = 1 / 2,
            three = 1 / 4,
            {print("No match! Using default"); 0}
)
## [1] "No match! Using default"
b
## [1] 0
```

The `switch()` statement can substitute for chained `if ... else` statements when all the conditions can be described by constant values or distinct values returned by the same test. The advantage is more concise and readable code. The equivalent of the first `switch()` example above when written using `if ... else` becomes longer. Given how terse code using `switch()` is, those not yet familiar with its use may find the more verbose style used below easier to understand. On the other hand, with numerous cases, a `switch()` statement is easier to read and understand.

```
my.object <- "two"
if (my.object == "one") {
  b <- 1
} else if (my.object == "two") {
  b <- 1 / 2
} else if (my.object == "four") {
  b <- 1 / 4
} else {
  b <- 0
}
b
## [1] 0.5
```

5.12 Consider another alternative approach, the use of a named vector to map values. In most of the examples above, the code for the cases is a constant value or an operation among constant values. Implement one of these examples using a named vector instead of a `switch()` statement.

5.6.2 Vectorised `ifelse()`

Vectorised *ifelse* is a peculiarity of the R language, but very useful for writing concise code that may execute faster than logically equivalent but not vectorised code. Vectorised conditional execution is coded by means of *function* `ifelse()` (written as a single word). This function takes three arguments: a `logical` vector usually the result of a test (parameter `test`), an expression to use for `TRUE` cases (parameter `yes`), and an expression to use for `FALSE` cases (parameter `no`). At each index position along the vectors, the value included in the returned vector is taken from `yes` if the corresponding member of the `test` logical vector is `TRUE` and from `no` if the corresponding member of `test` is `FALSE`. All three arguments can be any R statement returning the required vectors.

The flow chart for `ifelse()` is similar to that for `if ... else` shown on page 138 but applied in parallel to the individual members of vectors; e.g., the condition expression is evaluated at index position `1` controls which value will be present in the returned vector at index position `1`, and so on.

It is customary to pass arguments to `ifelse` by position. We give a first example with named arguments to clarify the use of the function.

```
my.test <- c(TRUE, FALSE, TRUE, TRUE)
ifelse(test = my.test, yes = 1, no = -1)
## [1]  1 -1  1  1
```

In practice, the most common idiom is to have as an argument passed to `test`, the result of a comparison calculated on the fly. As an example, the absolute values of the members of a vector are computed using `ifelse()` instead of with R function `abs()`.

```
nums <- -3:+3
ifelse(nums < 0, -nums, nums)
## [1] 3 2 1 0 1 2 3
```

In the case of `ifelse()`, the length of the returned value is determined by the length of the logical vector passed as an argument to its first formal parameter (named `test`)! A frequent mistake is to use a condition that returns a `logical` vector of length one, expecting that it will be recycled because arguments passed to the other formal parameters (named `yes` and `no`) are longer. However, no recycling will take place, resulting in a returned value of length one, with the remaining elements of the vectors passed to `yes` and `no` being discarded. Do try this by yourself, using logical vectors of different lengths. You can start with the examples below, making sure you understand why the returned values are what they are.

```
ifelse(TRUE, 1:5, -5:-1)
## [1] 1
ifelse(FALSE, 1:5, -5:-1)
## [1] -5
ifelse(c(TRUE, FALSE), 1:5, -5:-1)
## [1]  1 -4
ifelse(c(FALSE, TRUE), 1:5, -5:-1)
## [1] -5  2
ifelse(c(FALSE, TRUE), 1:5, 0)
## [1] 0 2
```

5.13 Some additional examples to play with, containing a few surprises. Study the examples below until you understand why returned values are what they are. In addition, create your own examples to test other possible cases. In other words, play with the code until you fully understand how `ifelse()` statements work.

```
a <- 1:10
ifelse(a > 5, 1, -1)
ifelse(a > 5, a + 1, a - 1)
ifelse(any(a > 5), a + 1, a - 1) # tricky
ifelse(logical(0), a + 1, a - 1) # even more tricky
ifelse(NA, a + 1, a - 1) # as expected
```

Hint: if you need to refresh your understanding of `logical` values and Boolean algebra see section 3.5 on page 49.

5.14 Using `ifelse()`, write a single statement to combine numbers from the two vectors `a` and `b` into a result vector `d`, based on whether the corresponding value in vector `c` is the character `"a"` or `"b"`. Then print vector `d` to make the result visible.

```
a <- -10:-1
b <- +1:10
c <- c(rep("a", 5), rep("b", 5))
# your code
```

If you do not understand how the three vectors are built, or you cannot guess the values they contain by reading the code, print them, and play with the arguments, until you understand what each parameter does. Also use `help(rep)` and/or `help(ifelse)` to access the documentation.

5.15 Continuing from the playground above, test the behaviour of `ifelse()` with `NA`, `NULL` and `logical()` passed as arguments to `test`. Also test the behaviour when only some members of a logical vector are not available (`NA`).

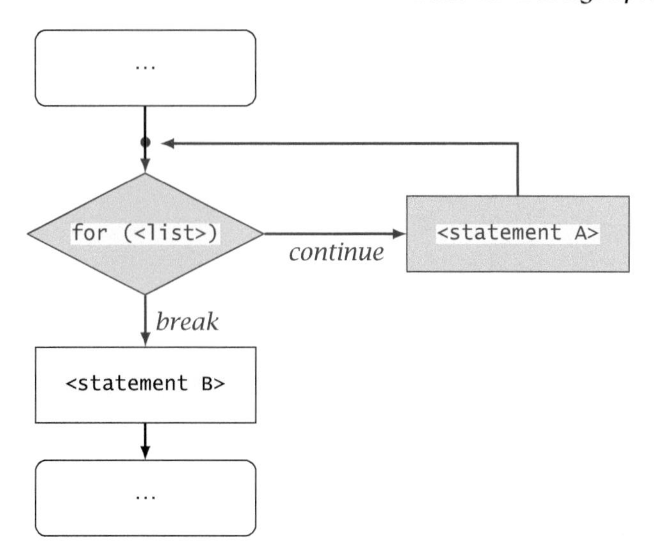

Figure 5.6
Flowchart for a `for` iteration loop.

5.7 Iteration

We give the name *iteration* to the process of repetitive execution of a program statement—e.g., *computed by iteration*. We use the same word, *iteration*, to name each one of these repetitions of the execution of a statement—e.g., *the second iteration*.

Iteration constructs make it possible to "decide" at run time the number of iterations, i.e., when execution breaks out of the loop and continues at the next statement in the script. Iteration can be used to apply the same computations to the different members of a vector or list (this section), but also to apply different functions to members of a vector, matrix, list, or data frame (section 5.10 on page 160).

In R, three types of iteration loops are available: `for`, `while` and `repeat` constructs. They differ in the origin of the values they iterate over, and in the type of test used to terminate iteration. When the same algorithm can be implemented with more than one of these constructs, using the least flexible of them usually results in easier to understand code.

In R, explicit loops as described in this section can in some cases be replaced by calls to *apply* functions (see section 5.8 on page 154) or with vectorised functions and operators (see page 30). The choice among these approaches affects readability and performance (see section 5.11 on page 162).

5.7.1 `for` loops

The most frequently used type of loop is a `for` loop. These loops work in R by "walking through" a list or vector of values to act upon (Figure 5.6). Within a loop, member values are available, sequentially, one at a time, through a variable that functions as a placeholder. The implicit test for the end of the vector or list takes place at the top of the construct before the loop statement is evaluated. The flow chart has the shape of a *loop* as the execution can be directed to an earlier position in the sequence of statements, allowing the same section of code to be evaluated multiple times, each time with a new value assigned to the placeholder variable.

In the diagram above, the argument to `for()` is shown as `<list>` but it can also be a `vector` of any mode. Objects of most classes derived from `list` or from an atomic vector can also fulfil the same role. The extraction operation with a numeric index must be supported by objects of the class passed as argument.

Similarly to `if` constructs, only one statement is controlled by `for`, however this statement can be a compound statement enclosed in braces `{ }` (see pages 132 and 138).

```
b <- 0 # variable needs to set to a valid numeric value!
for (a in 1:5) b <- b + a
b
## [1] 15
```

Here the statement `b <- b + a` is executed five times, with the placeholder variable `a` sequentially taking each of the values, 1, 2, 3, 4, and 5, the members of the anonymous vector `1:5`. The name used as a placeholder has to fulfil the same requirements as an ordinary R variable name. The list or vector following `in` can contain any valid R objects, as long as the code statements in the loop body can handle them.

In a `for` loop construct, even when it is a variable, the vector or list passed as argument cannot be modified by the code statement within the `for` loop.

A loop can be "unrolled" into a linear sequence of statements. Let's work through the `for` loop above.

```
b <- 0
# start of loop
# first iteration
a <- 1
b <- b + a
# second iteration
a <- 2
b <- b + a
# third iteration
a <- 3
b <- b + a
# fourth iteration
a <- 4
b <- b + a
# fifth iteration
a <- 5
b <- b + a
# end of loop
b
## [1] 15
```

The operation implemented in this example is a very frequent one, the sum of a vector, so base R provides a function optimised for efficiently computing it.

```
sum(1:5)
## [1] 15
```

It is important to note that a list or vector of length zero is a valid argument to `for`, that triggers no error, but skips the statements in the loop body.

```
b <- 0
for (a in numeric()) b <- b + a
print(b)
## [1] 0
```

By printing at each iteration variable `b`, the partial results at each iteration can be observed. Brackets are needed to form a compound statement from the two simple statements so that `print(b)` is also executed at each iteration.

```
a <- c(1, 4, 3, 6, 8)
for(x in a) {
  b <- x*2
  print(b)
  }
## [1] 2
## [1] 8
## [1] 6
## [1] 12
## [1] 16
```

The iteration constructs `for`, `while`, and `repeat` always silently return `NULL`, which is a different behaviour than that of `if`.

```
b <- for(x in a) x*2
x
## [1] 8
b
## NULL
```

Thus as shown in earlier examples of `for` loops, computed values need to be assigned to one or more variables within the loop so that they are not lost.

While in the examples above the code directly walked through the values in the vector, an alternative approach is to walk through a sequence of indices using the extraction operator `[ ]` to access the values in vectors or lists. This approach makes it possible to concurrently walk through more than one list or vector. In the example below, one member of vector `a` and of `b` are accessed in each iteration, `a` providing the input and `b` used to store the corresponding computed value.

```
b <- numeric() # an empty vector
for(i in seq(along.with = a)) {
  b[i] <- a[i]^2
}
b
## [1]  1 16  9 36 64
```

5.16 Adding calls to `print()` makes visible the values taken by variables `i`, `a`, and `b` at each iteration. Try to understand where these values come from at each iteration, by playing with the code and modifying it.

```
b <- numeric() # an empty vector
for(i in seq(along.with = a)) {
  b[i] <- a[i]^2
  print(i)
  print(a)
  print(b)
}
b
```

The same approach of adding calls to `print()` can be used for debugging any code that does not return the expected results.

Above I used `seq(along.with = a)` to build a numeric vector containing a sequence of the same length as vector `a`. Using this *idiom* ensures that a vector, in this example `a`, with length zero will be handled correctly, with `numeric(0)` assigned to `b`.

5.17 Run the examples below and explain why the two approaches are equivalent only when the length of `A` is one or more. Find the answer by assigning to `A`, vectors of different lengths, including zero (using `A <- numeric(0)`).

```
A <- -5:5 # assign different numeric vector to A
B <- numeric(length(A))
for(i in seq(along.with = A)) {
  B[i] <- A[i]^2
}
B

C <- numeric(length(A))
for(i in 1:length(A)) {
  C[i] <- A[i]^2
}
C
```

Using `seq(along.with = a)`, its equivalent `seq_along(a)`, as above creates a sequence of integers in `i`, that indexes all members of `a` in the "walk-through". There is no requirement in the R for this, and including only some of the valid indexes, or including them in arbitrary order is possible if needed, however, this is rarely the case. On exit from the loop, the iterator `i` remains accessible and contains its value at the last iteration.

Vectorisation usually results in the simplest and fastest code, as shown below (see section 5.11 on 162). However, not all `for` loops can be replaced by vectorised statements.

```
b <- a^2
b
## [1]  1 16  9 36 64
```

`for` loops as described above, in the absence of errors, have statically predictable behaviour. The compound statement in the loop will be executed once for each member of the vector or list. Special cases may require the alteration of the normal flow of execution in the loop. Two cases are easy to deal with, one is stopping iteration early with a call to `break()`, and another is jumping ahead to the next iteration with a call to `next()`. The example below shows the use of these

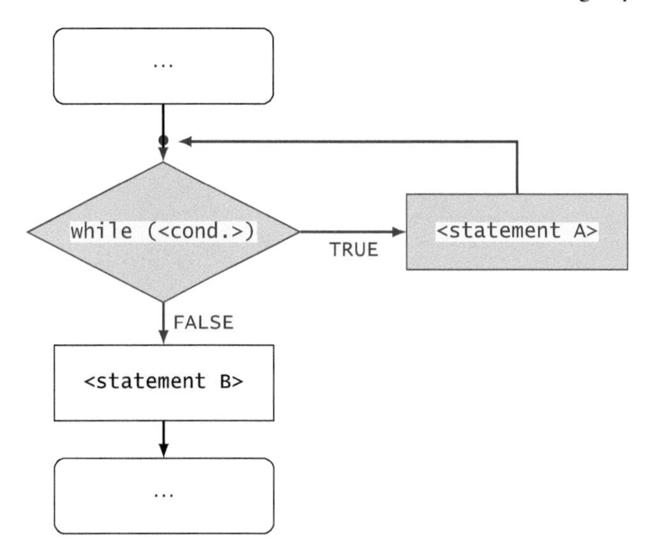

Figure 5.7
Flowchart for a `while` iteration loop.

two functions: we ignore negative values contained in `a`, and exit or break out of the loop when the accumulated sum `b` exceeds 100.

```
b <- 0
a <- -10:100
idxs <- seq_along(a)
for(i in idxs) {
  if (a[i] < 0) next()
  b <- b + a[i]
  if (b > 100) break()
}
b
## [1] 105
i
## [1] 25
a[i]
## [1] 14
```

Hint: if you find the code in the example above difficult to understand, insert `print()` statements and run it again inspecting how the values of `a`, `b`, `idxs` and `i` behave within the loop.

In `for` loops, the use of `break()` and `next()` should be reserved for exceptional conditions. When the `for` construct is not flexible enough for the computations being implemented, using a `while` or a `repeat` loop is preferable.

5.7.2 `while` loops

`while` loops are more flexible than `for` loops (Figure 5.7). Instead of walking through a list or vector, iteration is controlled by a logical condition of length one, just like in `if`. Differently to in an `if` construct, the controlled statement is executed repeatedly as long as the condition remains `TRUE`.

```
a <- 2
while (a < 50) {
  print(a)
  a <- a^2
}
## [1] 2
## [1] 4
## [1] 16
print(a)
## [1] 256
```

To ensure that a `while` loop is exited instead of circling for ever, the condition, `a < 50` in the example above, must depend on a value that is modified by the controlled statement, like `a` in this case.

5.18 Make sure that you understand why the final value of `a` is larger than 50.

The statements above can be simplified, by nesting the assignment inside a call to print.

```
a <- 2
print(a)
while (a < 50) print(a <- a^2)
```

In R, statements like `c <- 1:5` return *invisibly* (with no implicit call to `print()`) the value assigned. This makes possible *chained* assignments to several variables within a single statement like in the example below, as well as using an assignment statement as an argument to a function or operator.

```
a <- b <- c <- 1:5
a
```

5.19 Explain why a second `print(a)` has been added before `while()`. Hint: experiment if necessary.

As with `for` loops, we can use an index variable in a `while` loop to walk through vectors and lists. The difference is that we have to update the index values explicitly in our own code. The code example based on a `for` loop given on page 148 can be rewritten as a `while` loop.

```
a <- c(1, 4, 3, 6, 8)
b <- numeric() # an empty vector
i <- 1
while(i <= length(a)) {
  b[i] <- a[i]^2
  print(b)
  i <- i + 1
}
## [1] 1
## [1]  1 16
## [1]  1 16  9
## [1]  1 16  9 36
## [1]  1 16  9 36 64
b
## [1]  1 16  9 36 64
```

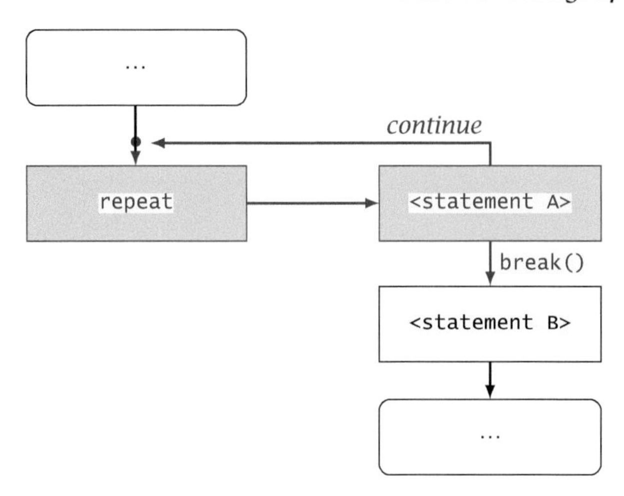

Figure 5.8
Flowchart for a `repeat` iteration loop.

`while` loops as described above will terminate when the condition tested is `FALSE`. In cases that require stopping iteration based on an additional test condition within the compound statement, we can call `break()` in the body of an `if` or `else` statement within the `while` statement. As in the case of `for` loops, it is good to use `break()` only for exceptional conditions.

5.7.3 `repeat` loops

The `repeat` construct is the most flexible as iteration only stops with a call to `break()`. One or more calls to `break()` can be located anywhere within the compound statement that forms the body of the loop (Figure 5.8).

```
a <- 2
repeat{
  print(a)
  if (a > 50) break()
  a <- a^2
}
## [1] 2
## [1] 4
## [1] 16
## [1] 256
```

5.20 Try to explain why the example above returns the values it does. Use the approach of adding `print()` statements, as described on page 148.

When `repeat` loop constructs contain more than one call to `break()`, each within a different `if` or `else` statement, indentation and/or comments can be used to highlight in the listing this infrequent use case .

5.21 Explain why a `repeat` construct is equivalent to a `while` construct with the test condition set equal to `logical` constant `TRUE`.

5.7.4 Nesting of loops

All the execution-flow control statements seen above can be nested, as syntactically they are themselves statements. I show an example with two `for` loops used to walk through rows and columns of a `matrix` constructed as follows.

```
A <- matrix(1:50, nrow = 10)
A
##       [,1] [,2] [,3] [,4] [,5]
##  [1,]    1   11   21   31   41
##  [2,]    2   12   22   32   42
##  [3,]    3   13   23   33   43
##  [4,]    4   14   24   34   44
##  [5,]    5   15   25   35   45
##  [6,]    6   16   26   36   46
##  [7,]    7   17   27   37   47
##  [8,]    8   18   28   38   48
##  [9,]    9   19   29   39   49
## [10,]   10   20   30   40   50
```

The nested loops below compute the sum for each row of the matrix. In the example below, the value of `i` changes for each iteration of the outer loop. The value of `j` changes for each iteration of the inner loop, and the inner loop is run in full for each iteration of the outer loop. The inner loop index `j` changes fastest.

```
row.sum <- numeric()
for (i in 1:nrow(A)) {
  row.sum[i] <- 0
  for (j in 1:ncol(A))
    row.sum[i] <- row.sum[i] + A[i, j]
}
print(row.sum)
##  [1] 105 110 115 120 125 130 135 140 145 150
```

The nested loops above work correctly with any two-dimensional matrix with at least one column and one row, but *crash* with an empty matrix (`matrix()` or `matrix(numeric())`). Thus it is good practice to enclose the `for` loop in an `if` statement as protection. For the example above, a suitable `logical` condition is `!is.null(dim(A)) && !any(dim(A) == 0`.

5.22 1) Modify the code in last chunk above so that it sums the values only in the first three columns of `A`, and 2) modify the same code so that it sums the values only in the last three rows of `A`.

Does the code you wrote work as expected when the number of rows in `A` is different from 10? and, also if the number of columns in `A` is different from 5? What would happen if `A` had fewer than three columns? Try to think first what to expect based on the code you wrote. Then create matrices of different sizes and test your code. After that, if necessary, try to improve the code, so that wrong results are never returned.

5.8 *Apply* Functions

Apply functions' role is similar to that of the iteration loops discussed above. One could say that apply functions "walk along" a vector, list or a dimension of a matrix or an array, calling a function with each member of the collection as argument. Notation is more concise than in `for` constructs. However, apply functions can be used only when the operations to be applied are *independent—i.e., the results from one iteration are not used in another iteration.*

Conceptually, `for`, `while` and `repeat` loops are interpreted as controlling a sequential evaluation of program statements. In contrast, R's *apply* functions are, conceptually, thought as evaluating a function in parallel for each of the different members of their input. So, while in loops the results of earlier iterations through a loop can be stored in variables and used in subsequent iterations, this is not possible in the case of *apply* functions.

The different *apply* functions in base R differ in the class of the values they accept for their `x` parameter, the class of the object they return and/or the class of the value returned by the applied function. `lapply()`, `vapply()` and `sapply()` expect a `vector` or `list` as an argument passed through `x`. `lapply()` returns a `list` or an `array`; and `vapply()` always *simplifies* its returned value into a vector, while `sapply()` does the simplification according to the argument passed to its `simplify` parameter. All these *apply* functions can be used to apply an R function that returns a value of the same or a different class as its argument. In the case of `apply()` and `lapply()` not even the length of the values returned for each member of the collection passed as an argument, needs to be consistent. Function `apply()` is used to apply a function to the elements along one dimension of an object that has two or more *dimensions* returning an array or a list or a vector depending on the size, and consistency in length and class among the values returned by the applied function.

5.8.1 Applying functions to vectors, lists and data frames

I exemplify the use of `lapply()`, `sapply()` and `vapply()`. Below, they are used to apply function `log()` to each member of a `numeric` vector. This is a function defined in R itself, but user-defined functions and functions imported from packages can be applied identically. How to define packages and define new functions are the subject of chapter 6 (on page 169).

The individual member objects in the list or vector passed as argument to parameter `x` of *apply* functions are passed as a positional argument to the first formal parameter of the applied function, i.e., only some R functions can be passed as an argument to `FUN`.

```
set.seed(123456) # so that vct1 does not change
vct1 <- runif(6) # A short vector as input to keep output short
str(vct1)
##  num [1:6] 0.798 0.754 0.391 0.342 0.361 ...
```

```
z <- lapply(X = vct1, FUN = log)
str(z)
## List of 6
##  $ : num -0.226
##  $ : num -0.283
##  $ : num -0.938
##  $ : num -1.07
##  $ : num -1.02
##  $ : num -1.62
```

The code above calls `log()` once with each of the six members of `vct1` as its first argument and collects the returned values into a `list`, hence the `l` in `lapply()`.

```
z <- sapply(X = vct1, FUN = log)
str(z)
##  num [1:6] -0.226 -0.283 -0.938 -1.074 -1.018 ...
```

The code above calls `log()` as in the previous example but collects the returned values into a vector, i.e., by default it *simplifies* the list into a `vector` or `matrix` when possible, hence the `s` in `sapply()`. Simplification can be skipped, in this case returning a list as `lapply()` above (returned value not shown).

```
z <- sapply(X = vct1, FUN = log, simplify = FALSE)
str(z)
```

`vapply()` always returns a vector (no example shown), hence the `v` in its name. The computed results are the same using `lapply()`, `sapply()` or `vapply()`, but the class and structure of the objects returned can differ, as well as how numbers are printed.

Function `log()` has a second parameter named `base` that can be passed and argument to override the default base (e) used to compute natural logarithms. Additional arguments like this can be passed by name, using the name of the parameter in the function passed as argument to `FUN`, in this case, `base`.

```
z <- sapply(X = vct1, FUN = log, base = 10)
str(z)
##  num [1:6] -0.0981 -0.1229 -0.4075 -0.4665 -0.4421 ...
```

Anonymous functions can be defined (see section 6.2 on page 169) and directly passed as an argument to `FUN` without the need of separately assigning them to a name.

```
z <- sapply(X = vct1, FUN = function(x) {log10(x + 1)})
str(z)
##  num [1:6] 0.255 0.244 0.143 0.128 0.134 ...
```

As explained in section 4.4 on page 94, class `data.frame` is derived from class `list`. The columns in a data frame are equivalent to members of a list, and functions can thus be applied to columns. The data frame `cars` from package 'datasets' contains data for speed and for stopping distance for cars stored in two columns or member variables, named `speed` and `dist`. The members of the returned `numeric` vector, containing the computed means, are named accordingly.

```
sapply(X = cars, FUN = mean)
## speed  dist
## 15.40 42.98
```

Here is a possible way of obtaining means and standard deviations of member vectors. The argument passed to `FUN.VALUE` provides a template for the type of the returned value and its organisation into rows and columns. Notice that the rows in the output are now named according to the names in `FUN.VALUE`.

A function that returns a numeric vector of length 2 containing mean and standard deviation can be defined by calling existing functions (see section 6.2 on page 169).

```
mean_and_sd <-
  function(x, na.rm = FALSE) {
    c(mean = mean(x, na.rm = na.rm),  sd = sd(x, na.rm = na.rm))
  }
```

and `vapply()` used to apply it to each member vector of the list. The argument passed to `FUN.VALUE` serves as a template indicating the values returned by function `mean_and_sd()`.

```
values <- vapply(X = cars,
                 FUN = mean_and_sd,
                 FUN.VALUE = c(mean = 0, sd = 0),
                 na.rm = TRUE)
class(values)
## [1] "matrix" "array"
values
##          speed     dist
## mean 15.400000 42.98000
## sd    5.287644 25.76938
```

5.23 Apply function `mean_and_sd()` defined above to the data frame `cars` from 'datasets'. The aim is to obtain the mean and standard deviation for each numeric column.

5.24 Obtain the summary of dataset `airquality` with function `summary()`, but in addition, write code with an *apply* function to count the number of non-missing values in each column. Hint: using `sum()` on a `logical` vector returns the count of `TRUE` values as `TRUE`, and `FALSE` are transparently converted into `numeric` 1 and 0, respectively, when `logical` values are used in arithmetic expressions.

In the examples above, the *apply* functions were used to "reduce" the data by applying summary functions. In the next code chunk, `lapply()` is used to construct the `list` of five vectors `ls1` using a vector of five numbers as argument passed to parameter `X`. As above, additional *named* arguments are relayed to each call of `rnorm()`.

```
set.seed(123456)
ls1 <- lapply(X = c(v1 = 2, v2 = 5, v3 = 3, v4 = 1, v5 = 4),
              FUN = rnorm, mean = 10, sd = 1)
str(ls1)
## List of 5
##  $ v1: num [1:2] 10.83 9.72
##  $ v2: num [1:5] 9.64 10.09 12.25 10.83 11.31
##  $ v3: num [1:3] 12.5 11.17 9.57
##  $ v4: num 9
##  $ v5: num [1:4] 8.89 9.94 11.17 11.05
```

In addition to functions returning pseudo-random draws from different probability distributions, constructors for objects of various classes can be used similarly.

5.8.2 Applying functions to matrices and arrays

Matrices and arrays have two or more dimensions, and contrary to data frames, they are not a special kind of one-dimensional lists. In R, the dimensions of a matrix, rows and columns, over which a function is applied are called *margins* (see section 3.11, and Figure 3.2 on page 71). The argument passed to parameter `MARGIN` determines *over* which margin the function will be applied. Arrays can have many dimensions (see Figure 3.3 on page 76), and consequently more margins. In the case of arrays with more than two dimensions, it is possible and can be useful to apply functions over multiple margins at once.

The individual *slices* of the matrix or array passed as argument to parameter `X` of *apply* functions are passed as a positional argument to the first formal parameter of the applied function, i.e., only some R functions can be passed as argument to `FUN`.

Matrix `mat1` constructed here will be used in examples. Adding names helps with understanding both here and when using matrices in real data analysis situations.

```
mat1 <- matrix(rnorm(6, mean = 10, sd = 1), ncol = 2)
mat1 <- round(mat1, digits = 1)
dimnames(mat1) <- # add row and column names
  list(paste("row", 1:nrow(mat1)), paste("col", 1:ncol(mat1)))
mat1
##       col 1 col 2
## row 1  10.1  11.7
## row 2   9.3  10.6
## row 3  10.9   9.2
```

Column (or row) means of matrices can be easily computed with `apply()`. However, in contrast to when using other *apply* functions, an argument must be passed to parameter `MARGIN`.

```
apply(mat1, MARGIN = 2, FUN = mean)
## col 1 col 2
##  10.1  10.5
```

5.25 Edit the example above so that it computes row means instead of column means.

5.26 As described above, we can pass arguments by name to the applied function. Can you guess why parameter names of *apply* functions are fully in uppercase, something very unusual for R coding style?

If the function applied returns a value of the same length as its input, then the dimensions of the value returned by `apply()` are the same as those of its input. Using the identity function `I()` that returns its argument unchanged, facilitates the comparison of output against input.

```
z <- apply(X = mat1, MARGIN = 2, FUN = I)
dim(z)
## [1] 3 2
```

```
z
##       col 1 col 2
## row 1  10.1  11.7
## row 2   9.3  10.6
## row 3  10.9   9.2
```

Passing `MARGIN = 1` as below instead of `MARGIN = 2` as above, rows and columns are transposed in the returned value!.

```
z <- apply(X = mat1, MARGIN = 1, FUN = I)
dim(z)
## [1] 2 3
z
##       row 1 row 2 row 3
## col 1  10.1   9.3  10.9
## col 2  11.7  10.6   9.2
```

The next, more realistic example, applies function `summary()` that returns a value usually shorter than its input, but longer than one. Both for column summaries (`MARGIN = 2`) and row summaries (`MARGIN = 1`), a matrix is returned. Each columns, a numeric vector in this example, contains the vector returned by a call to `summary()`. Column and row names from `mat1` are preserved, as well as the names in the value returned by `summary()`.

```
z <- apply(X = mat1, MARGIN = 2, FUN = summary)
z
##         col 1 col 2
## Min.      9.3  9.20
## 1st Qu.   9.7  9.90
## Median   10.1 10.60
## Mean     10.1 10.50
## 3rd Qu.  10.5 11.15
## Max.     10.9 11.70
```

```
z <- apply(X = mat1, MARGIN = 1, FUN = summary)
z
##         row 1  row 2  row 3
## Min.     10.1  9.300  9.200
## 1st Qu.  10.5  9.625  9.625
## Median   10.9  9.950 10.050
## Mean     10.9  9.950 10.050
## 3rd Qu.  11.3 10.275 10.475
## Max.     11.7 10.600 10.900
```

Binary operators in R are functions with two formal parameters which can be called using infix notation in expressions—i.e., `a + b`. By back-quoting their names they can be called using the same syntax as for ordinary functions, and consequently also passed to the `FUN` parameter of apply functions. A toy example, equivalent to the vectorised operation `vct1 + 5` follows. By enclosing operator `+` in back ticks (`) and passing by name a constant to its second formal parameter (`e2 = 5`) operator `+` behaves like an ordinary function. See section 6.2.3 on page 175).

```
set.seed(123456) # so that vct1 does not change
vct1 <- runif(10)
z <- sapply(X = vct1, FUN = `+`, e2 = 5)
str(z)
##  num [1:10] 5.8 5.75 5.39 5.34 5.36 ...
```

Table 5.1
R functions that can substitute for iteration loops. They accept vectors as arguments for their first parameter, except for `rowSums()`, `colSums()`, `rowMeans()`, and `colMeans()` which accept `matrix` objects. Only functions that return a value with the same dimensions as the argument passed as input are vectorised in the sense used in this book.

Function	Computation	Returned class, length
`sum()`	$\sum_{i=1}^{n} x_i$	`numeric`, 1
`rowSums()`	$\sum_{j=1}^{l} x_i$	`numeric`, n
`colSums()`	$\sum_{i=1}^{n} x_j$	`numeric`, l
`mean()`	$\sum_{i=1}^{n} x_i$	`numeric`, 1
`rowMeans()`	$\sum_{j=1}^{l} x_i / l$	`numeric`, n
`colMeans()`	$\sum_{i=1}^{n} x_j / n$	`numeric`, l
`prod()`	$\prod_{i=1}^{n} x_i$	`numeric`, 1
`cumsum()`	$\sum_{i=1}^{1} x_i, \cdots \sum_{i=1}^{j} x_i, \cdots \sum_{i=1}^{n} x_i$	`numeric`, $n_{\text{out}} = n_{\text{in}}$
`cumprod()`	$\prod_{i=1}^{1} x_i, \cdots \prod_{i=1}^{j} x_i, \cdots \prod_{i=1}^{n} x_i$	`numeric`, $n_{\text{out}} = n_{\text{in}}$
`cummax()`	cumulative maximum	`numeric`, $n_{\text{out}} = n_{\text{in}}$
`cummin()`	cumulative minimum	`numeric`, $n_{\text{out}} = n_{\text{in}}$
`runmed()`	running median	`numeric`, $n_{\text{out}} = n_{\text{in}}$
`diff()`	$x_2 - x_1, \cdots x_i - x_{i-1}, \cdots x_n - x_{n-1}$	`numeric`, $n_{\text{out}} = n_{\text{in}} - 1$
`diffinv()`	inverse of diff	`numeric`, $n_{\text{out}} = n_{\text{in}} + 1$
`factorial()`	$x!$	`numeric`, $n_{\text{out}} = n_{\text{in}}$
`rle()`	run-length encoding	`rle`, $n_{\text{out}} < n_{\text{in}}$
`inverse.rle()`	run-length decoding	`vector`, $n_{\text{out}} > n_{\text{in}}$

5.9 Functions that Replace Loops

R provides several functions that can be used to avoid writing iterative loops. The most frequently used are taken for granted: `mean()`, `var()` (variance), `sd()` (standard deviation), `max()`, and `min()`. Replacing code implementing an iterative algorithm by a single function call simplifies the script's code and can make it easier to understand. These functions are written in C and compiled, so even when iterative algorithms are used, they are fast (see section 5.11 on page 162). Table 5.1 lists several functions from base R that implement iterative algorithms. All these functions take a vector of arbitrary length as their first argument, except for `inverse.rle()`.

5.27 Build a `numeric` vector such as `x <- c(1, 9, 6, 4, 3)` and pass it as argument to the functions in Table 5.1. Do the corresponding computations manually for the functions your find most relevant, trying to understand what values they calculate.

5.10 The Multiple Faces of Loops

In this advanced section, I describe some uses of R loops that help with writing concise scrips. As these make heavy use of functions, if you are reading the book sequentially, you should skip this section and return to it after reading chapters 6 and 7.

In the same way as we can assign names to numeric, character and other types of objects, we can assign names to functions and expressions. We can also create lists of functions and/or expressions. The R language has a very consistent grammar, with all lists and vectors behaving in the same way. The implication of this is that we can assign different functions or expressions to a given name and, consequently, it is possible to write loops over lists of functions or expressions.

The next example, uses a *character vector of function names* together with function do.call() in the body of a for loop, to construct a numeric vector with members, named according to the function names, storing the computed values. Function do.call() accepts both character strings and function names as argument to its first parameter, and calls the corresponding function with arguments supplied as a list.

```
vct1 <- rnorm(10)
results <- numeric()
fun.names <- c("mean", "max", "min")
for (f.name in fun.names) {
  results[[f.name]] <- do.call(f.name, list(vct1))
}
results
##        mean        max        min
##   0.5453427  2.5026454 -1.1139499
```

When traversing a *list of functions* in a loop, the original names of the functions are not available as what is stored in the list are the definitions of the functions rather than their names. In this case, the function definitions are assigned to the placeholder variable (f in the chunk below) and the functions be called directly with (f()). The result is a numeric vector with anonymous members.

```
results <- numeric()
funs <- list(mean, max, min)
for (f in funs) {
  results <- c(results, f(x))
}
results
## [1] 8 8 8
```

A named list of functions makes it possible to gain full control of the naming of the results. It is possible to construct a numeric vector with named members with names matching the names given to the list members, which can be different to the names of the functions.

```
results <- numeric()
funs <- list(average = mean, maximum = max, minimum = min)
for (f in names(funs)) {
  results[[f]] <- funs[[f]](x)
```

```
}
results
## average maximum minimum
##       8       8       8
```

Next is an example using model formulas. In the this example, a loop is used to fit three models, obtaining a list of fitted models. It is not possible to pass to `anova()` this list of fitted models, as it expects each fitted model as a separate nameless argument to its ... parameter. It is possible to get around this problem using function `do.call()` to call `anova()`. Function `do.call()` passes the members of the list passed as its second argument as individual arguments to the function being called, using their names if present. `anova()` expects nameless arguments, so the names present in `results` have to be removed with a call to `unname()`.

```
my.data <- data.frame(x = 1:10, y = 1:10 + rnorm(10, 1, 0.1))
results <- list()
models <- list(linear = y ~ x, linear.orig = y ~ x - 1, quadratic = y ~ x + I(x^2))
for (m in names(models)) {
  results[[m]] <- lm(models[[m]], data = my.data)
}
str(results, max.level = 1)
## List of 3
##  $ linear     :List of 12
##   ..- attr(*, "class")= chr "lm"
##  $ linear.orig:List of 12
##   ..- attr(*, "class")= chr "lm"
##  $ quadratic  :List of 12
##   ..- attr(*, "class")= chr "lm"
do.call(anova, unname(results))
## Analysis of Variance Table
##
## Model 1: y ~ x
## Model 2: y ~ x - 1
## Model 3: y ~ x + I(x^2)
##   Res.Df     RSS Df Sum of Sq      F    Pr(>F)
## 1      8 0.05525
## 2      9 2.31266 -1   -2.2574 306.19 4.901e-07 ***
## 3      7 0.05161  2    2.2611 153.34 1.660e-06 ***
## ---
## Signif. codes:  0 '***' 0.001 '**' 0.01 '*' 0.05 '.' 0.1 ' ' 1
```

If the only aim is to pass `results` to `anova()` a `list` of nameless members can be constructed using positional indexing.

```
results <- list()
models <- list(y ~ x, y ~ x - 1, y ~ x + I(x^2))
for (i in seq(along.with = models)) {
  results[[i]] <- lm(models[[i]], data = my.data)
}
str(results, max.level = 1)
## List of 3
##  $ :List of 12
##   ..- attr(*, "class")= chr "lm"
##  $ :List of 12
##   ..- attr(*, "class")= chr "lm"
##  $ :List of 12
##   ..- attr(*, "class")= chr "lm"
```

```
do.call(anova, results)
## Analysis of Variance Table
##
## Model 1: y ~ x
## Model 2: y ~ x - 1
## Model 3: y ~ x + I(x^2)
##   Res.Df     RSS Df Sum of Sq      F    Pr(>F)
## 1      8 0.05525
## 2      9 2.31266 -1   -2.2574 306.19 4.901e-07 ***
## 3      7 0.05161  2    2.2611 153.34 1.660e-06 ***
## ---
## Signif. codes:  0 '***' 0.001 '**' 0.01 '*' 0.05 '.' 0.1 ' ' 1
```

5.11 Iteration When Performance Is Important

When working with large data sets, or many smaller data sets, one frequently needs to take performance into account. In R, explicit `for`, `while` and `repeat` are frequently considered to be slow. Vectorised operations are in general comparatively faster. As vectorisation (see page 30) usually also makes code simpler, it is good to use vectorisation whenever possible. Depending on the case, loops can be replaced using vectorised arithmetic operators, *apply* functions (see section 5.8 on page 154) and functions implementing frequently used operations (see section 5.9 on page 159). Improved performance needs to be balanced against the effort invested in writing faster code, as in most cases our own time is more valuable than computer running time. However, using vectorised operators and optimised functions becomes nearly effortless once one is familiar with them.

To demonstrate the magnitude of the differences in performance that can be expected, I used as a first case the computation of the differences between successive numbers in a vector, applied to vectors of lengths ranging from 10 to 100 million numbers (Figure 5.9). In relative terms, the difference in computation time was huge between loops and vectorisation for vectors of up to 1 000 numbers (near $\times 500$), but the total times were very short (5×10^{-3} s vs. 10×10^{-6} s). For these vectors, pre-allocation of a vector to collect the results made almost no difference and vectorisation with the extraction operator `[ ]` together with the minus arithmetic operator `-` was the fastest. There seems to be a significant overhead for explicit loops, as the running time was nearly independent of the length of these short vectors.

For vectors of 10 000 or more numbers there was only a very small advantage in using function `diff()` over using vectorised arithmetic and extraction operators. For `while` and `for` loops pre-allocation of the vector to collect results made an important difference ($\times 2$ to $\times 3$), larger in the case of `for`. However, vectorised operators and function `diff()` remained nearly $\times 10$ faster than the fastest explicit loop. For the longer vectors the time increased almost linearly with their length, with similar slopes for the different approaches. Because of the computation used for this example, *apply()* functions could not be used.

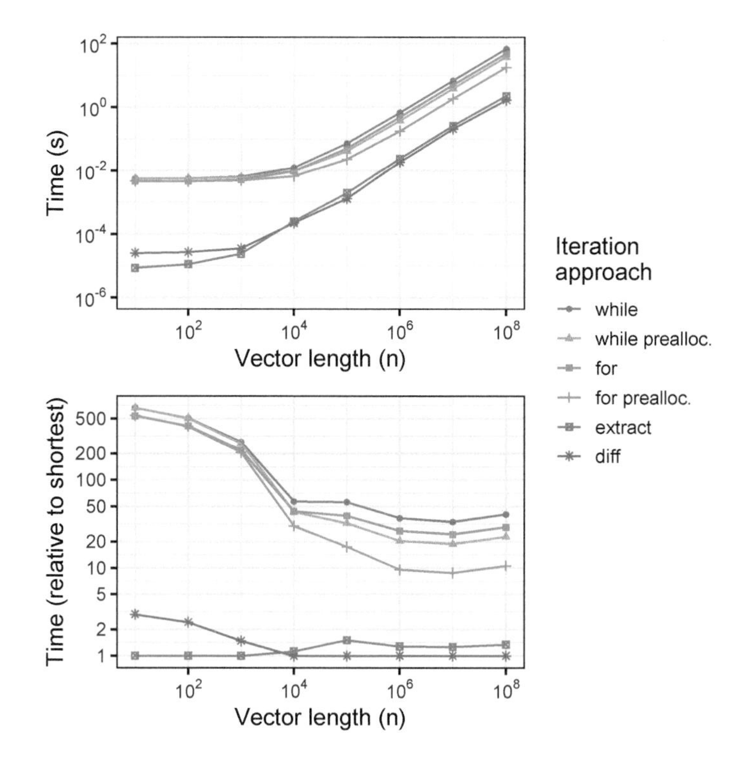

Figure 5.9
Benchmark results for different approaches to computing running differences in numeric (double) vectors of different lengths. The data in this figure were obtained in a computer with a 12-years old Xenon E3-1235 CPU with four cores, 32 GB of RAM, Windows 10 and R 4.3.1.

The chunks below show the code for the six approaches compared in Figure 5.9, where a is a numeric vector varying length constructed with function `rnorm()`.

```
b <- numeric() # do not pre-allocate memory
i <- 1
while (i < length(a)) {
  b[i] <- a[i+1] - a[i]
  i <- i + 1
}

b <- numeric(length(a)-1) # pre-allocate memory
i <- 1
while (i < length(a)) {
  b[i] <- a[i+1] - a[i]
  i <- i + 1
}

b <- numeric() # do not pre-allocate memory
for(i in seq(along.with = b)) {
  b[i] <- a[i+1] - a[i]
}
```

```
b <- numeric(length(a)-1) # pre-allocate memory
for(i in seq(along.with = b)) {
  b[i] <- a[i+1] - a[i]
}

# vectorised using extraction operators
b <- a[2:length(a)] - a[1:length(a)-1]

# vectorised function diff()
b <- diff(a)
```

In nested iteration loops, it is most important to vectorise, or otherwise enhance the performance of the innermost loop, as it is the one executed most frequently. The code for nested loops (used as an example in section 5.7.4 on page 153) can be edited to remove the explicit use of `for` loops. I assessed the performance of different approaches by collecting timings for square `matrix` objects with dimensions (rows $\times$ columns) ranging from 10×10, size $= 10^2$, to $10\,000 \times 10\,000$, size $= 10^8$ (Figure 5.10).

In this second case, pre-allocation of memory to `b` did not enhance performance in good agreement with the benchmarks for the first example as when largest its length was 10 000. The two nested loops always took the longest to run irrespective of the size of matrix `A`. A single loop over rows using a call to `sum()` for each row, improved performance compared to nested loops, most clearly for large matrices. This approach was out-performed by `apply()` only for small matrices, from which we can infer that `apply()` has a much smaller overhead than an explicit `for` loop. `rowSums()` was between $\times 5$ and $\times 20$ faster than the second fastest approach depending on the size of the matrix.

The chunks below show the code for the six approaches compared in Figure 5.10, where `A` was a numeric matrix constructed with function `rnorm()`.

The inner `for` loop can be replaced by function `sum()` which returns the sum of a vector. Within the loop, `A[i, ]` extracts whole rows, one at a time.

```
row.sum <- numeric()
for (i in 1:nrow(A)) {
  row.sum[i] <- 0
  for (j in 1:ncol(A))
    row.sum[i] <- row.sum[i] + A[i, j]
}
print(row.sum)

row.sum <- numeric(nrow(A)) # faster
for (i in 1:nrow(A)) {
  row.sum[i] <- sum(A[i, ])
}
```

The outer loop can be replaced by a call to `apply()` (see section 5.8 on page 154).

```
row.sum <- apply(A, MARGIN = 1, sum) # MARGIN=1 indicates rows
```

Calculating row sums is a frequent operation, thus, R provides a built-in function for this.

```
rowSums(A)
```

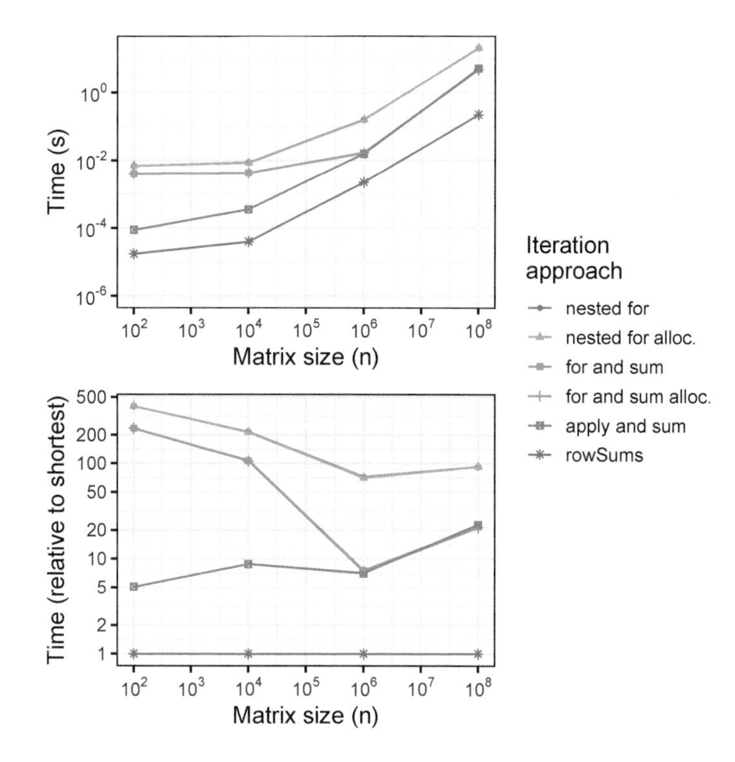

Figure 5.10
Benchmark results for different approaches to computing row sums of square numeric (double) matrices of different sizes. The data in this figure were obtained in a computer with a 12-years-old Xenon E3-1235 CPU with four cores, 32 GB of RAM, Windows 10, and R 4.3.1.

The simplest way of measuring the execution time of an R expression is to use function `system.time()`. Package 'microbenchmark', used for the benchmarks shown in Figures 5.9 and 5.10, provides finer time resolution.

As in these examples the computations in the body of the loop are very simple, the overhead of the iterative loops strongly affects the total computation time in these benchmarks. When the computations at each iteration are time consuming, the overhead of using explicit iteration loops gets diluted. Thus, removing the explicit use of iteration is most helpful, when it is easier to implement vectorised arithmetic or find optimised functions.

The timings in Figures 5.9 and 5.10 are only valid for the specific computer configuration, operating system and R version that I used. They provide only an approximate guide to what can be expected in different conditions. The scripts used are included in package 'learnrbook' in case readers wish to run them on their computers. As replication is used, the total run time for the script is relatively long.

You may be wondering: how do the faster approaches manage to avoid the overhead of iteration? Of course, they do not really avoid iteration, but the loops in functions written in C, C++, or FORTRAN are compiled into machine code as part of R itself or when packages binaries are created. In simpler words, the time required to convert and optimise the code written in these languages into machine code is spent during compilation, usually before we download and install R or packages. Instead, a loop coded in R is interpreted into machine code each time we source our script, and in some cases for each iteration in a loop. The R interpreter does some compilation into virtual machine code, as a preliminary stage which helps improve performance.

The examples in this section use numbers and arithmetic operations, but vectorisation and *apply* functions can be also used with vectors of other modes, such as vectors of `character` strings or `logical` values.

With modern computer processors, or CPUs, splitting the tasks across multiple cores for concurrent execution can enhance performance. To some extent this happens invisibly due to optimisations in the translation into machine code. Explicit approaches are available in package 'parallel' included in the R distribution and contributed packages such as 'future'. Parallelisation is also possible across interconnected computers. However, how to enhance performance based on parallel or distributed execution is beyond the scope of this book.

5.12 Object Names as Character Strings

In all assignment examples before this section, we have used object names included as literal character strings in the code expressions. In other words, the names are "decided" as part of the code, rather than at run time. In scripts or packages, the object name to be assigned may need to be decided at run time and, consequently, be available only as a character string stored in a variable. In this case, function `assign()` must be used instead of the operators <- or ->. The statements below demonstrate its use.

First using a `character` constant.

```
assign("a", 9.99)
a
## [1] 9.99
```

Next using a `character` value stored in a variable.

```
name.of.var <- "b"
assign(name.of.var, 9.99)
b
## [1] 9.99
```

The two toy examples above do not demonstrate why one may want to use `assign()`. Common situations where we may want to use character strings to store (future or existing) object names are 1) when we allow users to provide names for objects either interactively or as `character` data, 2) when in a loop we transverse a vector or list of object names, or 3) we construct at runtime object names from multiple character strings based on data or settings. A common case is when we import data from a text file and we want to name the object according to the name of the file on disk, or a character string read from the header at the top of the file.

Another case is when `character` values are the result of a computation.

```
for (i in 1:5) {
  assign(paste("square_of_", i, sep = ""), i^2)
}
ls(pattern = "square_of_*")
## [1] "square_of_1" "square_of_2" "square_of_3" "square_of_4" "square_of_5"
```

The complementary operation of *assigning* a name to an object is to *get* an object when we have available its name as a character string. The corresponding function is `get()`.

```
get("a")
## [1] 9.99
get("b")
## [1] 9.99
```

If we have available a character vector containing object names and we want to create a list containing these objects we can use function `mget()`. In the example below we use function `ls()` to obtain a character vector of object names matching a specific pattern and then collect all these objects into a list.

```
obj_names <- ls(pattern = "square_of_*")
obj_lst <- mget(obj_names)
str(obj_lst)
## List of 5
##  $ square_of_1: num 1
##  $ square_of_2: num 4
##  $ square_of_3: num 9
##  $ square_of_4: num 16
##  $ square_of_5: num 25
```

5.28 Think of possible uses of functions `assign()`, `get()` and `mget()` in scripts you use or could use to analyse your own data (or from other sources). Write a script to implement this, and iteratively test and revise this script until the result produced by the script matches your expectations.

5.13 Clean-Up

Sometimes we need to make sure that clean-up code is executed even if the execution of a script or function is aborted by the user or as a result of an error condition. A typical example is a script that temporarily sets a disk folder as the working directory or uses a file as temporary storage. Function `on.exit()` can be used to record that a user supplied expression needs to be executed when the current function, or a script, exits. Function `on.exit()` can also make code easier to read as it keeps creation and clean-up next to each other in the body of a function or in the listing of a script.

```
file.create("temp.file")
## [1] TRUE
on.exit(file.remove("temp.file"))
# code that makes use of the file goes here
```

Function `library()` attaches the namespace of the loaded packages and in some special cases one may want to detach them at the end of a script. We can use `detach()` similarly as with attached `data.frame` objects (see page 111). As an example, we detach the packages used in section 5.11. It is important to remember that the order in which they can be detached is determined by their interdependencies.

```
detach(package:patchwork)
detach(package:ggplot2)
detach(package:scales)
```

5.14 Further Reading

For further readings on the aspects of R discussed in the current chapter, I suggest the books *The Art of R Programming: A Tour of Statistical Software Design* (Matloff 2011) and *Advanced R* (Wickham 2019).

6

Base R: Adding New "Words"

> Computer Science is a science of abstraction—creating the right model for a problem and devising the appropriate mechanizable techniques to solve it.
>
> Alfred V. Aho and Jeffrey D. Ullman
> *Foundations of Computer Science*, 1992

6.1 Aims of This Chapter

In earlier chapters we have only used base R features. In this chapter you will learn how to expand the range of features available. I start by discussing how to define and use new functions, operators, and classes. What are their semantics and how they contribute to conciseness and reliability of computer scripts and programs. Later I focus on using existing packages to share extensions to R and touch briefly on how they work. I do not consider the important, but more advanced question of packaging functions and classes into new R packages. Instead I discuss how packages are installed and used.

6.2 Defining Functions and Operators

Abstraction can be defined as separating the fundamental properties from the accidental ones. Say obtaining the mean from a given vector of numbers is an actual operation. There can be many such operations on different numeric vectors, each one a specific case. When we describe an algorithm for computing the mean from any numeric vector, we formulate an abstraction of *mean*. In the same way, each time we separate operations from specific data we create a new abstraction. In this sense, functions are abstractions of operations or actions; they are like "verbs" describing actions separately from actors.

DOI: 10.1201/9781003404187-6

The main role of functions is that of providing an abstraction allowing us to avoid repeating blocks of code (groups of statements) applying the same operations on different data. The reasons to avoid repetition of similar blocks of code statements are that 1) if the algorithm or implementation needs to be revised—e.g., to fix a bug or error—it is best to make edits in a single place; 2) sooner or later pieces of repeated code can become different leading to inconsistencies and hard-to-track bugs; 3) abstraction and division of a problem into smaller chunks, greatly helps with keeping the code understandable to humans; 4) textual repetition makes the script file longer, and this makes debugging, commenting, etc., more tedious, and error prone; 5) with well-defined input and output, functions facilitate testing.

How does one, in practice, avoid repeating bits of code? One writes a function containing the statements that would need to be repeated, and later one *calls* ("uses") the function in their place. We have been calling R functions or operators in almost every example in this book; what we will next tackle is how to define new functions of our own.

The diagram in section 5.3 on page 132 describes a compound statement. A function is a code statement, simple or compound, that is partly isolated from the enclosing environment. The *function* abstraction relies on formal parameters working as placeholders for arguments within the function body. When the function is called (or "used") values are passed as arguments to the parameters, and used when executing the code within the function.

New functions and operators are defined using function `function()`, and saved like any other object in R by assignment to a variable name. In the example below, `x` and `y` are both formal parameters, or names used within the function for objects that will be supplied as *arguments* when the function is called.

Function `fun1()` has two formal parameters, `x` and `y`.

```
fun1 <- function(x, y){x * y}
```

When we call `fun1()` with `4` and `3` as arguments, the computation that takes place is `4 * 3` and the value returned is `12`. In this example, the returned value or result is printed, but it could have been assigned to a variable or used in further computations within the calling statement.

```
fun1(x = 4, y = 3)
## [1] 12
```

6.1 What is the computation that takes places in these function calls?

```
fun1(x = 10, y = 50)
fun1(x = 10, y = 50) * 3
```

Even though the statements within the function body do have access to the environment in which the function is called, it is safest to pass all input through the function parameters, and return all values to the caller. This ensures that the users of the function can treat it as a black box with no side effects.

In R, statements within the function usually do not affect directly any variable defined outside the function, the result from the computation is returned as a value. The diagram in Figure 6.1 describes a function that has no *side effects*, as

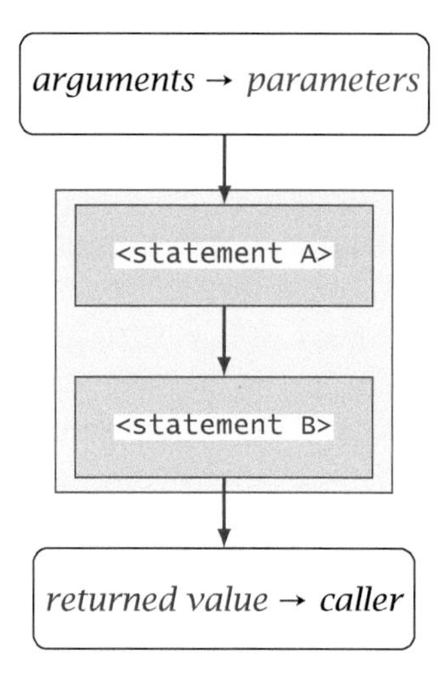

Figure 6.1
Diagram of function with no side effects, seen as a compound code statement receiving its input as arguments passed to its formal parameters and returning an object or value to the statement from where it was called or run. The body of the function is represented by the filled box.

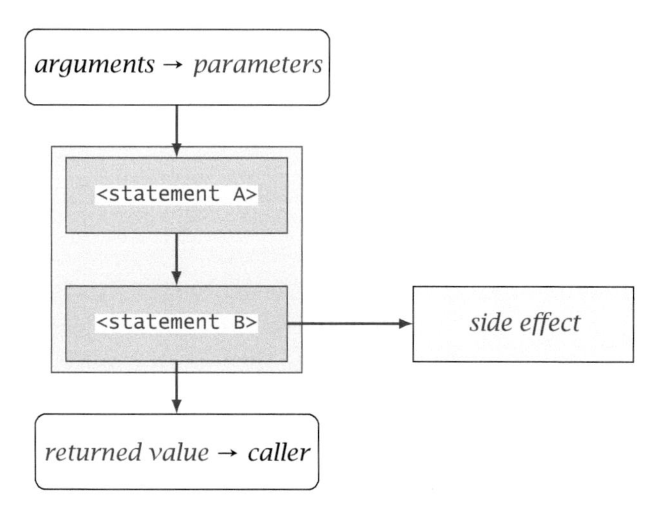

Figure 6.2
Diagram of function as a compound code statement receiving its input as arguments passed to its formal parameters and returning an object or value to the statement from where it was called or run. The body of the function is represented by the box filled in blue, while the side effect of the code in the function directly outside is represented by the box filled in yellow.

it does not affect its environment, it only returns a value to the caller. A value on which the caller has full control. The statement that calls the function "decides" what to do with the value received from the function.

When a function has a side effect, the caller is no longer in full control (Figure 6.2). Side effects can be actions that do not alter any object in the calling code, like when a call to `print()` displays text or numbers. Side effects can also be an assignment that modifies an object in the caller's environment, such as assigning a new value to a variable in the caller's environment, i.e., "outside the function".

A function can return only one object, so when multiple results are produced they need to be collected into a single object. In many cases, lists are used to collect all the values to be returned into one R object. For example, model fit functions like `lm()`, discussed in section 7.9 on page 200, return lists with multiple heterogeneous members, plus ancillary information stored in several attributes. In the case of `lm()` the returned object's class is `lm`, and its mode is `list`.

6.2 When function `return()` is called within a function, the flow of execution within the function stops and the argument passed to `return()` is the value returned by the function call. In contrast, if function `return()` is not explicitly called, the value returned by the function call is that returned by the last statement *executed* within the body of the function. Run these examples, and your own variations.

```
FN1 <- function(x) print("prn")
FN1("arg")
FN2 <- function(x){print("prn")
                   return(x)}
FN2("arg")
FN3 <- function(x){return(x)
                   print("prn")}
FN3("arg")
FN4 <- function(x){return()
                   print("prn")}
FN4("arg")
FN5 <- function(x){return(print(x))
                   print("prn")}
FN5("arg")
```

In base R, arguments to functions are passed by copy. This is something important to remember. If code in a function's body modifies the value of a parameter (the placeholder for an argument), its value outside the function is not affected, e.g., if the argument passed was a variable.

```
fn2 <- function(x){x <- 99}
a <- 1
fn2(a)
a
## [1] 1
```

In some other computer languages, arguments can be passed by reference, meaning that assignments to a formal parameter within the body of the function are back-referenced to the argument and modify it. It is possible to imitate such behaviour in R using some language trickery and consequently, occasionally functions in R use this approach.

Functions have their own *scope*. Any new variables created by normal assignment within the body of a function are visible only within the body of the function and are destroyed when the function returns from the call. In normal use, functions in R do not affect their environment through side effects.

Functions can be called without giving them a name. This is common when the function is simple and called only once. Anonymous functions are frequently used together with *apply functions*, as a definition passed directly as an argument to parameter FUN (see section 5.8 on page 154).

```
(function(x, y){x * y})(x = 4, y = 3)
## [1] 12
```

A new terse notation for defining functions was introduced in R 4.1.0, with \() as a synonym of function(). This is intended to make code concise, and especially useful for anonymous or lambda functions. However, I think this notation should be used sparingly, and possibly only at the R console. I have not used \() in code examples in the book, except for the one below.

```
(\(x, y){x * y})(x = 4, y = 3)
## [1] 12
```

6.2.1 Scope of names

Scoping in R is implemented using *environments* and *name spaces*. We can think of environments as having a boundary with asymmetric visibility. The code within a function runs in its own environment, in isolation from the calling environment in relation to assignments, but the values stored in objects in the calling environment can be retrieved. This protects from unintentional side effects by making difficult to overwrite object definitions in the calling environment. It is possible to override this protection with operator <<- or with function assign(). When used, assignment as side effects, can make the code much more difficult to read and debug, so its best to avoid them.

Parameters and local variables are not read-only, they behave like normal variables within the body of the function. However, assignments made using the operator <-, only affect a local copy that is destroyed when the function returns.

The visibility of names is determined by the *scoping rules* of a language. The clearest, but not the only situation when scoping rules matter, is when objects with the same name coexist. In such a situation, one will be accessible by its unqualified name and the other hidden but possibly accessible by qualifying the name with the namespace where it is defined.

As the R language has few reserved words for which no redefinition is allowed, we should take care not to accidentally reuse names that are part of the language. For example, pi is a constant defined in R with the value of the mathematical constant π. If we use the same name for one of our variables, the original definition is hidden and can no longer be normally accessed.

```
pi
## [1] 3.141593
pi <- "apple pie"
pi
## [1] "apple pie"
rm(pi)
pi
## [1] 3.141593
exists("pi")
## [1] TRUE
```

In the example above, the two variables are not defined in the same scope. In the example below, we assign a new value to a variable we have earlier created within the same scope, and consequently the second assignment overwrites, rather than hides, the existing definition.

```
my.pie <- "raspberry pie"
my.pie
## [1] "raspberry pie"
my.pie <- "apple pie"
my.pie
## [1] "apple pie"
rm(my.pie)
exists("my.pie")
## [1] FALSE
```

Name spaces play an important role in avoiding name clashes when contributed packages are attached (see section 6.4.4 on page 184).

Environments can be explicitly created with function `environment()`. However, `environment()` is rarely used in scripts while it can be useful within packages.

6.2.2 Ordinary functions

After the toy examples above, we will define a small but useful function: a function for calculating the standard error of the mean from a numeric vector. The standard error is given by $S_{\hat{x}} = \sqrt{S^2/n}$. We can translate this into the definition of an R function called SEM.

```
SEM <- function(x){sqrt(var(x) / length(x))}
```

As a test, we call `SEM()` with both `a` and `a.na` as argument.

```
a <- c(1, 2, 3, -5)
a.na <- c(a, NA)
SEM(x = a)
## [1] 1.796988
SEM(x = a.na)
## [1] NA
```

Our function `SEM(a)` never returns a wrong answer because NA values in its input always result in NA being returned. The downside is that unlike R's functions such as `var()`, `SEM()` does not support omitting NA values.

Adding `na.rm` as a second parameter and passing the argument it receives to the call to `var()` within the body of `SEM()` is not enough. To avoid returning wrong values, NA values should be also removed before counting the number of observations with `length()`. A good alternative is to define the function as follows.

```
sem <- function(x, na.rm = FALSE) {
  if (na.rm) {
    x <- na.omit(x)
  }
  sqrt(var(x)/length(x))
}

sem(x = a)
## [1] 1.796988
```

```
sem(x = a.na)
## [1] NA
sem(x = a.na, na.rm = TRUE)
## [1] 1.796988
```

R does not provide a function for standard error, so the function above is generally useful. Its user interface is consistent with that of functionally similar existing functions. We have added a new word to the R vocabulary available to us.

In the definition of `sem()` we set a default argument for parameter `na.omit` which is used unless the user explicitly passes an argument to this parameter.

6.3 Define your own function to calculate the mean in a similar way as `SEM()` was defined above. Hint: function `sum()` could be of help.

Within an expression, a function name followed by parentheses is interpreted as a call to the function, while the bare name of a function, returns its definition (similarly to any other R object). If the name is entered as a statement at the R console, its value is printed.

We first print (implicitly) the definition of our function from earlier in this section.

```
sem
## function(x, na.rm = FALSE) {
##   if (na.rm) {
##     x <- na.omit(x)
##   }
##   sqrt(var(x)/length(x))
## }
## <bytecode: 0x0000013b948be308>
```

Next, we print the definition of R's standard deviation function `sd()`.

```
sd
## function (x, na.rm = FALSE)
## sqrt(var(if (is.vector(x) || is.factor(x)) x else as.double(x),
##     na.rm = na.rm))
## <bytecode: 0x0000013b8f1afeb0>
## <environment: namespace:stats>
```

As can be seen at the end of the printouts, these functions written in the R language have been byte-compiled so that they execute faster. We can also see that the definition of `sd()` resides in `namespace:stats` because it has been attached from package 'stats'.

Functions that are part of the R language, but that are not coded using the R language, are called primitives and their full definition cannot be accessed through their name (c.f., `sem()` defined above and `sd`, with `list()` below).

```
list
## function (...)  .Primitive("list")
```

6.2.3 Operators

Operators are functions that use a different syntax for being called. If their name is enclosed in back ticks they can be called as ordinary functions. Binary operators like + have two formal parameters, and unary operators like unary – have only one formal parameter. The parameters of many binary R operators are named e1 and e2. This is just a convention, not enforced by the R language.

```
1 / 2
## [1] 0.5
`/`(1 , 2)
## [1] 0.5
`/`(e1 = 1 , e2 = 2)
## [1] 0.5
```

An important consequence of the possibility of calling operators using ordinary syntax is that operators can be used as arguments to *apply* functions in the same way as ordinary functions. When passing operator names as arguments to *apply* functions, we only need to enclose them in back ticks (see section 5.8 on page 154).

The name by itself and enclosed in back ticks allows us to access the definition of an operator.

```
`/`
## function (e1, e2)  .Primitive("/")
```

Defining a new operator. We will define a binary operator (taking two arguments) that subtracts from the numbers in a vector the mean of another vector. First, we need a suitable name, but we have less freedom as names of user-defined operators must be enclosed in percent signs. We will use %-mean% and as with any *special name*, we need to enclose it in quotation marks for the assignment.

```
"%-mean%" <- function(e1, e2) {
  e1 - mean(e2)
}
```

We can then use our new operator in a example.

```
10:15 %-mean% 1:20
## [1] -0.5  0.5  1.5  2.5  3.5  4.5
```

To print the definition, we enclose the name of our new operator in back ticks—i.e., we *back quote* the special name.

```
`%-mean%`
## function(e1, e2) {
##   e1 - mean(e2)
## }
```

6.3 Objects, Classes and Methods

New classes are normally defined within packages rather than in user scripts. To be really useful implementing a new class involves not only defining a class but also a set of specialised functions or *methods* that implement operations on objects belonging to the new class. Nevertheless, an understanding of how classes work is important even if only very occasionally a user will define a new method for an existing class within a script.

Classes are abstractions, but abstractions describing the shared properties of "types" or groups of similar objects. In this sense, classes are abstractions of "actors", they are like "nouns" in natural language. What we obtain with classes is the possibility of defining multiple versions of functions (or *methods*) sharing the same name but tailored to operate on objects belonging to different classes. We have already been using methods with multiple *specialisations* throughout the book, for example, `plot()` and `summary()`.

We start with a quotation from *S Poetry* (Burns 1998, page 13).

> The idea of object-oriented programming is simple, but carries a lot of weight. Here's the whole thing: if you told a group of people "dress for work," then you would expect each to put on clothes appropriate for that individual's job. Likewise it is possible for S[R] objects to get dressed appropriately depending on what class of object they are.

We say that specific methods are *dispatched* based on the class of the argument passed. This, together with the loose type checks of R, allows writing code that functions as expected on different types of objects, e.g., character and numeric vectors.

R has good support for the object-oriented programming paradigm, but as a system that has evolved over the years, currently R supports multiple approaches. The still most popular approach is called S3, and a more recent and powerful approach, with slower performance, is called S4. The general idea is that a name like "plot" can be used as a generic name and that the specific version of `plot()` called depends on the arguments of the call. Using computing terms we could say that the *generic* of `plot()` dispatches the original call to different specific versions of `plot()` based on the class of the arguments passed. S3 generic functions dispatch, by default, based only on the argument passed to a single parameter, the first one. S4 generic functions can dispatch the call based on the arguments passed to more than one parameter and the structure of the objects of a given class is known to the interpreter. In S3 functions, the specialisations of a generic are recognised/identified only by their name. And the class of an object by a character string stored as an attribute to the object (see section 4.6 on page 114 about attributes).

We first explore one of the methods already available in R. The definition of `mean` shows that it is the generic for a method.

```
mean
## function (x, ...)
## UseMethod("mean")
## <bytecode: 0x0000013b8b690df8>
## <environment: namespace:base>
```

We can find out which specialisations of a method are available in the current search path using `methods()`.

```
methods(mean)
##  [1] mean.Date         mean.default      mean.difftime     mean.POSIXct
##  [5] mean.POSIXlt      mean.quosure*     mean.vctrs_vctr* mean.yearmon*
##  [9] mean.yearqtr*     mean.zoo*
## see '?methods' for accessing help and source code
```

We can also use `methods()` to query all methods, including operators, defined for objects of a given class.

```
methods(class = "list")
## [1] all.equal     as.data.frame coerce        na.contiguous Ops
## [6] relist        sew           type.convert  within
## see '?methods' for accessing help and source code
```

S3 class information is stored as a character vector in an attribute named `"class"`. The most basic approach to the construction (= creation) of an object of a new S3 class, is to add the new class name to the `class` attribute of the object. As the implied class hierarchy is given by the order of the members of the character vector, the name of the new class must be added at the head of the vector. Even though this step can be done as shown here, in practice this step would normally take place within a *constructor* function and the new class, if defined within a package, would need to be registered. We show here this bare-bones example only to demonstrate how S3 classes are implemented in R.

```
a <- 123
class(a)
## [1] "numeric"
class(a) <- c("myclass", class(a))
class(a)
## [1] "myclass" "numeric"
```

Now we create a print method specific to `"myclass"` objects. Internally we are using function `sprintf()` and for the format template to work we need to pass a `numeric` value as an argument—i.e., obviously `sprintf()` does not "know" how to handle objects of the class we have just created!

```
print.myclass <- function(x) {
    sprintf("[myclass] %.0f", as.numeric(x))
}
```

Once a specialised method exists for a class, it will be used for objects of this class.

```
print(a)
## [1] "[myclass] 123"
print(as.numeric(a))
## [1] 123
```

Adding the name `"derivclass"` to the head of the `class` character vector, makes object `b` a member of both classes, `"myclass"` and `"derivclass"`, where `"derivclass"` is derived from `"myclass"`. As `"derivclass"` is at position `1`, it is for this object its *most derived class*.

```
b <- 456
class(b) <- c("derivclass", class(a))
```

A specialised `print()` method is not available for `"derivclass"`, the method for `"myclass"`, the next class name along the vector, is called.

```
print(b)
## [1] "[myclass] 456"
```

```
print(as.numeric(b))
## [1] 456
```

The S3 class system is "lightweight" in that it adds very little additional computation load, but it is rather "fragile" in that most of the responsibility for consistency and correctness of the design—e.g., not messing up dispatch by redefining functions or loading a package exporting functions with the same name, etc., is not checked by the R interpreter.

6.4 Packages

6.4.1 Sharing of R-language extensions

The most elegant way of adding new features or capabilities to R is through packages. A package can contain any, several or all of R function and operator definitions, data objects, classes, and their methods, plus the corresponding documentation. Some packages available through CRAN contain only one or two R objects while others contain hundreds of them. After loading and attaching a package, the objects that the package exports can be used as if they were part R itself.

Packages are, without doubt, the best mechanism for sharing extensions to R. However, in most situations, packages are also very useful for managing code that will be reused by a single person over time. R packages have strict rules about their contents, file structure, and documentation, which makes it possible among other things for the package documentation to be merged into R's help system when a package is loaded. With a few exceptions, packages can be written so that they will work on any computer where R runs.

In a "source package", the code written in R, and possibly in other programming languages, is contained in text files that are compressed together into a single archive file. In a "binary package" the source code is already processed into a form suitable for faster installation. Binary package files are specific to each major version of R, operating system, and computer architecture. In addition to being slower, package installation from sources can requires additional software, such as compilers. A compiler translates the text representation of a computer program written in C, C++, FORTRAN, etc., into machine code, i.e., instructions for the computer hardware. R code is compiled into instructions for a virtual machine, part of R, that does the final translation into machine code at runtime.

For distribution, a single compressed archive file is used for aech package. Packages can be shared as source- or binary-code files, sent for example through e-mail. However, the largest public repository of R packages is called CRAN (`https://cran.r-project.org/`), an acronym for Comprehensive R Archive Network. Packages available through CRAN are guaranteed to work, in the sense of not failing any tests built into the packages and not crashing or aborting prematurely. They are tested daily, as they may depend on other packages whose code will change when updated. The number of packages available through CRAN at the time of printing (2024-02-17) was 2.04×10^4.

A key repository for bioinformatics with R is Bioconductor (`https://www.bioconductor.org/`), containing packages that pass strict quality tests, adding an additional 3 400 packages. rOpenScience has established guidelines and a system for code peer review for R packages. These peer-reviewed packages are available through CRAN or other repositories and listed at the rOpenScience website (`https://ropensci.org/`). Occasionally, one may have, or want, to install packages or updates that are not yet in CRAN, either from the R Universe (`https://r-universe.dev/`) repositories, or from Git repositories (e.g., from GitHub).

A good way of learning how the extensions provided by a package work, is to experiment with them. When using a function we are not yet familiar with, looking at its help to check all its features expands our "toolbox". While documentation of exported objects is enforced, many packages include, in addition, comprehensive user guides or articles as *vignettes*. It is not unusual to decide which package to use from a set of alternatives based its documentation. In the case of packages adding extensive new functionality, they may be documented in depth in a book. Well-known examples are *Mixed-Effects Models in S and S-Plus* (Pinheiro and Bates 2000) and *ggplot2: Elegant Graphics for Data Analysis* (Wickham and Sievert 2016).

6.4.2 Download, installation and use

In R speak, "library" is the location where packages are installed. Packages are sets of functions, and data, specific for some particular purpose, that can be loaded into an R session to make them available so that they can be used in the same way as built-in R functions and data. Function `library()` is used to load and attach packages that are already installed in the local R library. In contrast, function `install.packages()` is used to install packages.

The instructions below assume that the user has access to repositories on the internet and enough user rights to install packages. This is rarely the case in organisations using strict security protocols. In such cases, the organisation may keep a mirror of CRAN in the intranet. The local/user's private R library can be kept in a folder where the user has writing and reading rights.

How to install or update a package from CRAN?

CRAN is the default repository for R packages. If you use RStudio or another IDE as a front end on any operating system or RGUI under MS-Windows, installation and updates can be done through a menu or GUI button. These menus use calls to `install.packages()` and `update.packages()` behind the scenes.

Alternatively, at the R command line, or in a script, `install.packages()` can be called with the name of the package as an argument. For example, to install package 'learnrbook' one can use

```
install.packages("learnrbook")
```

and to update already installed packages

```
update.packages()
```

How to install or update a package from GitHub?
Package 'remotes' makes it possible to install packages directly from GitHub, Bitbucket and other repositories based on Git. The code in the next chunk (not run here) can be used to install the latest, possibly, still under development, version of package 'learnrbook'.

```
remotes::install_github("aphalo/learnrbook-pkg")
```

Function pkg_install() from 'pak' can install packages, both from CRAN and Bioconductor repositories, and from Git repositories. The same function can be used to update specific already installed packages and dependencies.

```
pak::pkg_install("learnrbook") # from CRAN
pak::pkg_install("aphalo/learnrbook-pkg") # from GitHub
```

R packages can be installed either from sources, or from already built "binaries". Installing from sources, depending on the package, may require additional software to be available. This is because some R packages contain source code in other languages such as C, C++ or FORTRAN that needs to be compiled into machine code during installation. Under MS-Windows, the needed shell, commands, and compilers are not available as part of the operating system. Installing them is not difficult as they are available prepackaged in an installer under the name RTools (available from CRAN). MiKTEX) is usually needed to build the PDF of the package's manual.

Under MS-Windows, it is easier to install packages from binary .zip files than from .tar.gz source files. For OS X (Apple Mac) the situation is similar, with binaries available both for Intel and ARM (M1, M2 series) processors. Most, but not all, Linux distributions include in the default setup the tools needed for installation of R packages. Under Linux it is rather common to install packages from sources, although package binaries have recently become more easily available.

If the tools are available, packages can be easily installed from sources from within RStudio. However, binaries are for most packages also readily available. In CRAN, the binary for a new version of a package becomes available with a delay of one or two days compared to the source. For packages that need compilation, the installation from sources takes more time than installation from binaries.

6.4 Use help to look up the help page for install.packages(), and explore how to control whether the package is installed from a source or a binary file. Also explore, how to install a package from a file in a local disk instead of from a repository like CRAN.

Frequently the README file of a package includes instructions on how to install it from CRAN or another online repository. Exceptionally, packages may require additionally the installation of software outside R before their installation and/or use. When present, these rather exceptional requirements are always listed in the DESCRIPTION under SystemRequirements: and explained in more detail in the README file. In CRAN, each package has a home web page that can be easily found if one knows the name of the package, e.g., https://CRAN.R-project.org/package=learnrbook. Nowadays, it is common for the help for a package being also available as a web site, e.g., https://docs.r4photobiology.info/learnrbook/.

How to change the repository used to install packages?
Function `setRepositories()` can be used to enable other repositories in addition or instead of CRAN during an R session. In recent versions of R, the default list of repositories is taken from R option `"repos"` if defined. Consult `help("setRepositories")` for the details.

Alternatively, one can use function `pkg_install()` from package 'pak' as this function attempts to automatically set the correct repository based on the name of the package.

How to use an installed package?
To use the functions and other objects defined in a package, the package must first be loaded, and for the names of these objects to be visible in the user's workspace, the package needs to be attached. Function `library()` loads and attaches one package at a time. For example, to load and attach package 'learnrbook' we use.

```
library("learnrbook")
```

How to find the currently installed version of a package?
Function `packageVersion` returns the version as an object of class `"package_version"` that can not only be printed, but also meaningfully compared, e.g., to test for a minimum version requirement.

```
packageVersion(pkg="learnrbook")
## [1] '1.0.2.1'
```

As packages are contributed by independent authors, they should be cited in addition to citing R itself when they are used to obtain results or plots included in publications. R function `citation()` when called with the name of a package as its argument provides the reference that should be cited for the package, and without an explicit argument, the reference to cite for the version of R in use as shown below.

```
citation()
## To cite R in publications use:
##
##   R Core Team (2024). _R: A Language and Environment for Statistical
##   Computing_. R Foundation for Statistical Computing, Vienna, Austria.
##   <https://www.R-project.org/>.
##
## A BibTeX entry for LaTeX users is
##
##   @Manual{,
##     title = {R: A Language and Environment for Statistical Computing},
##     author = {{R Core Team}},
##     organization = {R Foundation for Statistical Computing},
##     address = {Vienna, Austria},
##     year = {2024},
##     url = {https://www.R-project.org/},
##   }
##
## We have invested a lot of time and effort in creating R, please cite it
## when using it for data analysis. See also 'citation("pkgname")' for
## citing R packages.
```

6.5 Look at the help page for function `citation()` for a discussion of why it is important that users cite R and packages when using them.

Conflicts among packages can easily arise, for example, when they use the same names for objects or functions. These are reported when the packages are attached (see section 6.4.4 on page 184 for a workaround). In addition, many packages use functions defined in packages in the R distribution itself or other independently developed packages by importing them. Updates to depended-upon packages can "break" (make non-functional) the dependent packages or parts of them. The rigourous testing by CRAN detects such problems in most cases when package revisions are submitted, forcing package maintainers to fix problems before distribution through CRAN is possible. However, if you use other repositories, I recommend that you make sure that revised (especially if under development) versions do work with your own code, before their use in "production" (important) data analyses.

6.4.3 Finding suitable packages

Due to the large number of contributed R packages, it can sometimes be difficult to find a suitable package for a task at hand. It is good to first check if the necessary capability is already built into base R. Base R plus the recommended packages (installed when R is installed) cover a lot of ground. Analysing data using almost any of the more common statistical methods does not require the use of contributed packages. Sometimes, contributed packages duplicate or extend the functionality in base R. When one considers the use of novel or specialised types of data analysis, the use of contributed packages can be unavoidable. Even in such cases, it is not unusual to have alternatives to choose from within the available contributed packages. Sometimes groups or suites of packages are designed to work well together.

The CRAN repository has a very broad scope and includes a section called "views". R views are web pages providing annotated lists of packages frequently used within a given field of research, engineering, or specific applications. These views are maintained by different expert editors. The R views can be found at `https://cran.r-project.org/web/views/`.

The Bioconductor repository specialises in bioinformatics with R. It also has a section with "views" and within it, descriptions of different data analysis workflows. The workflows are especially good as they reveal which sets of packages work well together. These views can be found at `https://www.bioconductor.org/packages/release/BiocViews.html`.

rOpenSci (Ram et al. 2019) fosters a culture that values open and reproducible research using shared data and reusable software. One aspect of this is making possible peer-review of R packages. rOpenSci does not keep a separate package repository for the peer-reviewed packages, they keep an index at `https://ropensci.org/packages/`. The packages included have become more diverse, but initially the main focus was on facilitating access to open data sources.

The CRAN repository keeps an archive of earlier versions of packages, on an individual package basis. This is also important for long-term reproducibility.

6.4.4 How packages work

R packages define all objects within a *namespace* with the same name as the package itself. Loading and attaching a package with `library()` makes visible only the exported objects. Attaching a package adds these objects to the search path so that they can be accessed without prepending the name of the namespace. Most packages do not export all the functions and objects defined in their code; some are kept internal, in most cases, to avoid making a commitment about their availability in future versions, which could constrain further development.

> Package namespaces can be detached and also unloaded with function `detach()` using a slightly different notation for the argument from that which we described for data frames in section 4.4.5 on page 110. This is very seldom needed, but one case I have come across is a package that redefines a generic function of a method of a package it imports, thus preventing the normal use of a third package that depends on the original definition of the generic.

When we reuse a name defined in a package, its definition in the package does not get overwritten, but instead, only hidden. These hidden objects remain accessible using the name *qualified* by prepending the name of the package followed by two colons, e.g., `base:mean()`.

If two packages define objects with the same name, then which one is visible depends on the order in which the packages were attached with `library()`. To avoid confusion in such cases, in scripts it is best to use the qualified names for calling objects defined with the same name in two packages. Using the qualified name for an object from an already attached package, is inconsequential for its interpretation by R, but can enhance the readability of the code.

> If one uses a qualified name for an object but does not attach the package with a call to `library()`, the package is only loaded. In other words, the names of the exported objects are not added to the search pass, but the code defining them is retrieved and available using qualified names.

Some functions that are part of R are collected into packages grouped by category: 'base', 'stats', 'datasets', etc., and can be called when needed using qualified names. We can find out the search order by calling `search()`, with the search starting at the `".GlobalEnv"` for statements evaluated at the R command line.

> 6.6 Namespaces isolate the names defined within them from those in other namespaces. This helps prevent name clashes, and makes it possible to access objects even when they are "hidden" by a different object with the same name.

```
class(cars)
head(cars, 3)
getAnywhere("cars")$where # defined in package
```

```
cars <- 1:10
class(cars)
head(cars, 3) # prints 'cars' defined in the global environment
rm(cars) # clean up
head(cars, 3)
getAnywhere("cars")$where # the first visible definition is in the global en-
vironemnt
```

In the playground above, I used a data frame object, but the same mechanisms apply to all R objects including functions. The situation when one of the definitions is a function and the other is not, is slightly different in that a call using parenthesis notation will distinguish between a function and an object of the same name that is not a function. Relying on this distinction is anyway very confusing and, thus, a bad idea.

```
mean
## function (x, ...)
## UseMethod("mean")
## <bytecode: 0x0000013b8b690df8>
## <environment: namespace:base>

mean <- mean(1:5)
mean
## [1] 3
mean(8:9)
## [1] 8.5

getAnywhere("mean")$where
## [1] ".GlobalEnv"      "package:base"     "namespace:base"
rm(mean)
getAnywhere("mean")$where
## [1] "package:base"    "namespace:base"
```

In this last example, `rm(mean)` removed the variable we had assigned a value to. Package namespaces protect the objects defined in the package from deletion or overwriting. This is different to defining a new object with the same name, which is allowed. The two statements below trigger errors and are not evaluated when typesetting the book.

```
datasets::cars <- "my car is green"
rm(datasets::cars)
```

The value returned by `getAnywhere()` has additional information than that in its member `where`. Do have a look at its help page with `help(getAnywhere)` for the details.

6.5 Further Reading

Several books describe in detail the different class systems available and how to use them in R. For an in-depth treatment of the subject please consult the books *Advanced R* (Wickham 2019) and *Extending R* (Chambers 2016).

The development of R packages is accessibly explained in the book *R Packages* (Wickham and Bryan 2023), using a practical approach and tools developed by the author and his collaborators. The book *Extending R* (Chambers 2016) has its focus on R itself, how it works, and how to develop extensions both with simple and challenging goals.

7

Base R: "Verbs" and "Nouns" for Statistics

> The purpose of computing is insight, not numbers.
>
> Richard W. Hamming
> *Numerical Methods for Scientists and Engineers*, 1987

7.1 Aims of This Chapter

This chapter aims to give the reader an introduction to the approach used in base R for the computation of statistical summaries, the fitting of models to observations and tests of hypothesis. This chapter does *not* explain data analysis methods, statistical principles or experimental designs. There are many good books on the use of R for different kinds of statistical analyses (see further reading on page 241) but most of them tend to focus on specific statistical methods rather than on the commonalities among them. Although base R's model fitting functions target specific statistical procedures, they use a common approach to model specification and for returning the computed estimates and test outcomes. This approach, also followed by many contributed extension packages, can be considered as part of the philosophy behind the R language. In this chapter, you will become familiar with the approaches used in R for calculating statistical summaries, generating (pseudo-)random numbers, sampling, fitting models, and carrying out tests of significance. We will use linear correlation, *t*-test, linear models, generalised linear models, non-linear models, and some simple multivariate methods as examples. The focus is on how to specify statistical models, contrasts and observations, how to access different components of the objects returned by the corresponding fit and summary functions, and how to use these extracted components in further computations or for customised printing and formatting.

DOI: 10.1201/9781003404187-7

Table 7.1
Frequently used simple statistical summaries and the corresponding R functions.

Function	Symbol	Formulation	Name
`mean()`	$\bar{x}$	$\sum x/n$	mean
`var()`	s^2	$\sum (x_i - \hat{x})^2/(n-1)$	sample variance
`sd()`	s	$\sqrt[2]{s^2}$	sample standard deviation
`median()`	M or $\tilde{x}$		median
`mad()`	MAD	median $\lvert x_i - \hat{x} \rvert$	median absolute deviation
`mode()`	MOD		mode
`max()`	x_{max}		maximum
`min()`	x_{min}		minimum
`range()`	x_{min}, x_{max}		range

7.2 Statistical Summaries

Being the main focus of the R language in data analysis and statistics, R provides functions both for simple and complex calculations, going from means and variances to fitting very complex models. Table 7.1 lists some frequently used functions. All these methods accept numeric vectors and/or matrices as arguments. In addition, function `quantile()` can be used to simultaneously compute multiple arbitrary quantiles for a vector of observations, and method `summary()` produces a summary that depends on the class of the argument passed to it. (See section 6.2.2 on page 174 for how to define your own functions.)

By default, if the argument contains `NA`s these functions return `NA`. The logic behind this is that if one value exists but is unknown, the true result of the computation is unknown (see page 33 for details on the role of `NA` in R). However, an additional parameter called `na.rm` allows us to override this default behaviour by requesting any `NA` in the input to be removed (or discarded) before calculating the summary,

```
x <- c(1:20, NA)
mean(x)
## [1] NA
mean(x, na.rm = TRUE)
## [1] 10.5
```

Function `mean()` can be used to compute the mean from all values, as in the example above, as well as trimmed means, i.e., means computed after discarding extreme values. The argument passed to parameter `trim` decides the fraction of the observations to discard at *each extreme* of the vector of values after ordering them from smallest to largest.

```
x <- c(1:20, 100)
mean(x)
## [1] 14.7619
mean(x, trim = 0.05)
## [1] 11
```

Table 7.2
Standard probability distributions in R. Partial list of base R functions related to probability distributions. The full list can be obtained by executing the command `help(Distributions)`.

Distribution	Symbol	Density	*P*-value	Quantiles	Draws
Normal	N	`dnorm()`	`pnorm()`	`qnorm()`	`rnorm()`
Student's	t	`dt()`	`pt()`	`qt()`	`rt()`
F	F	`df()`	`pf()`	`qf()`	`rf()`
binomial	B	`dbinom()`	`pbinom()`	`qbinom()`	`rbinom()`
multinomial	M	`dmultinom()`	`pmultinom()`	`qmultinom()`	`rmultinom()`
Poisson		`dpois()`	`ppois()`	`qpois()`	`rpois()`
X-squared	X^2	`dchisq()`	`pchisq()`	`qchisq()`	`rchisq()`
lognormal		`dlnorm()`	`plnorm()`	`qlnorm()`	`rlnorm()`
uniform		`dunif()`	`punif()`	`qunif()`	`runif()`

7.1 In contrast to the use of other functions, I do not provide examples of the use of all the functions listed in Table 7.1. Construct `numeric` vectors with artificial data or use real data to play with the remaining functions. Study the help pages to learn about the different parameters and their uses.

Other more advanced functions are also available in R, such as `boxplot.stats()` that computes the values needed to draw boxplots (see section 9.6.6 on page 328).

In many cases, you will want to compute statistical summaries by group or treatment in addition or instead of for a whole data set or vector. See section 4.4.2 on page 105 for details on how to compute summaries of data stored in data frames using base R functions, and section 8.7 on page 259 for alternative functions from contributed packages.

7.3 Standard Probability Distributions

Density functions, probability distribution functions, quantile functions, and functions for pseudo-random draws are available in R for several different standard (theoretical) probability distributions. Entering `help(Distributions)` at the R prompt will open a help page describing all the distributions available in base R. For each distribution, the different functions contain the same "root" in their names: `norm` for the normal distribution, `unif` for the uniform distribution, and so on. The "head" of the name indicates the type of values returned: "`d`" for density, "`q`" for quantile, "`r`" (pseudo-)random draws, and "`p`" for probability (Table 7.2).

Theoretical distributions are defined by mathematical functions that accept parameters that control the exact shape and location. In the case of the Normal distribution, these parameters are the *mean* (`mean`) controlling the location center and *(*standard deviation) (`sd`) controlling the spread away from the center of the distribution. The four different functions differ in which values are calculated (the unknowns) and which values are supplied as arguments (the known inputs).

In what follows, I use the Normal distribution as an example, but with differences in their parameters, the functions for other theoretical distributions follow a similar naming pattern.

7.3.1 Density from parameters

To obtain a single point from the distribution curve we pass a vector of length one as an argument for `x`.

```
dnorm(x = 1.5, mean = 1, sd = 0.5)
## [1] 0.4839414
```

To obtain multiple values we can pass a longer vector as an argument.

```
dnorm(x = seq(from = -1, to = 1, length.out = 5), mean = 1, sd = 0.5)
## [1] 0.0002676605 0.0088636968 0.1079819330 0.4839414490 0.7978845608
```

With 50 equally spaced values for x we can plot a line (`type = "l"`) that shows that the 50 generated data points give the illusion of a continuous curve. We also add a point showing the value for $x = 1.5$ calculated above.

```
vct1 <- seq(from = -1, to = 3, length.out = 50)

df1 <- data.frame(x = vct1,
                  y = dnorm(x = vct1, mean = 1, sd = 1))
plot(y~x, data = df1, type = "l", xlab = "z", ylab = "f(z)")
points(x = 2, y = dnorm(x = 2, mean = 1, sd = 1))
```

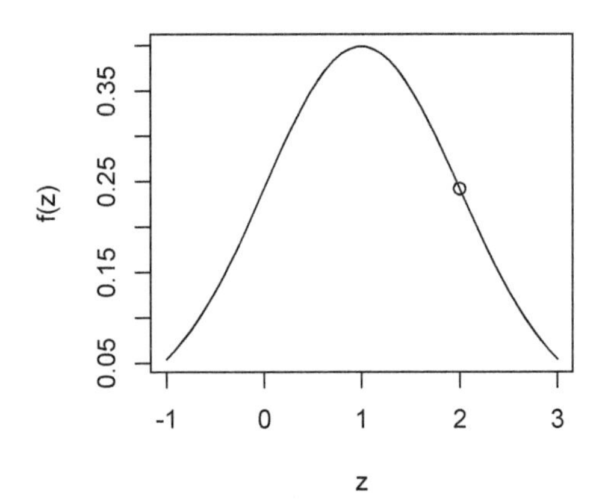

7.3.2 Probabilities from parameters and quantiles

With a known quantile value, it is possible to look up the corresponding P-value from the Normal distribution, i.e., the area under the curve, either to the right or to the left of a given value of `q` (by default, integrating the lower or left tail). When working with observations, the quantile, mean and standard deviation are in most cases computed from the same observations under the null hypothesis. In the example below, we use invented values for all parameters `q`, the quantile, `mean`, and `sd`, the standard deviation.

```
pnorm(q = 2, mean = 1, sd = 1)
## [1] 0.8413447
pnorm(q = 2, mean = 1, sd = 1, lower.tail = FALSE)
## [1] 0.1586553
pnorm(q = 2, mean = 1, sd = 4, lower.tail = FALSE)
## [1] 0.4012937
pnorm(q = c(2, 4), mean = 1, sd = 1, lower.tail = FALSE)
## [1] 0.158655254 0.001349898
```

In tests of significance, empirical z-values and t-values are computed by subtracting from the observed mean for one group or raw quantile, the "expected" mean (a hypothesised theoretical value, the mean of a control condition used as a reference, or the mean computed over all treatments under the assumption of no effect of treatments) and then dividing this difference by the standard deviation. Consequently, the p-values corresponding to these empirical z-values and t-values need to be looked up using `mean = 0` and `sd = 1` when calling `pnorm()` or `pt()` respectively. These frequently used values are the defaults.

7.3.3 Quantiles from parameters and probabilities

The reverse computation from that in the previous section is to obtain the quantile corresponding to a known P-value or area under one of the tails of the distribution curve. These quantiles are equivalent to the values in the tables of precalculated quantiles used in earlier times to assess significance with statistical tests.

```
qnorm(p = 0.01, mean = 0, sd = 1)
## [1] -2.326348
qnorm(p = 0.05, mean = 0, sd = 1)
## [1] -1.644854
qnorm(p = 0.05, mean = 0, sd = 1, lower.tail = FALSE)
## [1] 1.644854
```

Quantile functions like `qnorm()` and probability functions like `pnorm()` always do computations based on a single tail of the distribution, even though it is possible to specify which tail we are interested in. If we are interested in obtaining simultaneous quantiles for both tails, we need to do this manually. If we are aiming at quantiles for $P = 0.05$, we need to find the quantile for each tail based on $P/2 = 0.025$.

```
qnorm(p = 0.025, mean = 0, sd = 1)
## [1] -1.959964
qnorm(p = 0.025, mean = 0, sd = 1, lower.tail = FALSE)
## [1] 1.959964
```

When calculating a P-value from a quantile computed from observations in a test of significance, we need to first decide whether a two-sided or single-sided test is relevant, and in the case of a single sided test, which tail is of interest. In a two-sided test we need to multiply the returned P-value by 2. This works in the case of a symmetric distribution like the Normal, because the quantiles in the two tails differ only in sign. However, this is not the case for asymmetric distributions.

```
pnorm(q = 4, mean = 0, sd = 1) * 2
## [1] 1.999937
```

7.3.4 "Random" draws from a distribution

True random sequences can only be generated by physical processes. All "pseudo-random" sequences of numbers generated by computation are really deterministic although they share many properties with true random sequences (e.g., in relation to autocorrelation).

It is possible to compute not only pseudo-random draws from a uniform distribution but also from the Normal, t, F, and other distributions. In each case, the probability with which different values are "drawn" approximates the probabilities set by the corresponding theoretical distribution. Parameter `n` indicates the number of values to be drawn, or its equivalent, the length of the vector returned (see section 9.6.4 on page 324 for example plots).

```
rnorm(5)
## [1] -0.8248801  0.1201213 -0.4787266 -0.7134216  1.1264443
rnorm(n = 10, mean = 10, sd = 2)
##  [1] 12.394190  9.697729  9.212345 11.624844 12.194317 10.257707 10.082981
##  [8] 10.268540 10.792963  7.772915
```

7.2 Edit the examples in sections 7.3.2, 7.3.3, and 7.3.4 to do computations based on different distributions, such as Student's t, F or uniform.

It is impossible to generate truly random sequences of numbers by means of a deterministic process such as a mathematical computation. "Random numbers" as generated by R and other computer programs are *pseudo-random numbers*, long deterministic series of numbers that resemble random draws. Random number generation uses a *seed* value that determines where in the series the first value is fetched. The usual way of automatically setting the value of the seed is to take the milliseconds or a similar rapidly changing set of digits from the real-time clock of the computer. However, in cases when we wish to repeat a calculation using the same series of pseudo-random values, we can use `set.seed()` with an arbitrary integer as an argument to reset the generator to the same point in the underlying (deterministic) sequence.

7.3 Execute the statement `rnorm(3)` by itself several times, paying attention to the values obtained. Repeat the exercise, but now executing `set.seed(98765)` immediately before each call to `rnorm(3)`, again paying attention to the values obtained. Next execute `set.seed(98765)`, followed by `c(rnorm(3), rnorm(3))`, and then execute `set.seed(98765)`, followed by `rnorm(6)` and compare the output. Repeat the exercise using a different argument in the call to `set.seed()`. analyse the results and explain how `setseed()` affects the generation of pseudo-random numbers in R.

7.4 Observed Probability Distributions

It is common to estimate the value of the parameters for a standard distribution like Student's t or Normal distributions from observational data, assuming a priori the suitability of the distribution. If we compute the mean and standard deviation of a large sample, these two parameters define a specific Normal distribution curve. If we add the estimate of the degrees of freedom, $v = n - 1$, the three parameters define a specific t-distribution curve. Thus it is possible to use the functions described in section 7.3 on page 7.3, in statistical inference.

Package 'mixtools' provides tools for fitting and analysing *mixture models* such as the mix of two or more univariate Normal distributions. An example of its use could be to estimate mean and standard deviations for males and females in a dataset where the gender was not recorded at the time of observation.

It is also possible to describe the observed shape of the distribution, or empirical distribution, for a data set without relying on a standard distribution. The fitted empirical distribution can later be used to compute probabilities, quantiles, and random draws as from standard distributions. This also allows statistical inference, using methods such as the bootstrap or some additive models.

Function `density()` computes kernel density estimates, using different methods. A curve is used to describe the shape, and the bandwidth determines how flexible this curve is. The curve is a smoother that adapts to the observed shape of the distribution of observations. The object returned is a complex list that can be used to plot the estimated curve.

The code below estimates the empirical distribution for the waiting time in minutes between eruptions of the Old Faithful geyser at Yellowstone, a dataset from R.

```
d <- density(faithful$waiting, bw = "sj")
```

Using `str()` we can explore the structure of the object returned by function `density()`.

```
str(d)
## List of 8
##  $ x         : num [1:512] 35.5 35.6 35.8 35.9 36 ...
##  $ y         : num [1:512] 8.36e-06 9.89e-06 1.17e-05 1.38e-05 1.62e-05 ...
##  $ bw        : num 2.5
##  $ n         : int 272
##  $ old.coords: logi FALSE
##  $ call      : language density.default(x = faithful$waiting, bw = "sj")
##  $ data.name : chr "faithful$waiting"
##  $ has.na    : logi FALSE
##  - attr(*, "class")= chr "density"
```

The object saved as `d` is a `list` with seven members. The two numeric vectors, `x` and `y` describe the estimated probability distribution and produce the curve in the plot below. The numerical bandwidth estimated using method `"sj"` is in `bw`, and the length of vector `faithful$waiting`, the data used, is in `n`. Member `call` is the command used to call the function, the remaining two members have self-explanatory names. The returned object belongs to class `density`. The overall pattern is similar, but simpler than for the model fitting functions that we will see later in the chapter. The class name of the object is the same as the name of the function that created it, `call` provides a *trace* of how the object was created. Other members, facilitate computation of derived quantities and plotting. Being a list, the individual members can be extracted by name.

```
d$n
## [1] 272
```

As a `plot()` method is available for class `density` we can easily produce a plot of the estimated empirical density distribution. In this case, the fitted bimodal curve, with two maxima, is very different to the Normal.

```
plot(d)
```

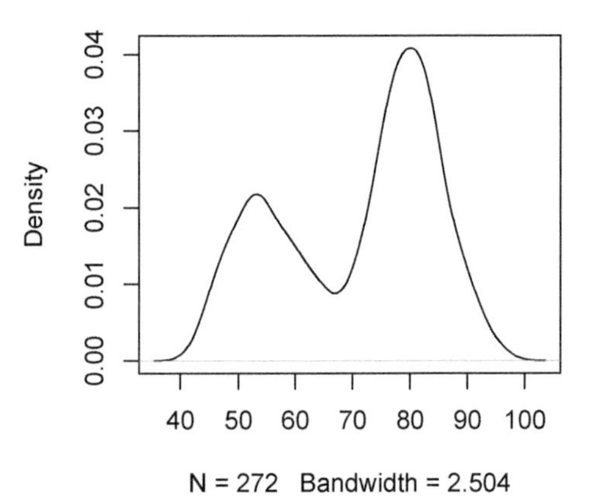

Observed probability distributions, especially empirical ones, nowadays play a central role in data visualisation including 1D and 2D empirical density plots based on the use of functions like `density()`, as well as traditional histograms (see section 9.6.5 on page 326 for examples of more elaborate and elegant plots).

7.5 "Random" Sampling

In addition to drawing values from a theoretical distribution, we can draw values from an existing set or collection of values. We call this operation (pseudo-)random sampling. The draws can be done either with replacement or without replacement. In the second case, all draws are taken from the whole set of values, making it possible for a given value to be drawn more than once. In the default case of not using replacement, subsequent draws are taken from the values remaining after removing the values chosen in earlier draws.

```
sample(x = LETTERS)
##  [1] "Z" "N" "Y" "R" "M" "E" "W" "J" "H" "G" "U" "O" "S" "T" "L" "F" "X" "P" "K"
```

```
## [20] "V" "D" "A" "B" "C" "I" "Q"
sample(x = LETTERS, size = 12)
##  [1] "M" "S" "L" "R" "B" "D" "Q" "W" "V" "N" "J" "P"
sample(x = LETTERS, size = 12, replace = TRUE)
##  [1] "K" "E" "V" "N" "A" "Q" "L" "C" "T" "L" "H" "U"
```

In practice, pseudo-random sampling is useful when we need to select subsets of observations. One such case is assigning treatments to experimental units in an experiment or selecting persons to interview in a survey. Another use is in bootstrapping to estimate variation in parameter estimates using empirical distributions.

How to sample random rows from a data frame?
As described in section 4.4 on page 94, data frames are commonly used to store one observation per row. To sample a subset of rows, we need to generate a random set of indices to use with the extraction operator (`[ ]`). Here we sample four rows from data frame `cars` included in R. These data consist of stopping distances for cars moving at different speeds as described in the documentation available by entering `help(cars)`).

```
cars[sample(x = 1:nrow(cars), size = 4), ]
##    speed dist
## 33    18   56
## 31    17   50
## 50    25   85
## 36    19   36
```

7.4 Consult the documentation of `sample()` and explain why the code below is equivalent to that in the example immediately above.

```
cars[sample(x = nrow(cars), size = 4), ]
```

7.6 Correlation

Both parametric (Pearson's) and non-parametric robust (Spearman's and Kendall's) methods for the estimation of the (linear) correlation between pairs of variables are available in base R. The different methods are selected by passing arguments to a single function. While Pearson's method is based on the actual values of the observations, non-parametric methods are based on the ordering or rank of the observations, and consequently less affected by observations with extreme values.

7.6.1 Pearson's r

Function `cor()` can be called with two vectors of the same length as arguments. In the case of the parametric Pearson method, we do not need to provide further arguments as this method is the default one. We use data set `cars`.

```
cor(x = cars$speed, y = cars$dist)
## [1] 0.8068949
```

It is also possible to pass a data frame (or a matrix) as the only argument. When the data frame (or matrix) contains only two columns, the returned correlation estimate is equivalent to that of passing the two columns individually as vectors. The object returned is a 2×2 `matrix` instead of a vector of length one.

```
cor(cars)
##           speed      dist
## speed 1.0000000 0.8068949
## dist  0.8068949 1.0000000
```

When the data frame or matrix contains more than two numeric vectors, the returned value is a matrix of estimates of pairwise correlations between columns. We here use `rnorm()` described above to create a long vector of pseudo-random values drawn from the Normal distribution and `matrix()` to convert it into a matrix with three columns (see page 70 for details about R matrices).

```
mat1 <- matrix(rnorm(54), ncol = 3,
               dimnames = list(rows = 1:18, cols = c("A", "B", "C")))
cor(mat1)
##            A         B          C
## A 1.00000000 0.1899797 0.07591003
## B 0.18997966 1.0000000 0.36800323
## C 0.07591003 0.3680032 1.00000000
```

7.5 Modify the code in the chunk immediately above constructing a matrix with six columns and then computing the correlations.

While `cor()` returns an estimate of r, the correlation coefficient, `cor.test()` computes, in addition, the t-value, P-value, and confidence interval for the estimate.

```
cor.test(x = cars$speed, y = cars$dist)
##
##  Pearson's product-moment correlation
##
## data:  cars$speed and cars$dist
## t = 9.464, df = 48, p-value = 1.49e-12
## alternative hypothesis: true correlation is not equal to 0
## 95 percent confidence interval:
##  0.6816422 0.8862036
## sample estimates:
##       cor
## 0.8068949
```

Above we passed two numeric vectors as arguments, one to parameter `x` and one to parameter `y`. Alternatively, we can pass a data frame as an argument to `data`, and a *model formula* to `formula`. The argument passed to `formula` determines which variables from `data` are used, and in which role. Briefly, the variable(s) to the left of the tilde ( ) are response variables, and those to the right are independent, or explanatory, variables. In the case of correlation, no assumption is made on cause and effect, and both variables appear to the right of the tilde. The code below is equivalent to that above. See section 7.13 on page 226 for details on the use of model formulas and section 7.8 on page 199 for examples of their use in model fitting.

```
cor.test(formula = ~ speed + dist, data = cars)
```

7.6 Functions `cor()` and `cor.test()` return R objects, that when using R interactively get automatically "printed" on the screen. One should be aware that `print()` methods do not necessarily display all the information contained in an R object. This is almost always the case for complex objects like those returned by R functions implementing statistical tests. As with any R object, we can save the result of an analysis into a variable. As described in section 4.3 on page 86 for lists, we can peek into the structure of an object with method `str()`. We can use `class()` and `attributes()` to extract further information. Run the code in the chunk below to discover what is actually returned by `cor()`.

```
MAT1 <- cor(cars)
class(MAT1)
attributes(MAT1)
str(MAT1)
```

Methods `class()`, `attributes()` and `str()` are very powerful tools that can be used when we are in doubt about the data contained in an object and/or how it is structured. Knowing the structure allows us to retrieve the data members directly from the object when predefined extractor methods are not available.

7.6.2 Kendall's τ and Spearman's ρ

We use the same functions as for Pearson's r but explicitly request the use of one of these methods by passing an argument.

```
cor(x = cars$speed, y = cars$dist, method = "kendall")
## [1] 0.6689901
cor(x = cars$speed, y = cars$dist, method = "spearman")
## [1] 0.8303568
```

Function `cor.test()`, described above, also allows the choice of method with the same syntax as shown for `cor()`.

7.7 Repeat the exercise in the playground immediately above, but now using non-parametric methods. How does the information stored in the returned `matrix` differ depending on the method, and how can we extract from the returned object information about the method used for the calculation of the correlation?

7.7 *t*-test

The *t*-test is based on Student's *t*-distribution. It can be applied to any parameter estimate for which its standard deviation is available, and the *t*-distribution is a plausible assumption. It is most frequently used to compare an estimate of the mean against a constant value, or the estimate of a difference between two means and a target difference, usually no difference. In R these can be computed manually using functions `mean()`, `sd()`, and `pt()` or with `t.test()`.

Although rarely presented in such a way, the t-test can be thought of as a special case of a linear model fit. Consistently with functions used to fit models to observations, we can use a *formula* to describe a t-test. A formula such as `y~x` is read as y is explained by x. We use *lhs* (left-hand-side) and *rhs* (right-hand-side) to signify all terms to the left and right of the tilde (`~`), respectively (`<lhs>~<rhs>`). (See section 7.13 on page 226 for a detailed discussion of model formulas, and section 7.8 on page 199 for examples of their use in model fitting.)

```
df1 <- data.frame(some.size = c(rnorm(10, mean = 2.5), rnorm(10, mean = 2.0)),
                  group = factor(rep(c("A", "B"), each = 10)))
```

The formula `some.size~1` is read as "the mean of variable `some.size` is explained by a constant value". The value estimated from observations, $\bar{x}$, is compared against the value of μ set as the null hypothesis, where μ is the *unknown* mean of the sampled population. By default, `t.test()` applies a two-sided test (`alternative = "two.sided"`) against `mu = 0`, but here we use `mu = 2` instead.

```
t.test(some.size ~ 1, mu = 2, data = df1)
##
##  One Sample t-test
##
## data:  some.size
## t = 1.078, df = 19, p-value = 0.2945
## alternative hypothesis: true mean is not equal to 2
## 95 percent confidence interval:
##  1.741200 2.808479
## sample estimates:
## mean of x
##   2.27484
```

The same test can be calculated step by step. In this case, this approach is not needed, but it is useful when we have a parameter estimate (not just mean) and its standard error available, as in model fits (see the advanced playground on page 206 for an example).

```
sem = sqrt(var(df1$some.size) / nrow(df1))
t.value = (mean(df1$some.size) - 2) / sem # Ho: mu = 2
p.value <- pt(t.value, df = nrow(df1) - 1, lower.tail = FALSE) * 2 # two tails
signif(c(t = t.value, df = nrow(df1) - 1, P = p.value), 4) # 4 digits
##       t      df       P
##  1.0780 19.0000  0.2945
```

The same function, with a different formula, tests for the difference between the means of two groups or treatments, $H_o : \mu_A - \mu_B = 0$. We read the formula `some.size~group` as "differences in `some.size` are explained by factor `group`". The difference between the means for the two groups is estimated and compared against the hypothesis. (In this case, the value of the argument passed to `mu`, zero by default, describes this difference.) By default, variances in the two groups are not to assumed equal,

```
t.test(some.size ~ group, data = df1)
##
##  Welch Two Sample t-test
##
```

Table 7.3
R functions implementing frequently used statistical tests. Student's t-test and correlation tests are described on pages 197 and 195, respectively.

Statistical test	Function name
Student's t-test (1 and 2 samples)	`t.test()`
Wilcoxon rank sum and signed rank tests	`wilcox.test`
Kolmogorov-Smirnov tests	`ks.test()`
Correlation tests (Pearson, Kendall, Spearman)	`cor.test()`
F-test to compare two variances	`var.test()`
Fisher's exact test for count data	`fisher.test()`
Pearson's Chi-squared (χ^2) test for count data	`chisq.test()`
Exact Binomial Test	`binom.test()`
Test of equal or given proportions	`prop.test()`

```
## data:  some.size by group
## t = 1.5864, df = 17.666, p-value = 0.1304
## alternative hypothesis: true difference in means between group A and group B ...
## 95 percent confidence interval:
##  -0.2538836  1.8108119
## sample estimates:
## mean in group A mean in group B
##        2.664072        1.885608
```

but with `var.equal = TRUE`, the variances in the populations from which observations in groups A and B were sampled are assumed equal, and pooled into a single estimate.

```
t.test(some.size ~ group, var.equal = TRUE, data = df1)
##
##  Two Sample t-test
##
## data:  some.size by group
## t = 1.5864, df = 18, p-value = 0.1301
## alternative hypothesis: true difference in means between group A and group B ...
## 95 percent confidence interval:
##  -0.2524857  1.8094140
## sample estimates:
## mean in group A mean in group B
##        2.664072        1.885608
```

The t-test serves as an example of how statistical tests are usually carried out in R. Table 7.3 lists R functions for frequently used statistical tests.

7.8 Model Fitting in R

The general approach to model fitting in R is to separate the actual fitting of a model from the inspection of the fitted model. A model fitting function minimally

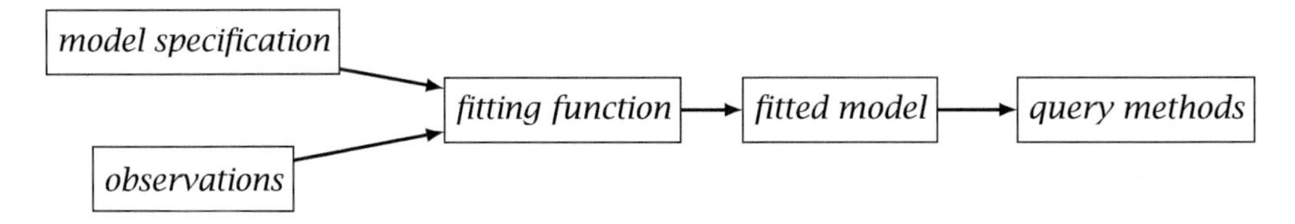

Figure 7.1
Model fitting in R is done in steps, and can be represented schematically as a flow of information.

requires a description of the model to fit, as a model `formula` and a data frame or vectors with the data or observations to which to fit the model. These functions in R return a model-fit object. This object contains the data, the model formula, the call, and the result of fitting the model. Several methods are available for querying it. The diagram in Figure 7.1 summarises the approach used in R for data analysis based on fitted models.

Models are described using model formulas such as `y~x` which we read as y is explained by x. We use *lhs* (left-hand-side) and *rhs* (right-hand-side) to signify all terms to the left and right of the tilde (`~`), respectively (`<lhs>~<rhs>`). Model formulas are used in different contexts: fitting of models, plotting, and tests like t-test. The syntax of model formulas is consistent throughout base R and numerous independently developed packages. However, their use is not universal, and several packages extend the basic syntax to allow the description of specific types of models. As most things in R, model formulas are objects and can be stored in variables. See section 7.13 on page 226 for a detailed discussion of model formulas.

Although there is some variation, especially for fitted model classes defined in extension packages, in most cases, the *query functions* bulked together in the rightmost box in the diagram in Figure 7.1 include `summary()`, `anova()` and `plot()`, with other methods such as `coef()`, `residuals()`, `fitted()`, `predict()`, `AIC()`, and `BIC()` usually also available. Additional methods may be available. However, as model fit objects are `list`-like, these and other values can be extracted and/or computed programmatically when needed. The examples in this chapter can be adapted to the fitting of types of models not described in this book.

Fitted model objects in R are self contained and include a copy of the data to which the model was fit, as well as residuals and possibly even intermediate results of computations. Although this can make the size of these objects large, it allows querying and even updating them in the absence of the data in the current R workspace.

7.9 Fitting Linear Models

Regression, analysis of variance (ANOVA) and analysis of covariance (ANCOVA) are all linear models, differing only on the type of explanatory variables included

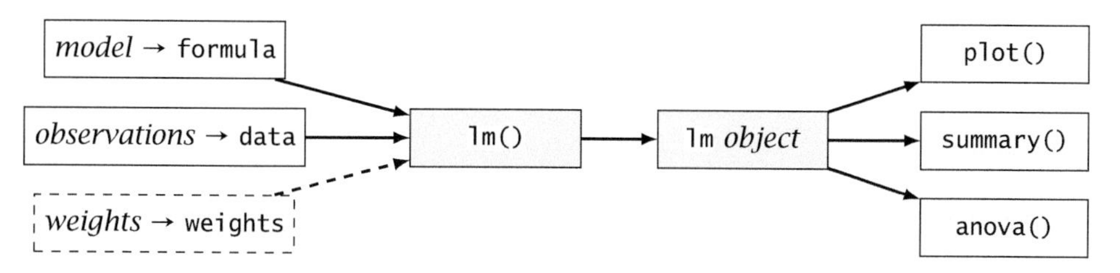

Figure 7.2
Linear model fitting in R is done in steps. The generic diagram from Figure 7.1 redrawn to show a linear model fit. Non-filled boxes are shared with the fitting of other types of models, and filled ones are specific to `lm()`. Only the three most frequently used query methods are shown, while both response and explanatory variables are under *observations*. Dashed boxes and arrows are optional as defaults are provided.

in the statistical model fitted. If in the fitted model all explanatory variables are continuous, i.e., `numeric`, vectors, the model is a regression model. If all explanatory variables are discrete, i.e., `factors`, the model is ANOVA. Finally, if the model contains but `numeric` variables and `factors` it is named ANCOVA. As in all cases the fitting approach is the same, based on ordinary least squares (OLS), in R, they are all implemented in function `lm()`.

There is another meaning of ANOVA, referring only to the tests of significance rather than to an approach to model fitting. Consequently, rather confusingly, results for tests of significance can both in the case of regression, ANOVA and ANCOVA, be presented in an ANOVA table. In this second, stricter meaning, ANOVA means a test of significance based on the ratios between pairs of variances.

If you do not clearly remember the difference between numeric vectors and factors, or how they can be created, please, revisit chapter 3 on page 23.

Figure 7.2 shows the steps needed to fit a linear model and extract the estimates and test results. The observations are stored in a data frame, one case or event per row, with values for both response and explanatory variables in variables or columns. The model formula is used to indicate which variables in the data frame are to be used and in which role: either response or explanatory, and when explanatory how they contribute to the estimated response. The object containing the results from the fit is queried to assess validity and make conclusions or predictions.

Weights are multiplicative factors used to alter the *weight* given to individual residuals when fitting a model to observations that are not equally informative. A frequent case is fitting a model to observations that are means of drastically different numbers of individual measurements. Some model fit functions compute the weights, but in most cases they are supplied as an argument to parameter `weights`. By default, `weights` have a value of `1` and thus do not affect the resulting model fit, when supplied or computed, the weights are saved to the model fit object.

7.9.1 Regression

The `cars` data set, containing two `numeric` variables, is used in the examples. A simple linear model $y = \alpha \cdot 1 + \beta \cdot x$ where y corresponds to stopping distance (`dist`) and x to initial speed (`speed`) is formulated in R as `dist ~ 1 + speed`. The fitted model object is assigned to variable `fm1` (a mnemonic for fitted-model one).

```
fm1 <- lm(dist ~ 1 + speed, data=cars)
class(fm1)
## [1] "lm"
```

The next step is diagnosis of the fit. Are assumptions of the linear model procedure used reasonably close to being fulfilled? In R it is most common to use plots to this end. We show here only one of the plots normally produced. This quantile vs. quantile plot is used to assess how much the distribution of the residuals deviates from the assumed Normal distribution.

```
plot(fm1, which = 2)
```

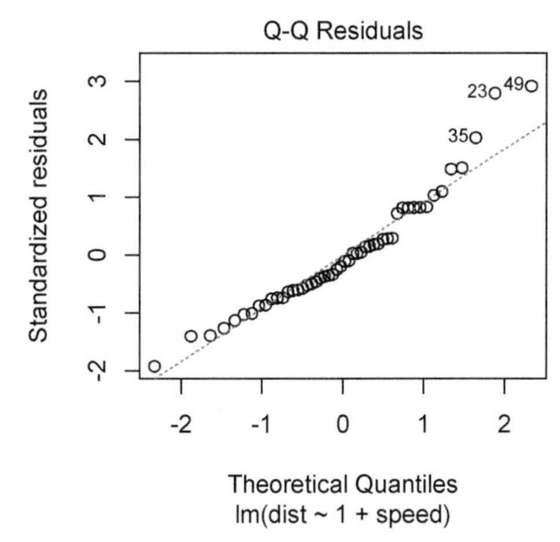

In the case of a regression, calling `summary()` with the fitted model object as argument is most useful as it provides a table of coefficient estimates and their errors. Remember that as is the case for most R functions, the value returned by `summary()` is printed when we call this method at the R prompt.

```
summary(fm1)
##
## Call:
## lm(formula = dist ~ 1 + speed, data = cars)
##
## Residuals:
##     Min      1Q  Median      3Q     Max
## -29.069  -9.525  -2.272   9.215  43.201
##
## Coefficients:
##             Estimate Std. Error t value Pr(>|t|)
## (Intercept) -17.5791     6.7584  -2.601   0.0123 *
## speed         3.9324     0.4155   9.464 1.49e-12 ***
## ---
## Signif. codes:  0 '***' 0.001 '**' 0.01 '*' 0.05 '.' 0.1 ' ' 1
```

```
##
## Residual standard error: 15.38 on 48 degrees of freedom
## Multiple R-squared:  0.6511, Adjusted R-squared:  0.6438
## F-statistic: 89.57 on 1 and 48 DF,  p-value: 1.49e-12
```

The summary is organised in sections. "Call:" shows `dist ~ 1 + speed` or the specification of the model fitted, plus the data used. "Residuals:" displays the extremes, quartiles and median of the residuals, or deviations between observations and the fitted line. "Coefficients:" contains estimates of the model parameters and their variation plus corresponding t-tests. In the last three lines, there is information on overall standard error and its degrees of freedom and overall coefficient of determination (R^2) and F-statistic.

Replacing α and β in $y = \alpha \cdot 1 + \beta \cdot x$ by the estimates for the intercept, $a = -17.6$, and slope, $b = 3.93$, we obtain an estimate for the regression line $y = -17.6 + 3.93x$. However, given the nature of the problem, we *know based on first principles* that the stopping distance must be zero when speed is zero. This suggests that we should not estimate the value of α but instead set $\alpha = 0$, or in other words, fit the model $y = \beta \cdot x$.

In R models, the intercept is included by default, so the model fitted above can be formulated as `dist ~ speed`—i.e., the missing `+ 1` does not change the model. To exclude the intercept, we need to specify the model as `dist ~ speed - 1` (or its equivalent `dist ~ speed + 0`), for a straight line passing through the origin ($x = 0$, $y = 0$). In the summary for this model there is an estimate for the slope but not for the intercept.

```
fm2 <- lm(dist ~ speed - 1, data = cars)
summary(fm2)
##
## Call:
## lm(formula = dist ~ speed - 1, data = cars)
##
## Residuals:
##     Min      1Q  Median      3Q     Max
## -26.183 -12.637  -5.455   4.590  50.181
##
## Coefficients:
##       Estimate Std. Error t value Pr(>|t|)
## speed   2.9091     0.1414   20.58   <2e-16 ***
## ---
## Signif. codes:  0 '***' 0.001 '**' 0.01 '*' 0.05 '.' 0.1 ' ' 1
##
## Residual standard error: 16.26 on 49 degrees of freedom
## Multiple R-squared:  0.8963, Adjusted R-squared:  0.8942
## F-statistic: 423.5 on 1 and 49 DF,  p-value: < 2.2e-16
```

The equation for `fm2` is $y = 2.91x$. From the residuals, it can be seen that it is inadequate, as the straight line does not follow the curvature of the cloud of observations.

7.8 You will now fit a second-degree polynomial, a different linear model: $y = \alpha \cdot 1 + \beta_1 \cdot x + \beta_2 \cdot x^2$. The function used is the same as for linear regression, `lm()`. We only need to alter the formulation of the model. The identity function `I()` is used to protect its argument from being interpreted as part of the model formula.

Instead, its argument is evaluated beforehand and the result is used as the, in this case second, explanatory variable.

```
fm3 <- lm(dist ~ speed + I(speed^2), data = cars)
plot(fm3, which = 3)
summary(fm3)
```

The "same" fit using an orthogonal polynomial can be specified using function `poly()`. Polynomials of different degrees can be obtained by supplying as the second argument to `poly()` the corresponding positive integer value. In this case, the different terms of the polynomial are bulked together in the summary.

```
fm3a <- lm(dist ~ poly(speed, 2), data = cars)
summary(fm3a)
```

It is possible to compare two model fits using `anova()`, testing whether one of the models describes the data better than the other. It is important in this case to take into consideration the nature of the difference between the model formulas, most importantly if they can be interpreted as nested—i.e., interpreted as a base model vs. the same model with additional terms.

```
anova(fm2, fm1)
```

Three or more models can also be compared in a single call to `anova()`. However, care is needed, as the order of the arguments matters.

```
anova(fm2, fm3, fm3a)
anova(fm2, fm3a, fm3)
```

Different criteria can be used to choose the "best" model: significance based on p-values or information criteria (AIC, BIC). AIC (Akaike's "An Information Criterion") and BIC ("Bayesian Information Criterion" = SBC, "Schwarz's Bayesian criterion") that penalise the resulting "goodness" based on the number of parameters in the fitted model. In the case of AIC and BIC, a smaller value is better, and values returned can be either positive or negative, in which case more negative is better. Estimates for both BIC and AIC are returned by `anova()`, and on their own by `BIC()` and `AIC()`

```
BIC(fm2, fm1, fm3, fm3a)
AIC(fm2, fm1, fm3, fm3a)
```

Once you have run the code in the chunks above, you will be able see that these three criteria do not necessarily agree on which is the "best" model. Find in the output P-value, BIC and AIC estimates, for the different models and conclude which model is favoured by each of the three criteria. In addition, you will notice that the two different formulations of the quadratic polynomial are equivalent.

Additional query methods give easy access to different aspects of fitted models: `vcov()` returns the variance-covariance matrix, `coef()` and its alias `coefficients()` return the estimates for the fitted model coefficients, `fitted()` and its alias `fitted.values()` extract the fitted values, and `resid()` and its alias `residuals()` the corresponding residuals (or deviations) (Figure 7.3). Less frequently used accessors are `getCall()`, `effects()`, `terms()`, `model.frame()`, and `model.matrix()`.

7.9 Familiarise yourself with these extraction and summary methods by reading their documentation and use them to explore `fm1` fitted above or model fits to other data of your interest.

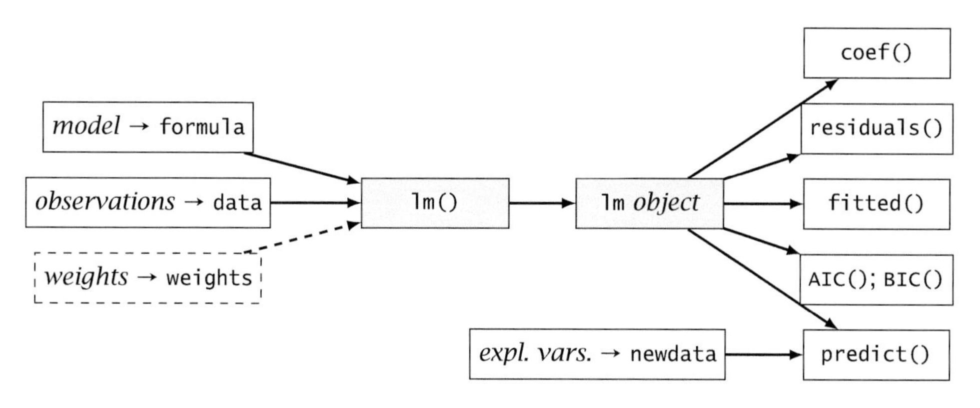

Figure 7.3
Diagram including additional methods used to query fitted model objects using linear models as an example. For other details see the legend of Figure 7.2.

It is usual to only look at the values returned by `anova()` and `summary()` as implicitly displayed by `print()`. However, both `anova()` and `summary()` return complex objects, derived from `list`, containing some members not displayed by `print()`. Access to members of these objects can be necessary to use them in further calculations or to print them in a format different to the default.

The class and structure of the objects returned by `summary()` depends on the class of the model fit object, i.e., `summary()` is a generic method with multiple specialisations.

```
class(summary(fm1))
## [1] "summary.lm"
```

One case where we need to extract individual members is when adding annotations to plots. Another case is when writing reports to include programmatically the computed values within the text. `str()` can be used to display the structure. Calling `str()` with `no.list = TRUE`, `give.attr = FALSE` and `vec.len = 2` as arguments restricts the output to an overview of the structure of `fm1`.

```
str(summary(fm1), no.list = TRUE, give.attr = FALSE, vec.len = 2)
##  $ call         : language lm(formula = dist ~ 1 + speed, data = cars)
##  $ terms        :Classes 'terms', 'formula'  language dist ~ 1 + speed
##  $ residuals    : Named num [1:50] 3.85 11.85 ...
##  $ coefficients : num [1:2, 1:4] -17.58 3.93 ...
##  $ aliased      : Named logi [1:2] FALSE FALSE
##  $ sigma        : num 15.4
##  $ df           : int [1:3] 2 48 2
##  $ r.squared    : num 0.651
##  $ adj.r.squared: num 0.644
##  $ fstatistic   : Named num [1:3] 89.6 1 ...
##  $ cov.unscaled : num [1:2, 1:2] 0.1931 -0.0112 ...
```

Extraction of members follows the usual R rules using `$`, `[ ]`, or `[[ ]]`.

```
summary(fm1)$adj.r.squared
## [1] 0.6438102
```

Estimates of `coefficients` are accompanied by estimates of their standard errors, *t*-values and *P*-values, while in the model object `fm1` these are not included.

```
coef(fm1)
## (Intercept)       speed
##  -17.579095    3.932409
str(fm1$coefficients)
##  Named num [1:2] -17.58 3.93
##  - attr(*, "names")= chr [1:2] "(Intercept)" "speed"
print(summary(fm1)$coefficients)
##               Estimate Std. Error   t value     Pr(>|t|)
## (Intercept) -17.579095  6.7584402 -2.601058 1.231882e-02
## speed         3.932409  0.4155128  9.463990 1.489836e-12
str(summary(fm1)$coefficients)
##  num [1:2, 1:4] -17.579 3.932 6.758 0.416 -2.601 ...
##  - attr(*, "dimnames")=List of 2
##   ..$ : chr [1:2] "(Intercept)" "speed"
##   ..$ : chr [1:4] "Estimate" "Std. Error" "t value" "Pr(>|t|)"
```

The class of the object returned by method `anova()` does not depend on the class of the model fit object, while its structure does depend.

```
anova(fm1)
## Analysis of Variance Table
##
## Response: dist
##           Df Sum Sq Mean Sq F value   Pr(>F)
## speed      1  21186 21185.5  89.567 1.49e-12 ***
## Residuals 48  11354   236.5
## ---
## Signif. codes:  0 '***' 0.001 '**' 0.01 '*' 0.05 '.' 0.1 ' ' 1

class(anova(fm1))
## [1] "anova"      "data.frame"

str(anova(fm1))
## Classes 'anova' and 'data.frame':	2 obs. of  5 variables:
##  $ Df     : int  1 48
##  $ Sum Sq : num  21185 11354
##  $ Mean Sq: num  21185 237
##  $ F value: num  89.6 NA
##  $ Pr(>F) : num  1.49e-12 NA
## - attr(*, "heading")= chr [1:2] "Analysis of Variance Table\n" "Response: dist"
```

As an example of the use of values extracted from the `summary.lm` object, I show how to test if the slope from a linear regression fit deviates significantly from a constant value different from the usual zero, which tests for the presence of an "effect" of the explanatory variable. When testing for deviations from a calibration by comparing two instruments or an instrument and a reference, a null hypothesis of one for the slope tests for deviations from the true readings. In some cases, when comparing the effectiveness of interventions we may be interested to test if a new approach surpasses that in current use by at least a specific margin. There exist practical situations where testing if a response exceeds a threshold is of interest.

A t-value can be computed for the slope as for the mean. When using `anova()` and `summary()` the null hypothesis is no effect or no response, i.e., slope = 0. The equivalent test with a null hypothesis of slope = 1 is easy to implement if we consider how we calculate a t-value (see section 7.7 on page 197). To compute the t-value we need an estimate for the slope and an estimate of its standard error. To look up the P-value, we need the degrees of freedom. All these values are available as members of the summary object of a fitted model.

```
est.slope.value <- summary(fm1)$coefficients["speed", "Estimate"]
est.slope.se <- summary(fm1)$coefficients["speed", "Std. Error"]
degrees.of.freedom <- summary(fm1)$df[2]
```

The estimate of the t-value, or quantile, is computed based on the difference between the estimate for the slope and a null hypothesis used as reference, and the standard error of the estimated slope. A probability is obtained based on the computed t-value, or quantile, and the t distribution with matching degrees of freedom with a call to `pt()` (see section 7.3 on page 189.) For a two-tail test we multiply by two the one-tail P estimate.

```
hyp.null <- 1
t.value <- (est.slope.value - hyp.null) / est.slope.se
p.value <- 2 * pt(q = t.value, df = degrees.of.freedom, lower.tail = FALSE)
cat("slope =", signif(est.slope.value, 3),
    "with s.e. =", signif(est.slope.se, 3),
    "\nt.value =", signif(t.value, 3),
    "and P-value =", signif(p.value, 3))
## slope = 3.93 with s.e. = 0.416
## t.value = 7.06 and P-value = 6.01e-09
```

This example is for a linear model fitted with function `lm()` but the same approach can be applied to other model fit procedures for which parameter estimates and their corresponding standard error estimates can be extracted or computed.

7.10 Check that the computations above after replacing `hyp.null <- 1` by `hyp.null <- 0` agree with the output of printing `summary()`.

Modify the example above so as to test whether the intercept is significantly larger than 5 feet, doing a one-sided test.

Method `predict()` uses the fitted model together with new data for the independent variables to compute predictions. As `predict()` accepts new data as input, it allows interpolation and extrapolation to values of the independent variables not present in the original data. In the case of fits of linear and some other models, method `predict()` returns, in addition to the prediction, estimates of the confidence and/or prediction intervals. The new data must be stored in a data frame with columns using the same names for the explanatory variables as in the data used for the fit, a response variable is not needed and additional columns are ignored. (The explanatory variables in the new data can be either continuous or factors, but they must match in this respect those in the original data.)

7.11 Predict using both `fm1` and `fm2` the distance required to stop cars moving at 0, 5, 10, 20, 30, and 40 mph. Study the help page for the `predict()` method for linear models (using `help(predict.lm)`). Explore the difference between `"prediction"` and `"confidence"` bands: why are they so different?

The objects returned by model fitting functions contain the full information, including the data to which the model was fit to. Their structure resembles a nested list. In most cases, the class of the objects returned by model fit functions agrees in name with the name of the function ("lm" in this example) but is not derived from "list". The query functions, either extract parts of the object or do additional calculations and formatting based on them. Different specialisations of these methods are called depending on the class of the model fit object. (See section 6.3 on page 176.)

```
class(fm1)
## [1] "lm"
names(fm1)
##  [1] "coefficients"  "residuals"     "effects"       "rank"
##  [5] "fitted.values" "assign"        "qr"            "df.residual"
##  [9] "xlevels"       "call"          "terms"         "model"
```

The structure of model fit objects is of interest only when the query or accessor functions do not provide the needed information and components have to be extracted using operator `[[ ]]`. Exploring these objects is also a way of learning how model fitting works in R. As with any other objects, `str()` shows the structure.

```
str(fm1, no.list = TRUE, give.attr = FALSE, vec.len = 2)
##  $ coefficients : Named num [1:2] -17.58 3.93
##  $ residuals    : Named num [1:50] 3.85 11.85 ...
##  $ effects      : Named num [1:50] -304 146 ...
##  $ rank         : int 2
##  $ fitted.values: Named num [1:50] -1.85 -1.85 ...
##  $ assign       : int [1:2] 0 1
##  $ qr           :List of 5
##   ..$ qr   : num [1:50, 1:2] -7.071 0.141 ...
##   ..$ qraux: num [1:2] 1.14 1.27
##   ..$ pivot: int [1:2] 1 2
##   ..$ tol  : num 1e-07
##   ..$ rank : int 2
##  $ df.residual  : int 48
##  $ xlevels      : Named list()
##  $ call         : language lm(formula = dist ~ 1 + speed, data = cars)
##  $ terms        :Classes 'terms', 'formula'  language dist ~ 1 + speed
##  $ model        :'data.frame': 50 obs. of  2 variables:
##   ..$ dist : num [1:50] 2 10 4 22 16 ...
##   ..$ speed: num [1:50] 4 4 7 7 8 ...
```

Member `call` contains the function call and arguments used to create object `fm1`.

```
str(fm1$call)
##  language lm(formula = dist ~ 1 + speed, data = cars)
```

7.9.2 Analysis of variance, ANOVA

In ANOVA, the explanatory variable is categorical, and in R, must be a `factor` or `ordered` factor (see section 3.12 on page 79). As a linear model, the fitting approach is the same as for linear and polynomial regression (Figure 7.2). The `InsectSprays` data set used in the next example gives insect counts in plots sprayed with different insecticides. In these data, `spray` is a factor with six levels.

What determines that this is an ANOVA is that `spray`, the explanatory variable, is a `factor`.

```
data(InsectSprays)
is.numeric(InsectSprays$spray)
## [1] FALSE
is.factor(InsectSprays$spray)
## [1] TRUE
levels(InsectSprays$spray)
## [1] "A" "B" "C" "D" "E" "F"
```

By using a factor instead of a numeric vector, a different model matrix is built from an equivalent formula.

```
fm4 <- lm(count ~ spray, data = InsectSprays)
```

Diagnostic plots are obtained in the same way as for linear regression. We show only the quantile-quantile plot for simplicity, but during data analysis it is very important to check all the diagnostics plots. As many of the residuals deviate from the one-to-one line we have to conclude the residuals do not follow the Normal distribution, and a different approach to model fitting should be used (see section 7.10 on page 217).

```
plot(fm4, which = 2)
```

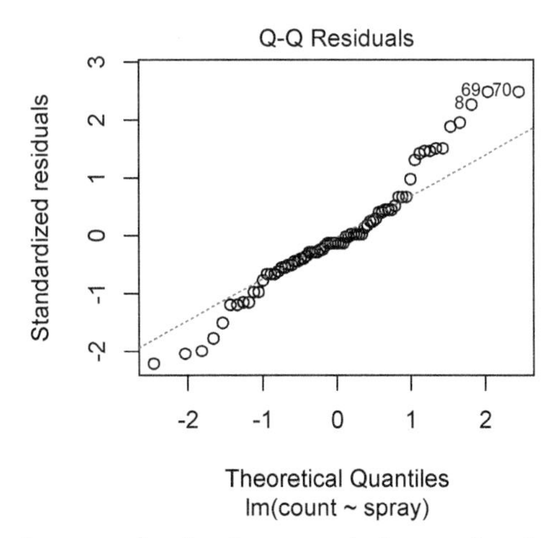

In ANOVA, most frequently the interest is in testing hypotheses with function `anova()`, which implements the F-test for the main effects of factors and their interactions. In this example, with a single explanatory variable, there is only one effect of interest, that of `sprays`.

```
anova(fm4)
## Analysis of Variance Table
##
## Response: count
##           Df Sum Sq Mean Sq F value    Pr(>F)
## spray      5 2668.8  533.77  34.702 < 2.2e-16 ***
```

```
## Residuals 66 1015.2   15.38
## ---
## Signif. codes:  0 '***' 0.001 '**' 0.01 '*' 0.05 '.' 0.1 ' ' 1
```

Function `summary()` can be used to extract parameter estimates informing of the size of the effects, but meaningfully only by using contrasts different to the default ones. Function `aov()` is a wrapper on `lm()` that returns an object that by default displays the output of `anova()` also with `summary()`, but even in this case it can be preferable to change the default contrasts (see `help(aov)`).

The contrasts used affect the estimates returned by `coef()` and `summary()` applied to an ANOVA model fit. The default used in R, `contr.treatment()`, is different to that used in S, `contr.helmert`. With `contr.treatment` the first level of the factor (assumed to be a control) is used as a reference for the estimation of coefficients for the remaining factor levels and testing of their significance. With `contr.helmert` the contrasts are of the second level with the first, the third with the average of the first two, and so on. These contrasts depend on the order of factor levels. Instead, `contr.sum` uses as reference the mean of all levels, i.e., using as a condition that the coefficient estimates add up to zero. Obviously what type of contrast is used changes what the coefficient estimates describe, and, consequently, how the p-values should be interpreted.

The approach used by default for model fits and ANOVA calculations varies among programs. There exist different so-called "types" of sums of squares, usually called I, II, and III. In orthogonal designs, the choice is of no consequence, but differences can be important for unbalanced designs, even leading to different conclusions. R's default, type I, is usually considered to suffer milder problems than type III, the default used by SPSS and SAS. In any case, for unbalanced data it is preferable to use the approach implemented in package 'nlme'.

The most straightforward way of setting a different default for contrasts for a whole series of model fits is by setting R option `contrasts`, which we here only print.

```
options("contrasts")
## $contrasts
##          unordered             ordered
## "contr.treatment"        "contr.poly"
```

The option is set to a named character vector of length two, with the first value, named `unordered` giving the name of the function used when the explanatory variable is an unordered `factor` (created with `factor()`) and the second value, named `ordered`, giving the name of the function used when the explanatory variable is an `ordered` factor (created with `ordered()`).

It is also possible to select the contrast to be used in the call to `aov()` or `lm()`.

```
fm4trea <- lm(count ~ spray, data = InsectSprays,
              contrasts = list(spray = contr.treatment))
fm4sum  <- lm(count ~ spray, data = InsectSprays,
              contrasts = list(spray = contr.sum))
```

In `fm4trea` we used `contr.treatment()`, thus contrasts for individual treatments are done against `Spray1` taking it as the control or reference, as can be inferred from the generated contrasts matrix. For this reason, there is no row for `spray1` in the summary table. Each of the rows `spray2` to `spray6` is a test comparing these treatments individually against `spray1`.

```
contr.treatment(length(levels(InsectSprays$spray)))
##   2 3 4 5 6
## 1 0 0 0 0 0
## 2 1 0 0 0 0
## 3 0 1 0 0 0
## 4 0 0 1 0 0
## 5 0 0 0 1 0
## 6 0 0 0 0 1
```

```
summary(fm4trea)
##
## Call:
## lm(formula = count ~ spray, data = InsectSprays, contrasts = list(spray = ...
##
## Residuals:
##    Min     1Q Median     3Q    Max
## -8.333 -1.958 -0.500  1.667  9.333
##
## Coefficients:
##             Estimate Std. Error t value Pr(>|t|)
## (Intercept)  14.5000     1.1322  12.807  < 2e-16 ***
## sprayB        0.8333     1.6011   0.520    0.604
## sprayC      -12.4167     1.6011  -7.755 7.27e-11 ***
## sprayD       -9.5833     1.6011  -5.985 9.82e-08 ***
## sprayE      -11.0000     1.6011  -6.870 2.75e-09 ***
## sprayF        2.1667     1.6011   1.353    0.181
## ---
## Signif. codes:  0 '***' 0.001 '**' 0.01 '*' 0.05 '.' 0.1 ' ' 1
##
## Residual standard error: 3.922 on 66 degrees of freedom
## Multiple R-squared:  0.7244, Adjusted R-squared:  0.7036
## F-statistic:  34.7 on 5 and 66 DF,  p-value: < 2.2e-16
```

In `fm4sum` we used `contr.sum()` with the sum constrained to be zero, thus estimates for the last treatment level are determined by the sum of the previous ones, and not tested for significance.

```
contr.sum(length(levels(InsectSprays$spray)))
##   [,1] [,2] [,3] [,4] [,5]
## 1    1    0    0    0    0
## 2    0    1    0    0    0
## 3    0    0    1    0    0
## 4    0    0    0    1    0
## 5    0    0    0    0    1
## 6   -1   -1   -1   -1   -1
```

```
summary(fm4sum)
##
## Call:
## lm(formula = count ~ spray, data = InsectSprays, contrasts = list(spray = ...
##
## Residuals:
##     Min      1Q Median      3Q     Max
## -8.333 -1.958 -0.500  1.667  9.333
##
## Coefficients:
##             Estimate Std. Error t value Pr(>|t|)
## (Intercept)   9.5000     0.4622  20.554  < 2e-16 ***
## spray1        5.0000     1.0335   4.838 8.22e-06 ***
## spray2        5.8333     1.0335   5.644 3.78e-07 ***
## spray3       -7.4167     1.0335  -7.176 7.87e-10 ***
## spray4       -4.5833     1.0335  -4.435 3.57e-05 ***
## spray5       -6.0000     1.0335  -5.805 2.00e-07 ***
## ---
## Signif. codes:  0 '***' 0.001 '**' 0.01 '*' 0.05 '.' 0.1 ' ' 1
##
## Residual standard error: 3.922 on 66 degrees of freedom
## Multiple R-squared:  0.7244, Adjusted R-squared:  0.7036
## F-statistic:  34.7 on 5 and 66 DF,  p-value: < 2.2e-16
```

7.12 Explore how taking the last level as reference in `contr.SAS()` instead of the first one as in `contr.treatment()` affects the estimates. Reorder the levels of factor `spray` so that the test using `contr.SAS()` becomes equivalent to that obtained above with `contr.treatment()`. Consider why `contr.poly()` is the default for ordered factors and when `contr.helmert()` could be most useful.

Contrasts, on the other hand, do not affect the table returned by `anova()` as this table does not deal with the effects of individual factor levels. The overall estimates shown at the bottom of the summary table remain unchanged. In other words, when using different contrasts what changes is how the total variation explained by the fitted model is partitioned into components to be tested for specific contributions to the overall model fit.

Post-hoc tests based on specific contrasts and multiple comparisons tests are most frequently applied after an ANOVA to test for differences among pairs of treatments or specific combinations of treatments. R function `Tukey.test()` implements Tukey's HSD (honestly significant difference) test for pairwise tests. Function `pairwise.t.test()` supports different correction methods for the P-values from simultaneous t-tests. Function `p.adjust()` applies adjustments to P-values and can be used when the test procedure does not apply them. The most comprehensive implementation of multiple comparisons is available in package 'multcomp'. Function `glht()` (general linear hypothesis testing) from this package supports different contrasts and adjustment methods.

Contrasts and their interpretation are discussed in detail by Venables and Ripley (2002) and Crawley (2012).

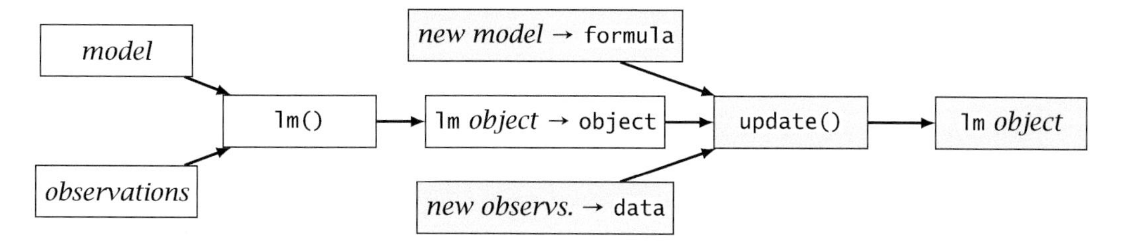

Figure 7.4
Diagram showing the steps for updating a fitted model (in filled boxes) together with the previous steps in unfilled boxes. Please, see Figure 7.2 for other details.

7.9.3 Analysis of covariance, ANCOVA

When a linear model includes both explanatory factors and continuous explanatory variables, we may call it *analysis of covariance* (ANCOVA). The formula syntax is the same for all linear models and, as mentioned in previous sections, what determines the type of analysis is the nature of the explanatory variable(s). As the formulation remains the same, no specific example is given. The main difficulty of ANCOVA is in the selection of the covariate and the interpretation of the results of the analysis, especially, when the covariate is not independent of the treatment described by the factor (e.g. Smith 1957).

7.9.4 Model update and selection

Model fit objects can be updated, i.e., modified, because they contain not only the results of the fit but also the data to which the model was fit (see page 208). Given that the call is also stored, all the information needed to recalculate the same fit is available. Method `update()` makes it possible to recalculate the fit with changes to the call, without passing again all the arguments to a new call to `lm()` (Figure 7.4). We can modify different arguments, including selecting part of the data by passing a new argument to formal parameter `subset`.

Method `update()` retrieves the call from the model fit object using `getCall()`, modifies it and, by default, evaluates it. The default `update()` method works as long as the model-fit object contains a member named `call` or if a specialisation of `getCall()` is available. Thus, method `update()` can be used with models fitted with other functions in addition to `lm()`.

For the next example, we recreate the model fit object `fm4` from page 209.

```
fm4 <- lm(count ~ spray, data = InsectSprays)
anova(fm4)
## Analysis of Variance Table
##
## Response: count
##           Df Sum Sq Mean Sq F value    Pr(>F)
## spray      5 2668.8  533.77  34.702 < 2.2e-16 ***
## Residuals 66 1015.2   15.38
## ---
## Signif. codes:  0 '***' 0.001 '**' 0.01 '*' 0.05 '.' 0.1 ' ' 1
fm4a <- update(fm4, formula = log10(count + 1) ~ spray)
```

```
anova(fm4a)
## Analysis of Variance Table
##
## Response: log10(count + 1)
##           Df Sum Sq Mean Sq F value    Pr(>F)
## spray      5 7.2649 1.45297  46.007 < 2.2e-16 ***
## Residuals 66 2.0844 0.03158
## ---
## Signif. codes:  0 '***' 0.001 '**' 0.01 '*' 0.05 '.' 0.1 ' ' 1
```

7.13 Print `fm4$call` and `fm4a$call`. These two calls differ in the argument to `formula`. What other members have been updated in `fm4a` compared to `fm4`?

In the chunk above we replaced the argument passed to `formula`. This is a frequent use, but, for example, to fit the same model to a subset of the data, we can pass a suitable argument to parameter `subset`.

```
fm4b <- update(fm4, subset = !spray %in% c("A", "B"))
anova(fm4b)
## Analysis of Variance Table
##
## Response: count
##           Df Sum Sq Mean Sq F value    Pr(>F)
## spray      3 1608.4  536.14  41.422 7.119e-13 ***
## Residuals 44  569.5   12.94
## ---
## Signif. codes:  0 '***' 0.001 '**' 0.01 '*' 0.05 '.' 0.1 ' ' 1
```

7.14 When having many treatments with long names, which is not the case here, instead of listing the factor levels for which to subset the data, it can be convenient to use regular expressions for pattern matching (see section 3.4 on page 46). Run the code below, and investigate why `anova(fm4b)` and `anova(fm4c)` produce the same ANOVA table printout, but the fit model objects are not identical. You can use `str()` to explore if any members differ between the two objects.

```
fm4c <- update(fm4, subset = !grepl("[AB]", spray))
anova(fm4c)
identical(fm4b, fm4c)
```

Method `update()` plays an additional role when the fitting is done by numerical approximation, as the previously computed estimates are used as the starting values for the numerical calculations required for fitting the updated model (see section 7.11 on page 220 as an example). This can drastically decrease computation time, or even easy the task of finding suitable starting values for parameter estimates by fitting increasingly more complex nested models.

Method `update()` used together with `AIC()` (or `anova()`) gives us the tools to compare nested models, and select one out of a group as shown above. When comparing several models doing the comparisons manually is tedious, and in scripts, in many cases difficult to write code that is flexible (or abstract) enough. Method `step()` automates stepwise selection of nested models such as the selection among polynomials of different degrees or which variables to retain in multiple regression. After fitting a model, method `step()` is used to update this model using an automatic stopping criterion (Figure 7.5).

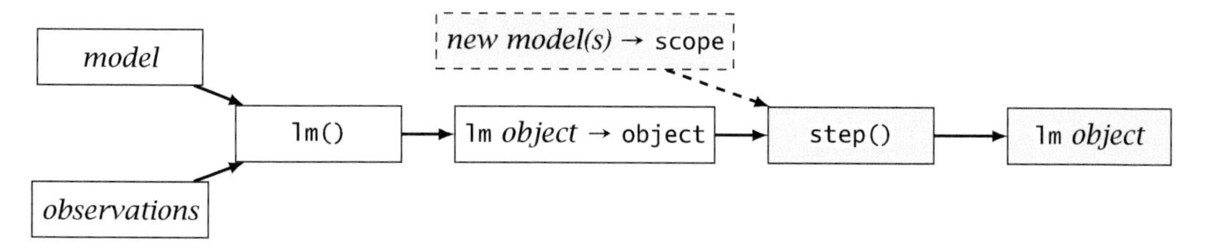

Figure 7.5
Diagram showing the steps used for stepwise model selection among nested models (in filled boxes) together with the previous steps in unfilled boxes. The range of models to select from can be set by the user. See Figure 7.2 for other details.

Stepwise model selection—either in the *forward* direction from simpler to more complex models, in the backward direction from more complex to simpler models or in both directions—is implemented in base R's *method* `step()` using Akaike's information criterion (AIC) as the selection criterion. Use of method `step()` from R is possible, for example, with `lm()` and `glm` fits. AIC is described on page 204.

For the next example, we use `fm3` from page 204, a linear model for a polynomial regression. If, as shown here, no models are passed through formal parameter `scope`, the previously fit model will be simplified, if possible. Method `step()` by default prints to the console a trace of the models tried and the corresponding AIC estimates.

```
fm3 <- lm(dist ~ speed + I(speed^2), data = cars)
fm3a <- step(fm3)
## Start:  AIC=274.88
## dist ~ speed + I(speed^2)
##
##              Df Sum of Sq   RSS    AIC
## - speed       1     46.42 10871 273.09
## <none>                    10825 274.88
## - I(speed^2)  1    528.81 11354 275.26
##
## Step:  AIC=273.09
## dist ~ I(speed^2)
##
##              Df Sum of Sq   RSS    AIC
## <none>                    10871 273.09
## - I(speed^2)  1     21668 32539 325.91
```

Method `summary()` reveals the differences between the original and updated models.

```
summary(fm3)
##
## Call:
## lm(formula = dist ~ speed + I(speed^2), data = cars)
##
## Residuals:
##     Min      1Q  Median      3Q     Max
## -28.720  -9.184  -3.188   4.628  45.152
##
## Coefficients:
```

```
##              Estimate Std. Error t value Pr(>|t|)
## (Intercept)  2.47014   14.81716   0.167    0.868
## speed        0.91329    2.03422   0.449    0.656
## I(speed^2)   0.09996    0.06597   1.515    0.136
##
## Residual standard error: 15.18 on 47 degrees of freedom
## Multiple R-squared:  0.6673, Adjusted R-squared:  0.6532
## F-statistic: 47.14 on 2 and 47 DF,  p-value: 5.852e-12
summary(fm3a)
##
## Call:
## lm(formula = dist ~ I(speed^2), data = cars)
##
## Residuals:
##     Min      1Q  Median      3Q     Max
## -28.448  -9.211  -3.594   5.076  45.862
##
## Coefficients:
##             Estimate Std. Error t value Pr(>|t|)
## (Intercept)  8.86005    4.08633   2.168   0.0351 *
## I(speed^2)   0.12897    0.01319   9.781  5.2e-13 ***
## ---
## Signif. codes:  0 '***' 0.001 '**' 0.01 '*' 0.05 '.' 0.1 ' ' 1
##
## Residual standard error: 15.05 on 48 degrees of freedom
## Multiple R-squared:  0.6659, Adjusted R-squared:  0.6589
## F-statistic: 95.67 on 1 and 48 DF,  p-value: 5.2e-13
```

If we pass a model with additional terms through parameter `scope` this model will be taken as the most complex model to be assessed. If, instead of one model, we pass two nested models in a list and name them `lower` and `upper`, they will delimit the scope of the stepwise search. In the next example, we see that first a backward search is done and term `speed` is removed because it removal decreases (= improves) AIC. Subsequently, a forward search is done unsuccessfully, as AIC increases.

```
fm3b <-
  step(fm3,
       scope = dist ~ speed + I(speed^2) + I(speed^3) + I(speed^4))
## Start:  AIC=274.88
## dist ~ speed + I(speed^2)
##
##              Df Sum of Sq   RSS    AIC
## - speed       1     46.42 10871 273.09
## <none>                    10825 274.88
## - I(speed^2)  1    528.81 11354 275.26
## + I(speed^4)  1    233.62 10591 275.79
## + I(speed^3)  1    190.35 10634 275.99
##
## Step:  AIC=273.09
## dist ~ I(speed^2)
##
##              Df Sum of Sq   RSS    AIC
## <none>                    10871 273.09
## + speed       1      46.4 10825 274.88
## + I(speed^3)  1       5.6 10866 275.07
## + I(speed^4)  1       0.0 10871 275.09
```

```
## - I(speed^2)  1   21667.8 32539 325.91
summary(fm3b)
##
## Call:
## lm(formula = dist ~ I(speed^2), data = cars)
##
## Residuals:
##     Min      1Q  Median      3Q     Max
## -28.448  -9.211  -3.594   5.076  45.862
##
## Coefficients:
##             Estimate Std. Error t value Pr(>|t|)
## (Intercept)  8.86005    4.08633   2.168   0.0351 *
## I(speed^2)   0.12897    0.01319   9.781  5.2e-13 ***
## ---
## Signif. codes:  0 '***' 0.001 '**' 0.01 '*' 0.05 '.' 0.1 ' ' 1
##
## Residual standard error: 15.05 on 48 degrees of freedom
## Multiple R-squared:  0.6659, Adjusted R-squared:  0.6589
## F-statistic: 95.67 on 1 and 48 DF,  p-value: 5.2e-13
```

7.15 Explain why the stepwise model selection in the code below differs from those in the two previous examples. Consult `help(step)` is necessary.

```
fm3c <-
  step(fm3,
       scope = list(lower = dist ~ speed,
                    upper = dist ~ speed + I(speed^2) + I(speed^3) + I(speed^4)))
summary(fm3c)
```

Functions `update()` and `step()` are *convenience functions* as they provide direct and/or simpler access to operations available through other functions or combined use of multiple functions.

7.10 Generalised Linear Models

Linear models make the assumption of normally distributed residuals. Generalised linear models, fitted with function `glm()`, are more flexible, and allow the assumed distribution to be selected as well as the link function (defaults are as in `lm()`). Figure 7.6 shows that the steps used to fit a model with `glm()` are the same as with `lm()` except that we can select the probability distribution assumed to describe the variation among observations. Frequently used probability distributions are binomial and Poisson (see `help(family)` for the variations and additional ones).

For count data, GLMs are preferred over LMs. In the example below, we fit the same model as above, but assuming a quasi-Poisson distribution instead of the Normal. An argument passed to `family` selects the assumed error distribution. The `InsectSprays` data set used in the next example, gives insect counts in plots sprayed with different insecticides. In these data, spray is a factor with six levels.

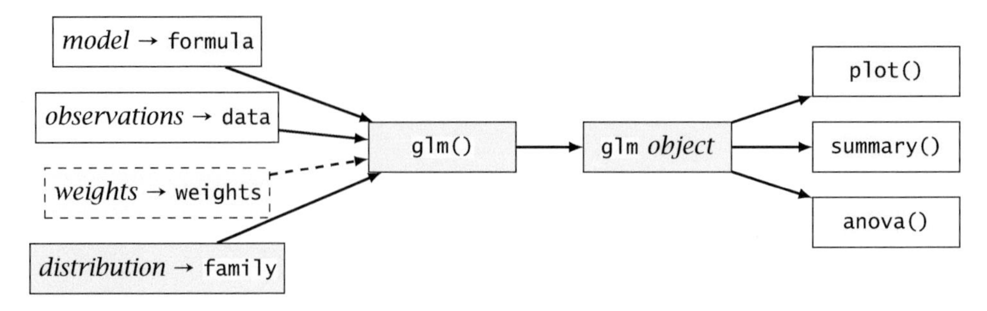

Figure 7.6
Generalised linear model fitting in R is done in steps similar to those used for linear models. Generic diagram from Figure 7.1 redrawn to show a generalised linear model fit. Non-filled boxes are shared with fitting of other types of models, and filled ones are specific to `glm()`. Only the three most frequently used query methods are shown, while both response and explanatory variables are under *observations*. Dashed boxes and arrows are optional as defaults are provided.

```
fm10 <- glm(count ~ spray, data = InsectSprays, family = quasipoisson)
```

Method `plot()` as for linear-model fits, produces diagnosis plots. We show, as before, the quantile-quantile plot of residuals. The Normal distribution assumed above in the linear model fit was not a good approximation (section 7.9.2 on page 208), as count data are known to follow a different distribution. This is clear by comparing the quantile-quantile plot for `fm4` (page 208) and the plot below for the model fit under the assumption of a Quasi-Poisson distribution.

```
plot(fm10, which = 2)
```

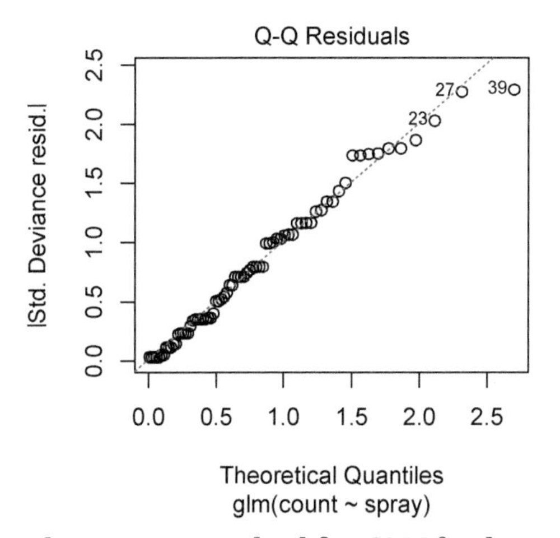

The printout from the `anova()` method for GLM fits has some differences to that for LM fits. In R versions previous to 4.4.0, no test statistics or P-values were computed unless requested by passing an argument to parameter `test`. In later versions of R, either a chi-squared test or an F-test are computed by default depending on whether the dispersion is fixed or free. We here use `"F"` as an argument to request an F-test.

```
anova(fm10, test = "F")
## Analysis of Deviance Table
##
## Model: quasipoisson, link: log
##
## Response: count
##
## Terms added sequentially (first to last)
##
##
##       Df Deviance Resid. Df Resid. Dev      F    Pr(>F)
## NULL                     71     409.04
## spray  5   310.71        66      98.33 41.216 < 2.2e-16 ***
## ---
## Signif. codes:  0 '***' 0.001 '**' 0.01 '*' 0.05 '.' 0.1 ' ' 1
```

We can extract different components similarly as described for linear models (see section 7.9 on page 200).

```
class(fm10)
## [1] "glm" "lm"
summary(fm10)
##
## Call:
## glm(formula = count ~ spray, family = quasipoisson, data = InsectSprays)
##
## Coefficients:
##             Estimate Std. Error t value Pr(>|t|)
## (Intercept)  2.67415    0.09309  28.728  < 2e-16 ***
## sprayB       0.05588    0.12984   0.430    0.668
## sprayC      -1.94018    0.26263  -7.388 3.30e-10 ***
## sprayD      -1.08152    0.18499  -5.847 1.70e-07 ***
## sprayE      -1.42139    0.21110  -6.733 4.82e-09 ***
## sprayF       0.13926    0.12729   1.094    0.278
## ---
## Signif. codes:  0 '***' 0.001 '**' 0.01 '*' 0.05 '.' 0.1 ' ' 1
##
## (Dispersion parameter for quasipoisson family taken to be 1.507713)
##
##     Null deviance: 409.041  on 71  degrees of freedom
## Residual deviance:  98.329  on 66  degrees of freedom
## AIC: NA
##
## Number of Fisher Scoring iterations: 5
head(residuals(fm10))
##          1          2          3          4          5          6
## -1.2524891 -2.1919537  1.3650439 -0.1320721 -0.1320721 -0.6768988
head(fitted(fm10))
##    1    2    3    4    5    6
## 14.5 14.5 14.5 14.5 14.5 14.5
```

If we use `str()` or `names()` we can see that there are some differences with respect to linear model fits. The returned object is of a different class and contains some members not present in linear models. Two of these have to do with the iterative approximation method used, `iter` contains the number of iterations used and `converged` the success or not in finding a solution.

```
class(fm10)
## [1] "glm" "lm"
names(fm10)
##  [1] "coefficients"      "residuals"         "fitted.values"
##  [4] "effects"           "R"                 "rank"
##  [7] "qr"                "family"            "linear.predictors"
## [10] "deviance"          "aic"               "null.deviance"
## [13] "iter"              "weights"           "prior.weights"
## [16] "df.residual"       "df.null"           "y"
## [19] "converged"         "boundary"          "model"
## [22] "call"              "formula"           "terms"
## [25] "data"              "offset"            "control"
## [28] "method"            "contrasts"         "xlevels"
fm10$converged
## [1] TRUE
fm10$iter
## [1] 5
```

Methods `update()` and `step()`, described for `lm()` in section 7.9.4 on page 213, can be also used with models fitted with `glm()`.

7.11 Non-Linear Regression

By *non-linear* it is meant non-linear *in the parameters* whose values are being estimated through fitting the model to observations. This is different from the shape of the function when plotted—i.e., polynomials of any degree are linear models. In contrast, the Michaelis-Menten equation used in chemistry and the Gompertz equation used to describe growth are models that are non-linear in their parameters.

While analytical algorithms exist for finding estimates for the parameters of linear models, in the case of non-linear models, the estimates are obtained by approximation. For analytical solutions, estimates can always be obtained (except in pathological cases affected by the limitations of floating point numbers described on page 27). For approximations obtained through iteration, cases when the algorithm fails to *converge* onto an answer are relatively common. Iterative algorithms attempt to improve an initial guess for the values of the parameters to be estimated, a guess frequently supplied by the user. In each iteration, the estimate obtained in the previous iteration is used as the starting value, and this process is repeated one time after another. The expectation is that after a finite number of iterations the algorithm will converge into a solution that "cannot" be improved further. In real life, we stop iteration when the improvement in the fit is smaller than a certain threshold, or when no convergence has been achieved after a certain maximum number of iterations. In the first case, we usually obtain good estimates; in the second case, we do not obtain usable estimates and need to look for different ways of obtaining them.

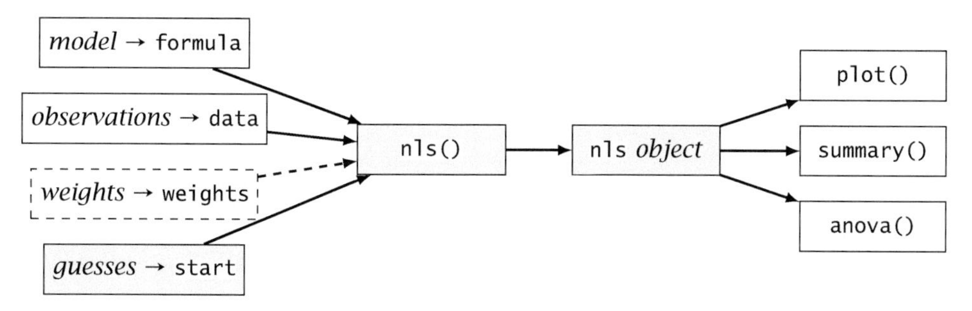

Figure 7.7
Non-linear model fitting in R is done in steps. Generic diagram from Figure 7.1 redrawn to show a non-linear model fit. Non-filled boxes are shared with fitting of other types of models, and filled ones are specific to `nls()`. Only the three most frequently used query methods are shown, while both response and explanatory variables are under *observations*. Dashed boxes and arrows are optional as defaults are provided.

When convergence fails, the first thing to do is to try different starting values and if this also fails, switch to a different computational algorithm. These steps usually help, but not always. Good starting values are in many cases crucial and in some cases "guesses" can be obtained using either graphical or analytical approximations.

Function `nls()` is R's workhorse for fitting non-linear models. The steps for its use are similar to those for LM and GLM (Figure 7.7). One difference is that starting values are needed, and another difference is in how the model to be fitted is specified: the user provides the names of the parameters and a model equation that includes in the *rhs* a call to an R function.

In cases when algorithms exist for "guessing" suitable starting values, R provides a mechanism for packaging the R function to be fitted together with the R function generating the starting values. These functions go by the name of *self-starting functions* and relieve the user from the burden of guessing and supplying suitable starting values. The self-starting functions available in R are `SSasymp()`, `SSasympOff()`, `SSasympOrig()`, `SSbiexp()`, `SSfol()`, `SSfpl()`, `SSgompertz()`, `SSlogis()`, `SSmicmen()`, and `SSweibull()`. Function `selfStart()` can be used to define new ones. All these functions can be used when fitting models with `nls` or `nlme`. The respective help pages give the details.

In calls to `nls()`, the rhs of the model `formula` is a function call. The names of its arguments if not present in `data` are assumed to be parameters to be fitted. Below, a named function

As example the Michaelis-Menten equation describing reaction kinetics in biochemistry and chemistry is fitted to the `Puromycin` data set. The mathematical formulation is given by

$$v = \frac{\mathrm{d}[P]}{\mathrm{d}t} = \frac{V_{\mathrm{max}}[S]}{K_{\mathrm{M}} + [S]} \tag{7.1}$$

and is implemented in R under the name `SSmicmen()` as a self-starting function.

```
data(Puromycin)
names(Puromycin)
## [1] "conc"  "rate"  "state"

fm21 <- nls(rate ~ SSmicmen(conc, Vm, K), data = Puromycin,
            subset = state == "treated")
```

As for other fitted models we use query methods (see section 7.9 on page 200).

```
class(fm21)
## [1] "nls"
summary(fm21)
##
## Formula: rate ~ SSmicmen(conc, Vm, K)
##
## Parameters:
##     Estimate Std. Error t value Pr(>|t|)
## Vm 2.127e+02  6.947e+00  30.615 3.24e-11 ***
## K  6.412e-02  8.281e-03   7.743 1.57e-05 ***
## ---
## Signif. codes:  0 '***' 0.001 '**' 0.01 '*' 0.05 '.' 0.1 ' ' 1
##
## Residual standard error: 10.93 on 10 degrees of freedom
##
## Number of iterations to convergence: 0
## Achieved convergence tolerance: 1.929e-06
residuals(fm21)
##  [1]  25.4339971  -3.5660029  -5.8109605   4.1890395 -11.3616075   4.6383925
##  [7]  -5.6846886 -12.6846886   0.1670798  10.1670798   6.0311723  -0.9688277
## attr(,"label")
## [1] "Residuals"
fitted(fm21)
##  [1]  50.5660  50.5660 102.8110 102.8110 134.3616 134.3616 164.6847 164.6847
##  [9] 190.8329 190.8329 200.9688 200.9688
## attr(,"label")
## [1] "Fitted values"
```

Methods `str()` and `names()` can reveal differences with respect to linear and generalised linear models. The fitted model object is of class `nls` contains additional members but lacks others. Two members are related to the iterative approximation method used, `control` containing nested members holding iteration settings, and `convInfo` (convergence information) with nested members with information on the outcome of the iterative algorithm.

```
class(fm21)
## [1] "nls"
names(fm21)
## [1] "m"           "convInfo"    "data"        "call"        "dataClasses"
## [6] "control"
```

```
fm21$convInfo
## $isConv
## [1] TRUE
##
## $finIter
## [1] 0
##
## $finTol
## [1] 1.928554e-06
##
## $stopCode
## [1] 0
##
## $stopMessage
## [1] "converged"
```

Method `update()`, described for `lm()` in section 7.9.4 on page 213, can be also used with models fitted with `nls()`. The use of previous estimates as guesses for updates is an important feature.

7.12 Splines and Local Regression

The name "spline" derives from the tool used by draftsmen to draw smooth curves. Originally, a spline of soft wood was used as a flexible guide to draw arbitrary curves. Later the wood splines were replaced by a rod of flexible metal, such as lead, encased in plastic or similar material but the original name persisted. In mathematics, splines are functions that describe smooth and flexible curves.

Most of the model fits given above as examples produce estimates for parameters that are interpretable in the real world, directly in the case of mechanistic models like the estimate of reaction constants or at least indicating broadly a relationship between two variables as in the case of linear regression. In the case of polynomials with degree higher than 2, parameter estimates no longer directly describe features of the data.

Splines take this a step farther and parameter estimates have no practical interest. The interest resides in the overall shape and position of the predicted curve. Splines consist of knots (or connection points) joined by straight or curved fitted lines, i.e., they are functions that are *piecewise*. The simplest splines, are piece-wise linear, given by chained straight line segments connecting knots.

In more complex splines, the segments are polynomials, frequently cubic polynomials, that fulfil certain constraints at the knots. For example, that the slope or first derivative is the same for the two connected curve "pieces" at the knot where they are connected. This constraint ensures that the curve is smooth. In some cases, similar constraints are imposed on higher order derivatives, for example, to the second derivative to ensure that the curve of the first derivative is also smooth at the knots.

Splines are used in free-hand drawing with computers to draw arbitrary smooth curves. They are also be used for interpolation, in which case observations, as-

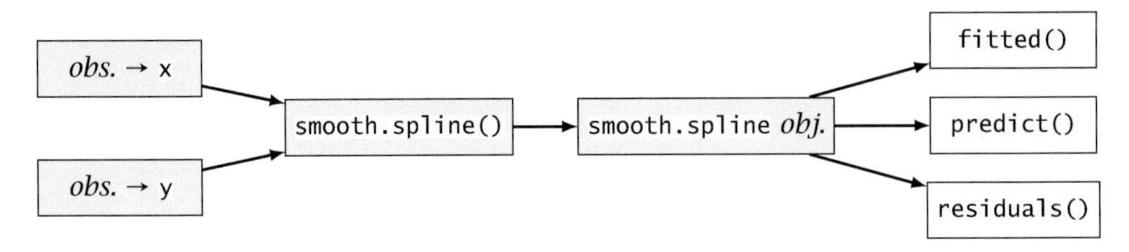

Figure 7.8
Fitting of smooth splines in R. Generic diagram from Figure 7.1 redrawn to show the fitting of splines. Non-filled boxes are shared with fitting of other types of models, and filled ones are specific to `smooth.spline()`. Only the three most frequently used query methods are shown, while response and explanatory variables are passed separately to *x* and *y*.

sumed to be error-free, become the knots of a spline used to approximate intermediate values. Finally, splines can be used as models to be fit to observations subject to random variation. In this case splines fulfil the role of smoothers, as a curve that broadly describes a relationship among variables.

Splines are frequently used as smooth curves in plots as described in section 9.6.3 on page 320. Function `spline()` is used for interpolation and function `smooth.spline()` for smoothing by fitting a cubic spline (a spline where the knots are connected by third degree polynomials). Function `smooth.spline()` has a different user interface than that we used for model fit functions described above, as it only accepts `numeric` vectors as arguments to parameters `x` and `y` (Figure 7.8). Additional parameters make it possible to override the defaults for number of knots and adjust the stiffness or tendency towards a straight line. The `plot()` method for splines, differently to the methods for other fit functions, produces a plot of the prediction. As no model formula is used, only one curve at a time is fitted and no statistical tests involving groups are possible. The most commonly used query functions are thus not the same as for linear and non-linear models.

```
fs1 <- smooth.spline(x = cars$speed, y = cars$dist)
print(fs1)
## Call:
## smooth.spline(x = cars$speed, y = cars$dist)
##
## Smoothing Parameter  spar= 0.7801305  lambda= 0.1112206 (11 iterations)
## Equivalent Degrees of Freedom (Df): 2.635278
## Penalized Criterion (RSS): 4187.776
## GCV: 244.1044
plot(fs1, type = "l")
points(x = cars$speed, y = cars$dist)
```

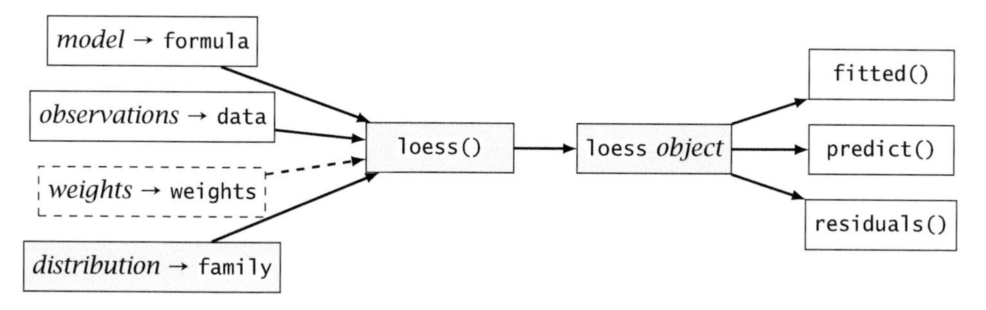

Figure 7.9
Loess model fitting in R is done in steps. Generic diagram from Figure 7.1 redrawn to show local polynomial regression model fitting. Non-filled boxes are shared with fitting of other types of models, and filled ones are specific to `loess()`. Only the three most frequently used query methods are shown, while both response and explanatory variables are under *observations*. Dashed boxes and arrows are optional as defaults are provided.

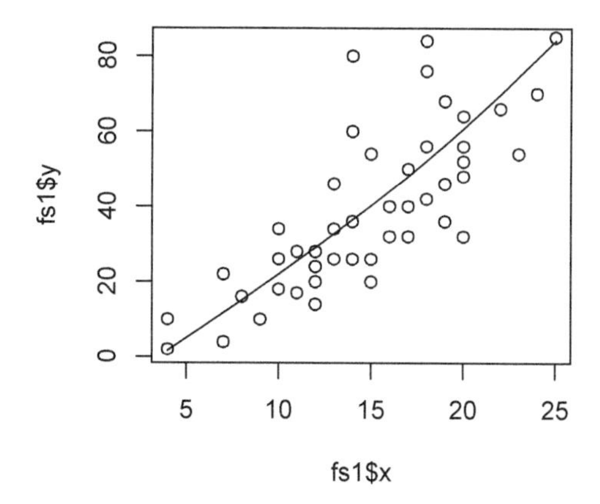

Function `loess` implements *local polynomial regression*. It fits a polynomial curve or surface (i.e., more than one explanatory variable can be included in the model formula) using local-weighted fitting. Its user interface is rather similar to that of `glm()` with `formula`, `family` and `data` formal parameters (Figure 7.9). Additional parameters control "stiffness" or the extent of the local data used for fitting (how much weight is given to observations as a function of their distance). The type of fit local or not used for individual explanatory variables can be controlled through parameter `parametric`.

```
floc <- loess(dist ~ speed, data = cars)
class(floc)
## [1] "loess"
summary(floc)
## Call:
## loess(formula = dist ~ speed, data = cars)
##
```

```
## Number of Observations: 50
## Equivalent Number of Parameters: 4.78
## Residual Standard Error: 15.29
## Trace of smoother matrix: 5.24  (exact)
##
## Control settings:
##    span     :  0.75
##    degree   :  2
##    family   :  gaussian
##    surface  :  interpolate    cell = 0.2
##    normalize:  TRUE
##   parametric:  FALSE
## drop.square:  FALSE
```

Function anova() can be used to compare two or more loess fits, but not on a single one.

Several modern approaches to data analysis, which do provide estimates of effects' significance and sizes, are based on the use of splines to describe the responses and even variance. Among them are additive models such as GAM and related methods (see Wood 2017) and functional data analysis (FDA) (Ramsay 2009). These methods are implemented in specialised extension packages and fall outside the scope of this book.

7.13 Model Formulas

Model formulas, such as y ~ x are widely used in R, both in model fitting as exemplified in previous sections of this chapter and in plotting when using base R plot() methods.

R is consistent and flexible in how it treats various objects, to an extent that can be surprising to those familiar with other computer languages. Model formulas are objects of class formula and mode call and can be manipulated and stored similarly to objects of other R classes.

```
class(y ~ x)
## [1] "formula"
mode(y ~ x)
## [1] "call"
```

Like any other R object formulas can be assigned to variables and be members of lists and vectors. Consequently, the first linear model fit example from page 202 can be rewritten as follows.

```
my.formula <- dist ~ 1 + speed
fm1 <- lm(my.formula, data=cars)
```

In some situations, e.g., calculation of correlations, models lacking a *lhs* term (a term on the left-hand side of ~) are used. At least one term must be present in the *rhs* of model formulas, as an expression ending in ~ is syntactically incomplete.

```
class(~ x + y)
## [1] "formula"
mode(~ x + y)
## [1] "call"
is.empty.model(~ x + y)
## [1] FALSE
```

Some details of R formulas can be important in advanced scripts. Two kinds of "emptiness" are possible for formulas. As with other classes, empty objects or vectors of length zero are valid and can be created with the class constructor. In the case of formulas, there is an additional kind of emptiness, a formula describing a model with no explanatory terms on its *rhs*.

An "empty" object of class `formula` can be created by a call to `formula()` with no arguments, similarly as a numeric vector of length zero is created by the call `numeric()`. The last, commented out, statement in the code below triggers an error as the argument passed to `is.empty.model()` is of length zero. (This behaviour is not consistent with `numeric` vectors of length zero; see for example the value returned by `is.finite(numeric())`.)

```
class(formula())
## [1] "formula"
mode(formula())
## [1] "list"
length(formula())
## [1] 0
# is.empty.model(formula())
```

A model formula describing a model with no explanatory terms on the rhs, is considered empty even if it is a valid object of class `formula` and, thus, not missing. While `y ~ 1` describes a model with only an intercept (estimating $a = \bar{x}$), `y ~ 0` or its equivalent `y ~ -1`, describes an empty model that cannot be fitted to data.

```
class(y ~ 0)
## [1] "formula"
mode(y ~ 0)
## [1] "call"
is.empty.model(y ~ 0)
## [1] TRUE
is.empty.model(y ~ 1)
## [1] FALSE
is.empty.model(y ~ x)
## [1] FALSE
```

The value returned by `length()` on a single formula is not always 1, the number of formulas in the vector of formulas, but instead the number of components in the formula. For longer vectors, it does return the number of member formulae. Because of this, it is better to store model formulas in objects of class `list` than in vectors, as `length()` consistently returns the expected value on lists.

```
length(formula())
## [1] 0
length(y ~ 0)
## [1] 3
length(y ~ 1)
## [1] 3
length(y ~ x)
## [1] 3
length(c(y ~ 1, y ~ x))
## [1] 2
length(list(y ~ 1))
## [1] 1
length(list(y ~ 1, y ~ x))
## [1] 2
```

As described above, `length()` applied to a single formula and to a list of formulas behaves differently. To call `length()` on each member of a list of formulas, we can use `sapply()` (see section 5.8 on page 154). As function `is.empty.model()` is not vectorised, we also have to use `sapply()` with a list of formulas.

```
sapply(list(y ~ 0, y ~ 1, y ~ x), length)
## [1] 3 3 3
sapply(list(y ~ 0, y ~ 1, y ~ x), is.empty.model)
## [1]  TRUE FALSE FALSE
```

In the examples in previous sections, we fitted simple models. More complex ones can be easily formulated using the same syntax. First of all, one can avoid using of operator `*` by explicitly defining all individual main effects and interactions using operators `+` and `:`. The syntax implemented in base R allows grouping by means of parentheses, so it is also possible to exclude some interactions by combining the use of `*` and parentheses.

The same symbols as for arithmetic operators are used for model formulas. Within a formula, symbols are interpreted according to formula syntax. When we mean an arithmetic operation that could be interpreted as being part of the model formula we need to "protect" it by means of the identity function `I()`. The next two examples define formulas for models with only one explanatory variable. With formulas like these, the explanatory variable will be computed on the fly when fitting the model to data. In the first case below, we need to explicitly protect the addition of the two variables into their sum, because otherwise they would be interpreted as two separate explanatory variables in the model. In the second case, `log()` cannot be interpreted as part of the model formula, and consequently does not require additional protection, neither does the expression passed as its argument.

```
y ~ I(x1 + x2)
y ~ log(x1 + x2)
```

R formula syntax allows alternative ways for specifying interaction terms. They allow "abbreviated" ways of entering formulas, which for complex experimental designs saves typing and can improve clarity. As seen above, operator `*` saves us from having to explicitly indicate all the interaction terms in a full factorial model.

```
y ~ x1 + x2 + x3 + x1:x2 + x1:x3 + x2:x3 + x1:x2:x3
```

Can be replaced by a concise equivalent.

```
y ~ x1 * x2 * x3
```

When the model to be specified does not include all possible interaction terms, we can combine the concise notation with parentheses. Below, equivalent formulas are shown using concise and verbose notation.

```
y ~ x1 + (x2 * x3)
y ~ x1 + x2 + x3 + x2:x3

y ~ x1 * (x2 + x3)
y ~ x1 + x2 + x3 + x1:x2 + x1:x3
```

The `^` operator provides a concise notation to limit the order of the interaction terms included in a formula.

```
y ~ (x1 + x2 + x3)^2
y ~ x1 + x2 + x3 + x1:x2 + x1:x3 + x2:x3
```

Operator `%in%` can also be used as a shortcut for including only some of all the possible interaction terms in a formula.

```
y ~ x1 + x2 + x1 %in% x2
```

7.16 Whether the two model formulas above are equivalent or not, can be investigated using function `terms()`.

```
terms(y ~ x1 + (x2 * x3))
terms(y ~ x1 * (x2 + x3))
terms(y ~ (x1 + x2 + x3)^2)
terms(y ~ x1 + x2 + x1 %in% x2)
```

7.17 For operator `^` to behave as expected, its first operand should be a formula with no interactions! Compare the result of expanding these two formulas with `terms()`.

```
y ~ (x1 + x2 + x3)^2
y ~ (x1 * x2 * x3)^2
```

7.18 Run the code examples below using the `npk` data set from R. They demonstrate the use of different model formulas in ANOVA. Use these examples plus your own variations on the same theme to build your understanding of the syntax of model formulas. Based on the terms displayed in the ANOVA tables, first work out what models are being fitted in each case. In a second step, write the mathematical formulation of each of the models. Finally, think how model choice may affect the conclusions from an analysis of variance.

```
data(npk)
anova(lm(yield ~ N * P * K, data = npk))
anova(lm(yield ~ (N + P + K)^2, data = npk))
anova(lm(yield ~ N + P + K + P %in% N + K %in% N, data = npk))
anova(lm(yield ~ N + P + K + N %in% P + K %in% P, data = npk))
```

Nesting of factors in experiments using hierarchical designs such as split-plots or repeated measures, results in the need to compute additional error terms, differing in their degrees of freedom. In a nested design with fixed effects, effects are tested based on different error terms depending on the design of an experiment, i.e., depending on the randomisation of the assignment of treatments to experimental units. In base-R model-formulas, nesting is described by explicit definition of error terms by means of `Error()` within the formula.

The syntax described above does not support complex statistical models as implemented in extension packages. For example, Nowadays, fitting linear mixed-effects (LME) models is the preferred approach for the analysis of data from experiments and surveys based on hierarchical designs. These methods are implemented in packages 'nlme' (Pinheiro and Bates 2000) and 'lme4' (Bates et al. 2015) that define extensions to the model formula syntax. The extensions make it possible to describe nesting and distinguish fixed and random effects. Packages implementing fitting of additive models have needed other extensions to the formula syntax. Additive model methods are described by Wood (2017) and Zuur (2012). Although the overall approach and syntax are followed in most contributed packages, different packages have extended the formula syntax in different ways. These extensions fall outside the scope of this book.

R will accept any syntactically correct model formula, even when the results of the fit are not interpretable. It is *the responsibility of the user to ensure that models are meaningful.* The most common, and dangerous, mistake is specifying for factorial experiments, models that are missing lower-order terms.

Fitting models like those below to data from an experiment based on a three-way factorial design should be avoided. In both cases, simpler terms are missing, while higher-order interaction(s) that include the missing term are included in the model. Such models are not interpretable, as the variation from the missing term(s) ends being "disguised" within the remaining terms, distorting their apparent significance and parameter estimates.

```
y ~ A + B + A:B + A:C + B:C
y ~ A + B + C + A:B + A:C + A:B:C
```

In contrast to those above, the models below are interpretable, even if not "full" models (not including all possible interactions).

```
y ~ A + B + C + A:B + A:C + B:C
y ~ (A + B + C)^2
y ~ A + B + C + B:C
y ~ A + B * C
```

Manipulation of model formulas. Because this is a book about the R language, it is pertinent to describe how formulas can be manipulated. Formulas, as any other R objects, can be saved in variables including lists. Why is this useful? For example, if we want to fit several different models to the same data, we can write a `for` loop that walks through a list of model formulas (see section 5.10 on page 160). Obviously, user-defined functions can accept formulas as arguments as `lm()` and other model-fitting functions do. In addition, it is relatively simple for user code to programmatically create and edit R formulas, in the same way as functions `update()` and `step()` do under the hood.

A conversion constructor is available under the name `as.formula()`. It is useful when formulas are input interactively by the user or read from text files. With `as.formula()` we can convert a character string into a formula.

```
as.formula("y ~ x")
## y ~ x
```

As there are many functions for the manipulation of `character` strings available in base R and through extension packages, it is easiest to build model formulas as strings. We can use functions like `paste()` to assemble a formula as text, and then use `as.formula()` to convert it to an object of class `formula`, usable for fitting a model.

```
paste("y", "x", sep = "~") |> as.formula()
## y ~ x
```

For the reverse operation of converting a formula into a string, we have available methods `as.character()` and `format()`. The first of these methods returns a character vector containing the components of the formula as individual strings, while `format()` returns a single character string with the formula formatted for printing.

```
as.character(y ~ x)
## [1] "~" "y" "x"
format(y ~ x)
## [1] "y ~ x"
```

This conversion makes it possible to edit a formula as a character string.

```
format(y ~ x) |> gsub("x", "x + z", x = _) |> as.formula()
## y ~ x + z
```

It is also possible to *edit* formula objects with method `update()`. In the replacement formula, a dot can replace either the left-hand side (lhs) or the right-hand side (rhs) of the existing formula. We can also remove terms as can be seen below. In some cases, the dot corresponding to the lhs can be omitted, but including it makes the syntax clearer.

```
my.formula <- y ~ x1 + x2
update(my.formula, . ~ . + x3)
## y ~ x1 + x2 + x3
update(my.formula, . ~ . - x1)
## y ~ x2
update(my.formula, . ~ x3)
## y ~ x3
update(my.formula, z ~ .)
## z ~ x1 + x2
update(my.formula, . + z ~ .)
## y + z ~ x1 + x2
```

As R provides high-level functions for model selection editing model formulas is not very frequently needed for model fitting.

A model matrix of dummy coefficients is used in the actual computations. This matrix can be derived from a model formula, a contrast name, and the data for the explanatory variables using function `model.matrix()`.

7.14 Time Series

Longitudinal data consist of repeated measurements, usually done over time, on the same experimental units. Longitudinal data, when replicated on several experimental units at each time point, are called repeated measurements, while when not replicated, they are called time series. Base R provides special support for the analysis of time series data, while repeated measurements can be analysed with nested linear models, mixed-effects models, and additive models.

Time series data are data collected in such a way that there is only one observation, possibly of multiple variables, available at each point in time. This brief section introduces only the most basic aspects of time-series analysis. In most cases, time steps are of uniform duration and occur regularly, which simplifies data handling and storage. R not only provides methods for the analysis and manipulation of time-series, but also a specialised class for their storage, "ts". Regular time steps allow more compact storage—e.g., a ts object does not need to store time values for each observation but instead a combination of two of start time, step size and end time. When analysing time-series data, it is frequently necessary to convert time data between one of the special R classes and character strings, and to operate on dates and times (see section 8.8 on page 267).

By now, you surely guessed that to create an object of class "ts" one needs to use a constructor called ts() or a conversion constructor called as.ts() and that you can look up the arguments they accept by consulting help using help(ts).

```
my.ts <- ts(1:10, start = 2019, deltat = 1/12)
```

The print() method for ts objects is special, and adjusts the printout according to the time step or deltat of the series.

```
print(my.ts)
##      Jan Feb Mar Apr May Jun Jul Aug Sep Oct
## 2019   1   2   3   4   5   6   7   8   9  10
```

The structure of the ts object is simple. Its mode is numeric but its class is ts. It is similar to a numeric vector with the addition of one attribute named tsp describing the time steps, as a numeric vector of length 3, giving start and end time and the size of the steps.

```
mode(my.ts)
## [1] "numeric"
class(my.ts)
## [1] "ts"
is.ts(my.ts)
## [1] TRUE
str(my.ts)
##  Time-Series [1:10] from 2019 to 2020: 1 2 3 4 5 6 7 8 9 10
attributes(my.ts)
## $tsp
## [1] 2019.00 2019.75   12.00
##
## $class
## [1] "ts"
```

Data set `nottem`, included in R, contains meteorological data for Nottingham. The annual cycle of mean air temperatures (in degrees Fahrenheit) as well as variation among years are clear when data are plotted.

Reexpression of the temperatures in the time-series from degrees Fahrenheit into degrees Celsius can be achieved as in `numeric` vectors using vectorised arithmetic and recycling.

```
nottem.celcius <- (nottem - 32) * 5 / 9
```

```
is.ts(nottem.celcius)
## [1] TRUE
```

```
plot(nottem.celcius)
```

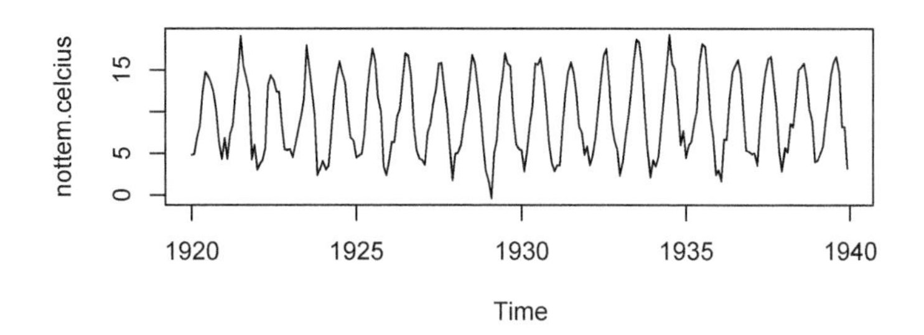

7.19 Explore the structure of the `nottem.celcius` object (or the `nottem` object), and consider how and why it differs or not from that of the object `my.ts` that we created above. Similarly explore time series `ausres`, another of the data sets included in R.

```
str(nottem.celcius)
attributes(nottem.celcius)
```

Many time series of observations display cyclic variation at different frequencies. Outdoors, air temperature varies cyclically between day and night and throughout the year. Superimposed on these regular cycles there can be faster random variation and long-term trends. One approach to the analysis of time series data is to estimate the separate contribution of these components.

An efficient approach to time series decomposition, based on LOESS (see section 7.12 on 223), is STL (Seasonal and Trend decomposition using Loess). A seasonal window of 7 months, the minimum accepted, allows the extraction of the annual cycles and a long-term trend leaving as a remainder some unexplained variation. In the plot, is important to be aware that the scale limits in the different panels are different, and re-set for each plot.

```
nottem.stl <- stl(nottem.celcius, s.window = 7)
plot(nottem.stl)
```

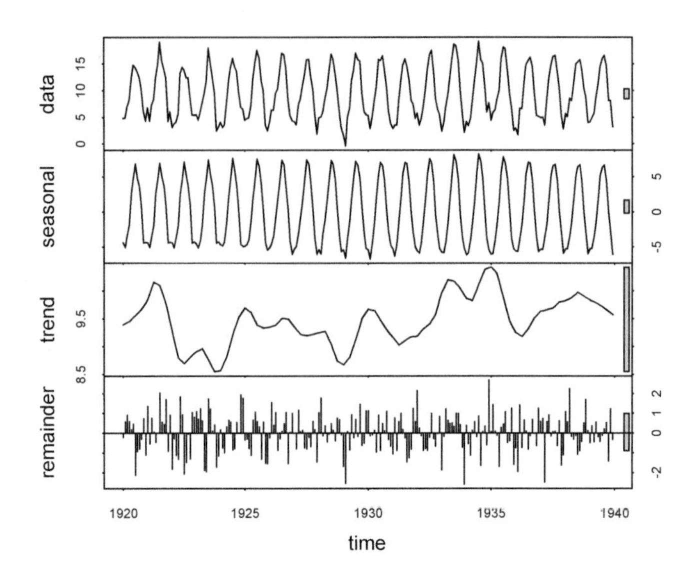

It is interesting to explore the class and structure of the object returned by `stl()`, as we may want to extract components. We can see that the structure of this object is rather similar to model-fit objects of classes `lm` and `glm`.

```
class(nottem.stl)
## [1] "stl"
str(nottem.stl, no.list = TRUE, give.attr = FALSE, vec.len = 2)
##  $ time.series: Time-Series [1:240, 1:3] from 1920 to 1940: -4.4 -5.08 ...
##  $ weights    : num [1:240] 1 1 1 1 1 ...
##  $ call       : language stl(x = nottem.celcius, s.window = 7)
##  $ win        : Named num [1:3] 7 23 13
##  $ deg        : Named int [1:3] 0 1 1
##  $ jump       : Named num [1:3] 1 3 2
##  $ inner      : int 2
##  $ outer      : int 0
```

As with other fit methods, method `summary()` is available. However, this method for class `stl` returns unchanged the `stl` object received as an argument and displays a summary. In other words, it behaves similarly to `print()` methods with respect to the returned object, but produces a different printout than `print()` as its side effect.

```
summary(nottem.stl)
##  Call:
##  stl(x = nottem.celcius, s.window = 7)
##
##  Time.series components:
##     seasonal               trend                remainder
##  Min.   :-6.693714   Min.   : 8.548340   Min.   :-2.5950749
##  1st Qu.:-4.413237   1st Qu.: 9.201837   1st Qu.:-0.6907277
##  Median :-0.650109   Median : 9.456694   Median : 0.0593786
##  Mean   : 0.001867   Mean   : 9.462835   Mean   : 0.0017326
##  3rd Qu.: 4.595458   3rd Qu.: 9.779625   3rd Qu.: 0.6445627
##  Max.   : 8.215818   Max.   :10.424848   Max.   : 2.6914745
##  IQR:
```

```
##      STL.seasonal STL.trend STL.remainder data
##      9.0087       0.5778    1.3353        8.5833
##    % 105.0          6.7      15.6          100.0
##
##  Weights: all == 1
##
##  Other components: List of 5
##  $ win  : Named num [1:3] 7 23 13
##  $ deg  : Named int [1:3] 0 1 1
##  $ jump : Named num [1:3] 1 3 2
##  $ inner: int 2
##  $ outer: int 0
```

7.20 Consult `help(stl)` and `help(plot.stl)` and create different plots and decompositions by passing different arguments to the formal parameters of these methods.

Method `print()` shows the different components. Extract the seasonal component and plot is on its own against time.

In the Nottingham temperature time series, the period of the variation is clearly annual, but for many other time series an interesting feature to characterise is autocorrelation and its periodicity. Function `acf()` computes and plots the autocorrelation function (ACF) vs. the lag. The time series has monthly data, while the scale for lag in the plot below is in years. The autocorrelation is one at zero lag, and slightly less with a lag of one year, while its is negative between winter and summer temperatures.

```
acf(nottem)
```

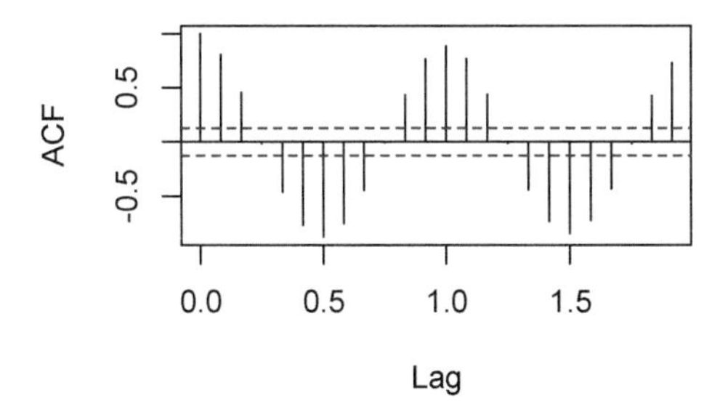

More advanced time-series analysis and forecasting methods are implemented in contributed packages and fall outside the scope of this book. The textbook *Forecasting: Principles and Practice* (Hyndman and Athanasopoulos 2021) is comprehensive, starting with an introduction to time series and continuing all the way to the description of modern forecasting methods, using R throughout.

7.15 Multivariate Statistics

All the methods presented above are univariate, as even if in some cases we considered multiple explanatory variables on the rhs of model formulas, the lhs contained at most one response variable. There are many different multivariate methods available, and a few of them are implemented in base R functions. The current section does not describe these methods in depth, it only provides a few simple examples for some of the frequently used ones.

7.15.1 Multivariate analysis of variance

Multivariate methods take into account several response variables simultaneously, as part of a single analysis. In practice, it is usual to use contributed packages for multivariate data analysis in R, except for simple cases. We will look first at *multivariate* ANOVA or MANOVA. In the same way as `aov()` is a wrapper that uses internally `lm()`, `manova()` is a wrapper that uses internally `aov()`.

Multivariate model formulas in base R require the use of column binding (`cbind()`) on the left-hand side (lhs) of the model formula. The well-known `iris` data set, containing size measurements for flowers of three species of *Iris*, is used in the examples below.

```
mmf2 <- manova(cbind(Petal.Length, Petal.Width) ~  Species, data = iris)
anova(mmf2)
## Analysis of Variance Table
##
##              Df  Pillai approx F num Df den Df    Pr(>F)
## (Intercept)   1 0.98786   5939.2      2    146 < 2.2e-16 ***
## Species       2 1.04645     80.7      4    294 < 2.2e-16 ***
## Residuals   147
## ---
## Signif. codes:  0 '***' 0.001 '**' 0.01 '*' 0.05 '.' 0.1 ' ' 1
summary(mmf2)
##            Df Pillai approx F num Df den Df    Pr(>F)
## Species     2 1.0465   80.661      4    294 < 2.2e-16 ***
## Residuals 147
## ---
## Signif. codes:  0 '***' 0.001 '**' 0.01 '*' 0.05 '.' 0.1 ' ' 1
```

7.21 Modify the example above to use `aov()` instead of `manova()` and save the result to a variable named `mmf3`. Use `class()`, `attributes()`, `names()`, `str()` and extraction of members to explore objects `mmf1`, `mmf2` and `mmf3`. Are they different?

7.15.2 Principal components analysis

Principal components analysis (PCA) is used to simplify a data set by combining variables with similar and "mirror" behaviour into principal components. At a later stage, we frequently try to interpret these components in relation to known and/or assumed independent variables. Base R's function `prcomp()` computes the principal components and accepts additional arguments for centring and scaling.

```
pc <- prcomp(iris[c("Sepal.Length", "Sepal.Width",
                    "Petal.Length", "Petal.Width")],
             center = TRUE,
             scale = TRUE)
```

By printing the returned object, we can see the loadings of each variable in the principal components `PC1` to `PC4`.

```
class(pc)
## [1] "prcomp"
pc
## Standard deviations (1, .., p=4):
## [1] 1.7083611 0.9560494 0.3830886 0.1439265
##
## Rotation (n x k) = (4 x 4):
##                     PC1         PC2        PC3        PC4
## Sepal.Length  0.5210659 -0.37741762  0.7195664  0.2612863
## Sepal.Width  -0.2693474 -0.92329566 -0.2443818 -0.1235096
## Petal.Length  0.5804131 -0.02449161 -0.1421264 -0.8014492
## Petal.Width   0.5648565 -0.06694199 -0.6342727  0.5235971
```

In the summary, the rows "Proportion of Variance" and "Cumulative Proportion" are most informative of the contribution of each principal component (PC) to explaining the variation among observations.

```
summary(pc)
## Importance of components:
##                           PC1    PC2     PC3     PC4
## Standard deviation     1.7084 0.9560 0.38309 0.14393
## Proportion of Variance 0.7296 0.2285 0.03669 0.00518
## Cumulative Proportion  0.7296 0.9581 0.99482 1.00000
```

Method `plot()` generates a bar plot of variances corresponding to the different components.

```
plot(pc)
```

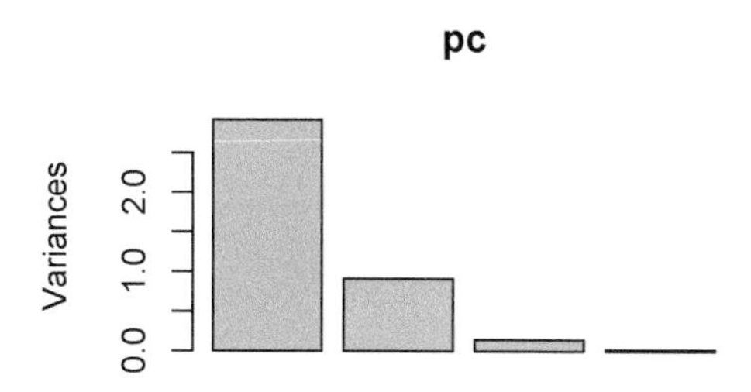

Method `biplot()` produces a plot with one principal component (PC) on each axis, plus arrows for the loadings.

```
biplot(pc)
```

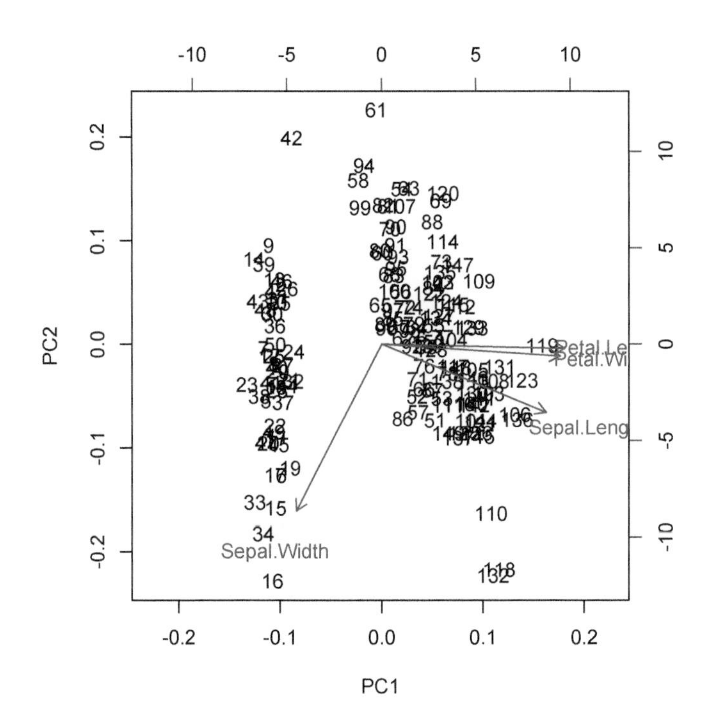

Visually more elaborate plots of the principal components and their loadings can be obtained using package ‘ggplot’ described in chapter 9 on page 271. Package ‘ggfortify’ extends ‘ggplot’ so as to make it easy to plot principal components and their loadings.

7.22 For growth and morphological data, a log-transformation can be suitable given that variance is frequently proportional to the magnitude of the values measured. We leave it as an exercise to repeat the above analysis using transformed values for the dimensions of petals and sepals. How much does the use of transformations change the outcome of the analysis?

7.23 As for other fitted models, the object returned by function `prcomp()` is list-like with multiple components and belongs to a class of the same name as the function, not derived from class `"list"`.

```
class(pc)
str(pc, max.level = 1)
```

7.15.3 Multidimensional scaling

The aim of multidimensional scaling (MDS) is to visualise in 2D space the similarity between pairs of observations. The values for the observed variable(s) are used to compute a measure of distance among pairs of observations. The nature of the data will influence what distance metric is most informative. For MDS we start with a matrix of distances among observations. We will use, for the next examples, distances in kilometres between geographic locations in Europe from the data set `eurodist`.

```
loc <- cmdscale(eurodist)
```

We can see that the returned object `loc` is a `matrix`, with names for one of the dimensions.

```
class(loc)
## [1] "matrix" "array"
dim(loc)
## [1] 21  2
head(loc, 8)
##                   [,1]       [,2]
## Athens     2290.274680  1798.8029
## Barcelona  -825.382790   546.8115
## Brussels     59.183341  -367.0814
## Calais      -82.845973  -429.9147
## Cherbourg  -352.499435  -290.9084
## Cologne     293.689633  -405.3119
## Copenhagen  681.931545 -1108.6448
## Geneva       -9.423364   240.4060
```

To make the code easier to read, two vectors are first extracted from the matrix and named `x` and `y`. We force aspect to equality so that distances on both axes are comparable.

```
x <- loc[, 1]
y <- -loc[, 2] # change sign so North is at the top
plot(x, y, type = "n", asp = 1,
     main = "cmdscale(eurodist)")
text(x, y, rownames(loc), cex = 0.6)
```

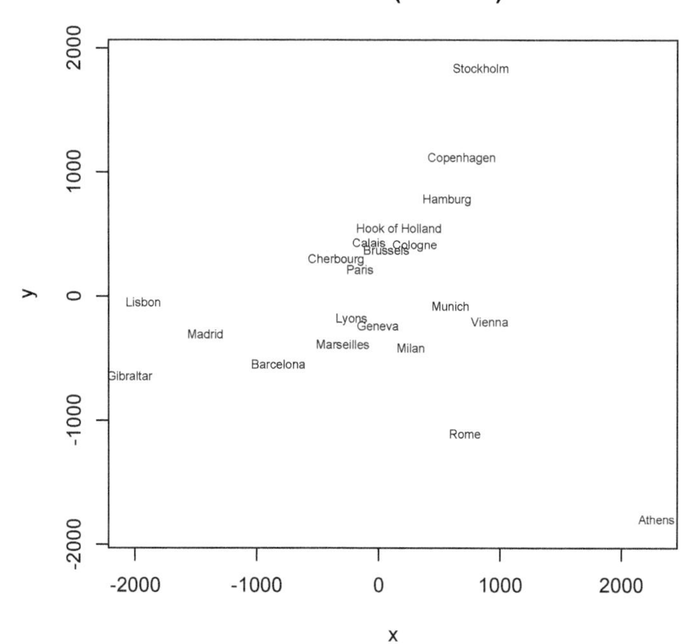

7.24 Find data on the mean annual temperature, mean annual rainfall and mean number of sunny days at each of the locations in the `eurodist` data set. Next, compute suitable distance metrics, for example, using function `dist`. Finally, use MDS to visualise how similar the locations are with respect to each of the three variables. Devise a measure of distance that takes into account the three climate variables and use MDS to find how distant the different locations are.

7.15.4 Cluster analysis

In cluster analysis, the aim is to group observations into discrete groups with maximal internal homogeneity and maximum group-to-group differences. In the next example, we use function `hclust()` from the base-R package 'stats'. We use, as above, the `eurodist` data which directly provides distances. In other cases, a matrix of distances between pairs of observations needs to be first calculated with function `dist` which supports several methods.

```
hc <- hclust(eurodist)
print(hc)
##
## Call:
## hclust(d = eurodist)
##
## Cluster method   : complete
## Number of objects: 21
```

```
plot(hc)
```

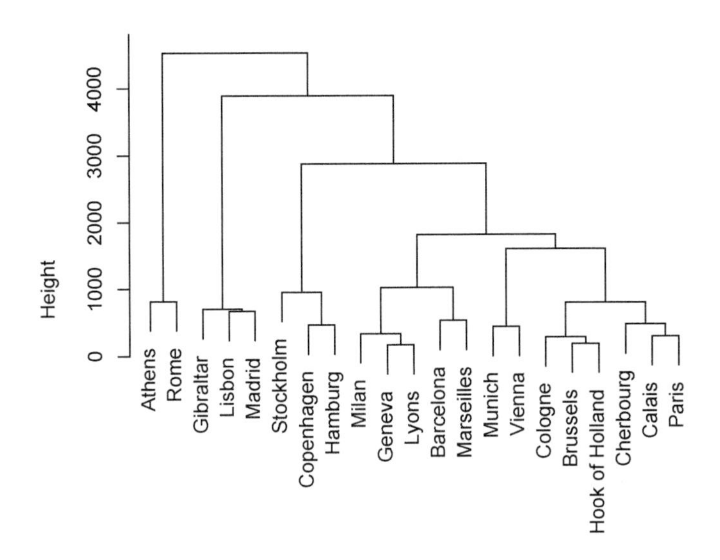

We can use `cutree()` to limit the number of clusters by directly passing as an argument the desired number of clusters or the height at which to cut the tree.

```
cutree(hc, k = 5)
##          Athens       Barcelona        Brussels          Calais       Cherbourg
##               1               2               3               3               3
##         Cologne      Copenhagen          Geneva       Gibraltar         Hamburg
##               3               4               2               5               4
## Hook of Holland          Lisbon           Lyons          Madrid      Marseilles
##               3               5               2               5               2
##           Milan          Munich           Paris            Rome       Stockholm
##               2               3               3               1               4
##          Vienna
##               3
```

The object returned by `hclust()` contains details of the result of the clustering, which allows further manipulation and plotting.

```
str(hc)
## List of 7
##  $ merge      : int [1:20, 1:2] -8 -3 -6 -4 -16 -17 -5 -7 -2 -12 ...
##  $ height     : num [1:20] 158 172 269 280 328 428 460 460 521 668 ...
##  $ order      : int [1:21] 1 19 9 12 14 20 7 10 16 8 ...
##  $ labels     : chr [1:21] "Athens" "Barcelona" "Brussels" "Calais" ...
##  $ method     : chr "complete"
##  $ call       : language hclust(d = eurodist)
##  $ dist.method: NULL
##  - attr(*, "class")= chr "hclust"
```

7.16 Further Reading

R and its extension packages provide implementations of most known statistical methods. For some methods, alternative implementations exist in different packages. The present chapter only attempts to show how some of the most frequently used implementations are used, as this knowledge is frequently taken for granted in specialised books, several of which I list here. Two recent text books on statistics, following a modern approach, and using R for examples, are *OpenIntro Statistics* (Diez et al. 2019) and *Modern Statistics for Modern Biology* (Holmes and Huber 2019). They differ in the subjects emphasised, with the second one focusing more on genetic and molecular biology. Three examples of books introducing statistical computations in R are *Introductory Statistics with R* (Dalgaard 2008), *A Handbook of Statistical Analyses Using R* (Everitt and Hothorn 2010) and *A Beginner's Guide to R* (Zuur et al. 2009). The book *Biometry for Forestry and Environmental Data with Examples in R* (Mehtätalo and Lappi 2020) presents both the statistical theory and code examples. The comprehensive *The R Book* (Crawley 2012) and the classic reference *Modern Applied Statistics with S* (Venables and Ripley 2002) both present statistical theory in parallel with the R code examples. More specific books are also available from which a few suggestions for further reading are *An Introduction to Applied Multivariate Analysis with R* (Everitt and Hothorn 2011), *Linear Models with R* (Faraway 2004), *Extending the Linear Model with R: Generalized Linear, Mixed Ef-*

fects and Nonparametric Regression Models (Faraway 2006), *Forecasting: Principles and Practice* (Hyndman and Athanasopoulos 2021), *An Introduction to Statistical Learning: with Applications in R* (James et al. 2013), *Mixed-Effects Models in S and S-Plus* (Pinheiro and Bates 2000) and *Generalized Additive Models* (Wood 2017).

8

R Extensions: Data Wrangling

> Essentially everything in S[R], for instance, a call to a function, is an S[R] object. One viewpoint is that S[R] has self-knowledge. This self-awareness makes a lot of things possible in S[R] that are not in other languages.
>
> Patrick J. Burns
> *S Poetry*, 1998

8.1 Aims of This Chapter

Base R and the recommended extension packages (installed by default) include many functions for manipulating data. The R distribution supplies a complete set of functions and operators that allow all the usual data manipulation operations. These functions have stable and well-described behaviour, so in my view, they should be preferred unless some of their limitations justify the use of alternatives defined in contributed packages. In the present chapter, I describe the new syntax introduced by the most popular contributed R extension packages aiming at changing (usually improving one aspect at the expense of another) in various ways how we can manipulate data in R. These independently developed packages extend the R language not only by adding new "words" to it but by supporting new ways of meaningfully connecting "words"—i.e., providing new "grammars" for data manipulation. While at the current stage of development of base R not breaking existing code has been the priority, several of the still "young" packages in the 'tidyverse' have prioritised experimentation with enhanced features over backwards compatibility. The development of 'tidyverse' packages seems to have initially emphasised users' convenience more than encouraging safe/error-free user code. The design of package 'data.table' has prioritised performance at the expense of easy of use. I do not describe in depth these new approaches but instead only briefly compare them to base R highlighting the most important differences.

DOI: 10.1201/9781003404187-8

8.2 Introduction

By reading previous chapters, you have already become familiar with base R classes, methods, functions, and operators for storing and manipulating data. Most of these had been originally designed to perform optimally on rather small data sets (see Matloff 2011). The performance of these functions has improved significantly over the years and random-access memory in computers has become cheaper, making constraints imposed by the original design of R less limiting. On the other hand, the size of data sets has also increased.

Some contributed packages have aimed at improving performance by relying on different compromises between usability, speed, and reliability than used for base R. Package 'data.table' is the best example of an alternative implementation of data storage and manipulation that maximises the speed of processing for large data sets using a new semantics and requiring a new syntax. We could say that package 'data.table' is based on a theoretical abstraction, or "grammar of data", that is different from that in the R language. The compromise in this case has been the use of a less intuitive syntax, and by defaulting to passing arguments by reference instead of by copy, increasing the "responsibility" of the programmer or data analyst with respect to not overwriting or corrupting data. This focus on performance has made obvious the performance bottlenecks present in base R, which have been subsequently alleviated while maintaining backwards compatibility for users' code.

Another recent development is the 'tidyverse', which is a formidable effort to redefine how data analysis operations are expressed in R code and scripts. In many ways, it is also a new abstraction, or "grammar of data". With respect to its implementation, it can also be seen as a new language built on-top of the R language. It is still young and evolving, and the developers from Posit still remain relentless about fixing what they consider earlier misguided decisions in the design of the packages comprising the 'tidyverse'. This is a wise decision for the future, but can be annoying to occasional users who may not be aware of the changes that have taken place between uses. As a user I highly value long-term stability and backwards compatibility of software. Older systems like base R provide this, but their long development history shows up as occasional inconsistencies and quirks. The 'tidyverse' as a paradigm is nowadays popular among data analysts while among users for whom data analysis is not the main focus, it is more common to make use of only individual packages as the need arises, e.g., using the new grammar only for some stages of the data analysis workflow.

When a computation included a chain of sequential operations, until R 4.1.0, using base R by itself we could either store the returned value in a temporary variable at each step in the computation, or nest multiple function calls. The first approach is verbose, but allows readable scripts, especially if the names used for temporary variables are wisely chosen. The second approach becomes very difficult to read as soon as there is more than one nesting level. Attempts to find an alternative syntax have borrowed the concept of data *pipes* from Unix shells (Kernigham and Plauger 1981). Interestingly, that it has been possible to write packages that define the operators needed to "add" this new syntax to R is a testimony to its flexibility and extensibility. Two packages, 'magrittr' and 'wrapr', define operators for pipe-based syntax. In year 2021, a pipe operator was added to the R language itself and more recently its features enhanced.

In much of my work I emphasise reproducibility and reliability, preferring base R over extension packages, except for plotting, whenever practical. For run once and delete or quick-and-dirty data analyses, I tend to use the *tidyverse*. However, with modern computers and some understanding of what are the performance bottlenecks in R code, I have rarely found it worthwhile the effort needed for improved performance by using extension packages. The benefit to effort balance will be different for those readers who analyse huge data sets.

The definition of the *tidyverse* is rather vague, as package 'tidyverse' loads and attaches a set of packages of which most but not all follow a consistent design and support this new grammar. The packages that are attached by package 'tidyverse' has changed over time. Package 'tidyverse', however, defines a function that lists them.

```
tidyverse::tidyverse_packages()
##  [1] "broom"         "conflicted"    "cli"           "dbplyr"
##  [5] "dplyr"         "dtplyr"        "forcats"       "ggplot2"
##  [9] "googledrive"   "googlesheets4" "haven"         "hms"
## [13] "httr"          "jsonlite"      "lubridate"     "magrittr"
## [17] "modelr"        "pillar"        "purrr"         "ragg"
## [21] "readr"         "readxl"        "reprex"        "rlang"
## [25] "rstudioapi"    "rvest"         "stringr"       "tibble"
## [29] "tidyr"         "xml2"          "tidyverse"
```

In this chapter, you will become familiar with packages 'tibble', 'dplyr', 'tidyr', and 'lubridate'. Package 'ggplot2' will be described in chapter 9 as it implements the grammar of graphics and has little in common with other members of the 'tidyverse'. As many of the functions in the *tidyverse* can be substituted by existing base R functions, recognising similarities and differences between them has become important since both approaches are now in common use, and frequently even coexist within R scripts.

In any design, there is a tension between opposing goals. In software for data analysis, a key pair of opposed goals are usability, including concise but expressive code, and avoidance of ambiguity. Base R function `subset()` has an unusual syntax, as it evaluates the expression passed as the second argument within the namespace of the data frame passed as its first argument (see section 4.4.5 on page 110). This saves typing, enhancing usability, at the expense of increasing the risk of bugs, as by reading the call to subset, it is not obvious which names are resolved in the environment of the call to `subset()` and which ones within its first argument—i.e., as column names in the data frame. In addition, changes elsewhere in a script can change how a call to subset is interpreted. In reality, subset is a wrapper function built on top of the extraction operator `[ ]` (see section 3.10 on page 64). It is a convenience function, mostly intended to be used at the console, rather than in scripts or package code. To extract columns or rows from a data frame, it is always safer to use the `[ , ]` or `[[ ]]` operators at the expense of some verbosity.

Package 'dplyr', and much of the 'tidyverse', relies on a similar approach as subset to enhance convenience at the expense of ambiguity. Package 'dplyr' has undergone quite drastic changes during its development history with respect to how to handle the dilemma caused by "guessing" of the environment where names should be looked up. There is no easy answer; a simplified syntax leads to ambiguity, and a fully specified syntax is verbose. Recent versions of the package introduced a terse syntax to achieve a concise way of specifying where to look up names. I do appreciate the advantages of the grammar of data that is implemented in the 'tidyverse'. However, the actual implementation, can result in ambiguities and subtleties that are even more difficult to deal by inexperienced or occasional users than those caused by inconsistencies in base R. My opinion is that for code that needs to be highly reliable and produce reproducible results in the future, we should for the time being prefer base R constructs. For code that is to be used once, or for which reproducibility can depend on the use of a specific (old or soon to become old) version of packages like 'dplyr', or which is not a burden to thoroughly test and update regularly, the conciseness and power of the new syntax can be an advantage.

Package 'poorman' re-implements many of the functions in 'dplyr' and a few from 'tidyr' using pure R code instead of compiled C++ code and with no dependencies on other extension packages. This light-weight approach can be useful when R's data frames rather than tibbles are preferred or when the possible enhanced performance with large data sets is not needed.

8.3 Packages Used in This Chapter

```
install.packages(learnrbook::pkgs_ch_data)
```

To run the examples included in this chapter, you need first to load and attach some packages from the library (see section 6.4 on page 179 for details on the use of packages).

```
library(learnrbook)
library(tibble)
library(magrittr)
library(wrapr)
library(stringr)
library(dplyr)
library(tidyr)
library(lubridate)
```

8.4 Replacements for `data.frame`

8.4.1 Package 'data.table'

The function call semantics of the R language is that arguments are passed to functions by copy (see section 6.2 on page 169). Functions and methods from package 'data.table' pass arguments by reference, avoiding making copies. In base R, if the arguments are modified within the code of a function, these changes are local to the function. However, any assignments within the functions and methods defined in package 'data.table' modify the variables passed as their arguments.

If implemented naively, the copy semantics used in base R would impose a huge toll on performance. However, R in most situations only makes a copy in memory if and when the value changes. Consequently, for modern versions of R, which are good at avoiding unnecessary copying of objects, the normal R semantics has only a moderate negative impact on performance. However, this impact can still be a problem as modification is detected at the object level, and consequently, R can make copies of large objects such as a whole data frame when only values in a single column or even just an attribute have changed.

Passing arguments by reference, as in 'data.table', simplifies the needed tests for delayed copying and by avoiding the need to copy arguments, achieves the best possible performance. This is a specialised package but extremely useful when dealing with very large data sets. Writing user code, such as scripts, with 'data.table' requires a good understanding of the pass-by-reference semantics. Obviously, package 'data.table' makes no attempt at backwards compatibility with base-R `data.frame`.

In contrast to the design of package 'data.table', the focus of the 'tidyverse' is not only performance. The design of this grammar has also considered usability. Design compromises have been resolved differently than in base R or 'data.table' and in some cases code written using base R can significantly outperform the 'tidyverse' and vice versa. There exist packages that implement a translation layer from the syntax of the 'tidyverse' into that of 'data.table' or relational database queries.

8.4.2 Package 'tibble'

Package 'tibble' aimed at enhanced performance, like 'data.table', but not at the expense of usability. The `tibble()` constructor supports semantics that allow more concise code compared to the `data.frame()` constructor. The `print()` method for tibbles displays them concisely and provides additional information. With small data sets, differences in performance are in most cases irrelevant. Early on, package `tibble()` was consistently faster than base R data frames, but the performance of R has improved over the years. Nowadays, there is no clear winner. The decision to use package `tibble()` depends mostly on whether one uses the other packages from the 'tidyverse', mainly 'dplyr' and 'tidyr', or not.

The authors of package 'tibble' describe their `tbl` class as nearly backwards compatible with data.frame and make it a derived class. This backwards compatibility is only partial so in some situations data frames and tibbles are not equivalent.

The class and methods that package 'tibble' defines lift some of the restrictions imposed by the design of base R data frames at the cost of creating some incompatibilities due to changed (improved) syntax for member extraction. Tibbles simplify the creation of "columns" of class list and remove support for columns of class matrix. Handling of attributes is also different, with no row names added by default. There are also differences in default behaviour of both constructors and methods.

Although, objects of class tbl can be passed as arguments to functions that expect data frames as input, these functions are not guaranteed to work correctly with tibbles as a result of the differences in behaviour of some methods and operators.

It is easy to write code that will work correctly both with data frames and tibbles by avoiding constructs that behave differently. However, code that is syntactically correct according to the R language may fail to work as expected if a tibble is used in place of a data frame. Only functions tested to work correctly with both tibbles and data frames can be relied upon as compatible.

That it has been possible to define tibbles as objects of a class derived from data.frame reveals one of the drawbacks of the simple implementation of S3 object classes in R. Allowing this is problematic because the promise of compatibility implicit in a derived class is not always fulfilled. An independently developed method designed for data frames will not necessarily work correctly with tibbles, but in the absence of a specialised method for tibbles it will be used (dispatched) when the generic method is called with a tibble as argument.

One should be aware that although the constructor tibble() and conversion function as_tibble(), as well as the test is_tibble() use the name tibble, the class attribute is named tbl. This is inconsistent with base R conventions, as it is the use of an underscore instead of a dot in the name of these methods.

```
my.tb <- tibble(numbers = 1:3)
is_tibble(my.tb)
## [1] TRUE
inherits(my.tb, "tibble")
## [1] FALSE
class(my.tb)
## [1] "tbl_df"     "tbl"        "data.frame"
```

Furthermore, to support tibbles based on different underlying data sources such as data.table objects or databases, a further derived class is needed. In our example, as our tibble has an underlying data.frame class, the most derived class of my.tb is tbl_df.

Function show_classes(), defined below, concisely reports the class of the object passed as an argument and of its members (*apply* functions are described in section 5.8 on page 154).

```
show_classes <- function(x) {
  cat(
    paste(paste(class(x)[1],
```

```
    "containing:"),
    paste(names(x),
          sapply(x, class), collapse = ", ", sep = ": "),
    sep = "\n")
    )
}
```

The `tibble()` constructor by default does not convert character data into factors, while the `data.frame()` constructor did before R version 4.0.0. The default can be overridden through an argument passed to these constructors, and in the case of `data.frame()` also by setting an R option. This new behaviour extends to function `read.table()` and its wrappers (see section 10.6 on page 388).

```
my.df <- data.frame(codes = c("A", "B", "C"), numbers = -1:1, integers = 1L:3L)
is.data.frame(my.df)
## [1] TRUE
is_tibble(my.df)
## [1] FALSE
show_classes(my.df)
## data.frame containing:
## codes: character, numbers: integer, integers: integer
```

Tibbles are, or pretend to be (see above), data frames—or more formally class `tibble` is derived from class `data.frame`. However, data frames are not tibbles.

```
my.tb <- tibble(codes = c("A", "B", "C"), numbers = -1:1, integers = 1L:3L)
is.data.frame(my.tb)
## [1] TRUE
is_tibble(my.tb)
## [1] TRUE
show_classes(my.tb)
## tbl_df containing:
## codes: character, numbers: integer, integers: integer
```

The `print()` method for tibbles differs from that for data frames in that it outputs a header with the text "A tibble:" followed by the dimensions (number of rows × number of columns), adds under each column name an abbreviation of its class and instead of printing all rows and columns, a limited number of them are displayed. In addition, individual values are formatted more compactly and using colour to highlight, for example, negative numbers in red.

```
print(my.df)
##   codes numbers integers
## 1     A      -1        1
## 2     B       0        2
## 3     C       1        3
print(my.tb)
## # A tibble: 3 x 3
##   codes numbers integers
##   <chr>   <int>    <int>
## 1 A          -1        1
## 2 B           0        2
## 3 C           1        3
```

The default number of rows printed depends on R option `tibble.print_max` that can be set with a call to `options()`. This option plays for tibbles a similar role as option `max.print` plays for base R `print()` methods.

```
options(tibble.print_max = 3, tibble.print_min = 3)
```

8.1 Print methods for tibbles and data frames differ in their behaviour when not all columns fit in a printed line. 1) Construct a data frame and an equivalent tibble with at least 50 rows and then test how the output looks when they are printed. 2) Construct a data frame and an equivalent tibble with more columns than will fit in the width of the R console and then test how the output looks when they are printed.

Data frames can be converted into tibbles with `as_tibble()`.

```
my_conv.tb <- as_tibble(my.df)
is.data.frame(my_conv.tb)
## [1] TRUE
is_tibble(my_conv.tb)
## [1] TRUE
show_classes(my_conv.tb)
## tbl_df containing:
## codes: character, numbers: integer, integers: integer
```

Tibbles can be converted into “real” data.frames with `as.data.frame()`.

```
my_conv.df <- as.data.frame(my.tb)
is.data.frame(my_conv.df)
## [1] TRUE
is_tibble(my_conv.df)
## [1] FALSE
show_classes(my_conv.df)
## data.frame containing:
## codes: character, numbers: integer, integers: integer
```

When dealing with tibbles, column- and row binding should be done with functions `bind_rows()` and `bind_cols()` from ‘dplyr’, not with functions `rbind()` and `cbind()` from R. See explanation below.

Not all conversion functions work consistently when converting from a derived class into its parent. The reason for this is disagreement between authors on what the *correct* behaviour is based on logic and theory. You are not likely to be hit by this problem frequently, but it can be difficult to diagnose.

We have already seen that calling `as.data.frame()` on a tibble strips the derived class attributes, returning a data frame. We will look at the whole character vector stored in the `"class"` attribute to demonstrate the difference. We also test the two objects for equality, in two different ways. Using the operator `==` tests for equivalent objects. Objects that contain the same data. Using `identical()` tests that objects are exactly the same, including attributes such as `"class"`, which we retrieve using `class()`.

```
class(my.tb)
## [1] "tbl_df"     "tbl"        "data.frame"
class(my_conv.df)
## [1] "data.frame"
my.tb == my_conv.df
##      codes numbers integers
## [1,]  TRUE    TRUE     TRUE
## [2,]  TRUE    TRUE     TRUE
## [3,]  TRUE    TRUE     TRUE
identical(my.tb, my_conv.df)
## [1] FALSE
```

Now we derive from a tibble, and then attempt a conversion back into a tibble.

```
my.xtb <- my.tb
class(my.xtb) <- c("xtb", class(my.xtb))
class(my.xtb)
## [1] "xtb"        "tbl_df"     "tbl"        "data.frame"
my_conv_x.tb <- as_tibble(my.xtb)
class(my_conv_x.tb)
## [1] "tbl_df"     "tbl"        "data.frame"
my.xtb == my_conv_x.tb
##      codes numbers integers
## [1,]  TRUE    TRUE     TRUE
## [2,]  TRUE    TRUE     TRUE
## [3,]  TRUE    TRUE     TRUE
identical(my.xtb, my_conv_x.tb)
## [1] FALSE
```

The two viewpoints on conversion functions are as follows. If the argument passed to a conversion function is an object of a derived class, 1) it should be returned after stripping the derived class, or 2) it should be returned as is, without stripping the derived class. Base `R` follows, as far as I have been able to work out, approach 1). Some packages in the 'tidyverse' sometimes follow, or have followed in the past, approach 2). If in doubt about the behaviour of some function, then you will need to do a test similar to the one used above.

As tibbles have been defined as a class derived from `data.frame`, if methods have not been explicitly defined for tibbles, the methods defined for data frames are called, and these are likely to return a data frame rather than a tibble. Even a frequent operation like column binding is affected, at least at the time of writing.

```
class(my.df)
## [1] "data.frame"
class(my.tb)
## [1] "tbl_df"     "tbl"        "data.frame"

class(cbind(my.df, my.tb))
## [1] "data.frame"
class(cbind(my.tb, my.df))
## [1] "data.frame"
```

```
class(cbind(my.df, added = -3:-1))
## [1] "data.frame"
class(cbind(my.tb, added = -3:-1))
## [1] "data.frame"
identical(cbind(my.tb, added = -3:-1), cbind(my.df, added = -3:-1))
## [1] TRUE
```

There are additional important differences between the constructors `tibble()` and `data.frame()`. One of them is that in a call to `tibble()`, member variables ("columns") being defined can be used in the definition of subsequent member variables.

```
tibble(a = 1:5, b = 5:1, c = a + b, d = letters[a + 1])
## # A tibble: 5 x 4
##       a     b     c d
##   <int> <int> <int> <chr>
## 1     1     5     6 b
## 2     2     4     6 c
## 3     3     3     6 d
## # i 2 more rows
```

8.2 What is the behaviour if you replace `tibble()` by `data.frame()` in the statement above?

While objects passed directly as arguments to the `data.frame()` constructor to be included as "columns" can be factors, vectors or matrices (with the same number of rows as the data frame), arguments passed to the `tibble()` constructor can be factors, vectors or lists (with the same number of members as rows in the tibble). As we saw in section 4.4 on page 94, base R's data frames can contain columns of classes `list` and `matrix`. The difference is in the need to use `I()`, the identity function, to protect these variables during construction and assignment to true `data.frame` objects as otherwise list members and matrix columns will be assigned to multiple individual columns in the data frame.

```
tibble(a = 1:5, b = 5:1, c = list("a", 2, 3, 4, 5))
## # A tibble: 5 x 3
##       a     b c
##   <int> <int> <list>
## 1     1     5 <chr [1]>
## 2     2     4 <dbl [1]>
## 3     3     3 <dbl [1]>
## # i 2 more rows
```

A list of lists or a list of vectors can be directly passed to the constructor.

```
tibble(a = 1:5, b = 5:1, c = list("a", 1:2, 0:3, letters[1:3], letters[3:1]))
## # A tibble: 5 x 3
##       a     b c
##   <int> <int> <list>
## 1     1     5 <chr [1]>
## 2     2     4 <int [2]>
## 3     3     3 <int [4]>
## # i 2 more rows
```

8.5 Data Pipes

The first obvious difference between scripts using 'tidyverse' packages is the frequent use of *pipes*. This is, however, mostly a question of preferences, as pipes can be as well used with base R functions. In addition, since version 4.0.0, R has a native pipe operator |>, described in section 5.5 on page 134. Here I describe other earlier implementations of pipes, and the differences among these and R's pipe operator.

8.5.1 'magrittr'

A set of operators for constructing pipes of R functions is implemented in package 'magrittr'. It preceded the native R pipe by several years. The pipe operator defined in package 'magrittr', %>%, is imported and re-exported by package 'dplyr', which in turn defines functions that work well in data pipes.

Operator %>% plays a similar role as R's |>.

```
data.in <- 1:10

data.in %>% sqrt() %>% sum() -> data0.out
```

The value passed can be made explicit using a dot as placeholder passed as an argument by name and by position to the function on the *rhs* of the %>% operator. Thus . in 'magrittr' plays a similar but not identical role as _ in base R pipes.

```
data.in %>% sqrt(x = .) %>% sum(.) -> data1.out
all.equal(data0.out, data1.out)
## [1] TRUE
```

R's native pipe operator requires, consistently with R in all other situations, that functions that are to be evaluated use the parenthesis syntax, while 'magrittr' allows the parentheses to be missing when the piped argument is the only one passed to the function call on *rhs*.

```
data.in %>% sqrt %>% sum -> data5.out
all.equal(data0.out, data5.out)
## [1] TRUE
```

Package 'magrittr' provides additional pipe operators, such as "tee" (%T>%) to create a branch in the pipe, and %<>% to apply the pipe by reference. These operators are much less frequently used than %>%.

8.5.2 'wrapr'

The %.>%, or "dot-pipe", operator from package 'wrapr', allows expressions both on the rhs and lhs, and *enforces the use of the dot* (.), as placeholder for the piped object. Given the popularity of 'dplyr' the pipe operator from 'magrittr' has been the most used.

Rewritten using the dot-pipe operator, the pipe in the previous chunk becomes

```
data.in %.>% sqrt(.) %.>% sum(.) -> data2.out
all.equal(data0.out, data2.out)
```

```
## [1] TRUE
```

However, as operator %>% from 'magrittr' recognises the . placeholder without enforcing its use, the code below where %.>% is replaced by %>% returns the same value as that above.

```
data.in %>% sqrt(.) %>% sum(.) -> data3.out
all.equal(data0.out, data3.out)
## [1] TRUE
```

To use operator |> from R, we need to edit the code using (_) as placeholder and passing it as an argument to parameters by name in the function calls on the *rhs*.

```
data.in |> sqrt(x = _) |> sum(x = _) -> data4.out
all.equal(data0.out, data4.out)
## [1] TRUE
```

We can, in this case, simply use no placeholder, and pass the arguments by position to the first parameter of the functions.

```
data.in |> sqrt() |> sum() -> data4.out
all.equal(data0.out, data4.out)
## [1] TRUE
```

The dot-pipe operator %.>% from 'wrapr' allows us to use the placeholder . in expressions on the *rhs* of operators in addition to in function calls.

```
data.in %.>% (.^2) -> data7.out
```

In contrast, operator %>% does not support expressions, only function call syntax on the *rhs*, forcing calling of operators with parenthesis syntax

```
data.in %>% `^`(e1 = ., e2 = 2) -> data9.out
all.equal(data7.out, data9.out)
## [1] TRUE
```

In conclusion, R syntax for expressions is preserved when using the dot-pipe operator from 'wrapr', with the only caveat that because of the higher precedence of the %.>% operator, we need to "protect" bare expressions containing other operators by enclosing them in parentheses. In the examples above, we showed a simple expression so that it could be easily converted into a function call. The %.>% operator supports also more complex expressions, even with multiple uses of the placeholder.

```
data.in %.>% (.^2 + sqrt(. + 1))
##  [1]   2.414214   5.732051  11.000000  18.236068  27.449490  38.645751
##  [7]  51.828427  67.000000  84.162278 103.316625
```

8.5.3 Comparing pipes

Under-the-hood, the implementations of operators |>, %>% and %.>% are different, with |> expected to have the best performance, followed by %.>% and %>% being slowest. As implementations evolve, performance may vary among versions. However, |> being part of R is likely to remain the fastest.

Being part of the R language, |> will remain available and most likely also backwards compatible, while packages could be abandoned or redesigned by their maintainers. For this reason, it is preferable to use the |> in scripts or code expected to be reused, unless compatibility with R versions earlier than 4.2.0 is needed. Elsewhere in the book I have used R's pipe operator |>.

Pipes can be used with any R function, but how elegant can be their use depends on the order of formal parameters. This is especially the case when passing arguments implicitly to the first parameter of the function on the *rhs*. Several of the functions and methods defined in 'tidyr', 'dplyr', and a few other packages from the 'tidyverse' fit this need.

Writing a series of statements and saving intermediate results in temporary variables makes debugging easiest. Debugging pipes is not as easy, as this usually requires splitting them, with one approach being the insertion of calls to `print()`. This is possible, because `print()` returns its input invisibly in addition to displaying it.

```
data.in |> print() |> sqrt() |> print() |> sum() |> print() -> data10.out
##  [1]  1  2  3  4  5  6  7  8  9 10
##  [1] 1.000000 1.414214 1.732051 2.000000 2.236068 2.449490 2.645751 2.828427
##  [9] 3.000000 3.162278
## [1] 22.46828
data10.out
## [1] 22.46828
```

Debugging nested function calls is difficult. So, in general, it is preferable to use pipes instead of deeply nested function calls. However, it is best to avoid very long pipes. Normally, while writing scripts or analysing data it is important to check the correctness of intermediate results, so saving them to variables can save time and effort.

The design of R's native pipes has benefited from the experience gathered by earlier implementations and being now part of the language, we can expect it to become the reference one once its implementation is stable. The designers of the three implementations have to some extent disagreed in their design decisions. Consequently, some differences are more than aesthetic. R pipes are simpler, easier to use and expected to be fastest. Those from 'magrittr' are the most feature rich, but not as safe to use, and purportedly given a more complex implementation, the slowest. Package 'wrapr' is an attempt to enhance pipes compared to 'magrittr' focusing in syntactic simplicity and performance. R's |> operator has been enhanced since its addition in R only two years ago. These enhancements have all been backwards compatible.

The syntax of operators |> and %>% is not identical. With R's |>, (as of R 4.3.0) the placeholder _ can be only passed to parameters by name, while with operator %>% from 'magrittr' the placeholder `.` can be used to pass arguments both by name and by position. With operator `%.>%` the use of the placeholder `.` is mandatory, and it can be passed by name or by position to the function call on the *rhs*. Other differences are deeper like those related to the use of the extraction operator in the *rhs* or support or not for expressions that are not explicit function calls.

In the case of R, the |> pipe is conceptually a substitution with no alteration of the syntax or evaluation order. This avoids *surprising* the user and simplifies implementation. In other words, R pipes are an alternative way of writing nested function calls. Quoting R documentation:

> Currently, pipe operations are implemented as syntax transformations. So an expression written as `x |> f(y)` is parsed as `f(x, y)`. It is worth emphasising that while the code in a pipeline is written sequentially, regular R semantics for evaluation apply and so piped expressions will be evaluated only when first used in the rhs expression.

While frequently the different pipe operators can substitute for each other by adjusting the syntax, in some cases the differences among them in the order and timing of evaluation of the terms needs to be taken into account.

In some situations, operator %>% from package 'magrittr' can behave unexpectedly. One example is the use of `assign()` in a pipe. With R's operator |> assignment takes place as expected.

```
data.in |> assign(x = "data6.out", value = _)
all.equal(data.in, data6.out)
## [1] TRUE
```

Named arguments are also supported with the dot-pipe operator from 'wrapr'.

```
data.in %.>% assign(x = "data7.out", value = .)
all.equal(data.in, data7.out)
## [1] TRUE
```

However, the pipe operator (%>%) from package 'magrittr' silently and unexpectedly fails to assign the value to the name.

```
data.in %>% assign(x = "data8.out", value = .)
if (exists("data8.out")) {
  all.equal(data.in, data8.out)
} else {
  print("'data8.out' not found!")
}
## [1] "'data8.out' not found!"
```

Although there are usually alternatives to get the computations done correctly, unexpected silent behaviour can be confusing.

8.6 Reshaping with 'tidyr'

Data stored in table-like formats can be arranged in different ways, wide and long (Figure 8.1). In base R, most model fitting functions, and the `plot()` method using (model) formulas, expect data to be arranged in "long form" so that each row in a data frame corresponds to a single observation (or measurement) event on a subject. Each column corresponds to a different measured feature, or ancillary information like the time of measurement, or a factor describing a classification of subjects according to treatments or features of the experimental design (e.g., blocks). Covariates measured on the same subject at an earlier point in time may also be

Subject	Height January	Height February	Height March
A	5	7	9
B	4	7	9
C	6	7	8

Subject	Date	Height
A	January	5
B	January	4
C	January	6
A	February	7
B	February	7
C	February	7
A	March	9
B	March	9
C	March	8

Figure 8.1
Wide (left) and long (right) data formats.

stored in a column. Data arranged in *long form* has been nicknamed as "tidy" and this is reflected in the name given to the 'tidyverse' suite of packages. However, this longitudinal arrangement of data has been the preferred format since the inception of S and R. Data in which columns correspond to measurement events are described as being in a *wide form.*

It is rather frequently used when observations in each subject are repeated in time. In this case, there is one row per subject and one column for each combination of response variable and time of measurement. Real-world data at the time of acquisition are rather frequently stored in the wide format, or even in ad-hoc non-rectangular formats, so in many cases the first task in data analysis is to reshape the data. Package 'tidyr' provides functions for reshaping data from wide to long form and *vice versa.*

Package 'tidyr' replaced package 'reshape2', which in turn replaced package 'reshape', while additionally the functions implemented in 'tidyr' have been recently replaced by new ones with different syntax and name. If a data analyst uses these functions every day, the cost involved is frequently tolerable or even desirable given the improvements. However, for R users in applied fields, to whom this book is targeted, in the long run using function `reshape()` from base R can be better, even when its syntax is not as straightforward (see section 4.5 on page 112). This does not detract from the advantages of using a clear workflow as emphasised by the proponents of the *tidyverse.* Here I only want to emphasise that using some of the packages from the 'tidyverse' as with any software with an evolving user interface can have in some cases a cost that needs to be taken into consideration.

I use in examples below the `iris` data set included in base R. Some operations on R `data.frame` objects with functions and operators from the 'tidyverse' packages will return `data.frame` objects while others will return tibbles—i.e., `"tb"` objects. Consequently, it is safer to first convert into tibbles the data frames we will work with.

```
iris.tb <- as_tibble(iris)
```

Function `pivot_longer()` from 'tidyr' converts data from wide form into long form (or "tidy"). We use it here to obtain a long-form tibble. By comparing `iris.tb` with `long_iris.tb` we can appreciate how `pivot_longer()` reshaped its input.

```
long_iris.tb <-
  pivot_longer(iris.tb,
               cols = -Species,
               names_to = "part",
               values_to = "dimension")
long_iris.tb
## # A tibble: 600 x 3
##   Species part         dimension
##   <fct>   <chr>            <dbl>
## 1 setosa  Sepal.Length       5.1
## 2 setosa  Sepal.Width        3.5
## 3 setosa  Petal.Length       1.4
## # i 597 more rows
```

Differently to base R, in most functions from the 'tidyverse' packages we can use bare column names preceded by a minus sign to signify "all other columns".

Function `pivot_wider()` does not directly implement the exact inverse operation of `pivot_longer()`. With multiple rows with shared codes, i.e., replication, in our case within each species and flower part, the returned tibble has columns that are lists of vectors. We need to expand these columns with function `unnest()` in a second step.

```
wide_iris.tb <-
  pivot_wider(long_iris.tb,
              names_from = "part",
              values_from = "dimension",
              values_fn = list) |>
  unnest(cols = -Species)
wide_iris.tb
## # A tibble: 150 x 5
##   Species Sepal.Length Sepal.Width Petal.Length Petal.Width
##   <fct>          <dbl>       <dbl>        <dbl>       <dbl>
## 1 setosa           5.1         3.5          1.4         0.2
## 2 setosa           4.9         3            1.4         0.2
## 3 setosa           4.7         3.2          1.3         0.2
## # i 147 more rows
```

8.3 Is `wide_iris.tb` equal to `iris.tb`, the tibble we converted into long shape and back into wide shape? Run the comparisons below, and print the tibbles to find out.

```
identical(iris.tb, wide_iris.tb)
all.equal(iris.tb, wide_iris.tb)
all.equal(iris.tb, wide_iris.tb[ , colnames(iris.tb)])
```

What has changed? Would it matter if our code used indexing with a numeric vector to extract columns? or if it used column names as character strings?

Starting from version 1.0.0 of 'tidyr', functions `gather()` and `spread()` are deprecated and replaced by functions `pivot_longer()` and `pivot_wider()`. These new functions, described above, use a different syntax than the old ones.

8.4 Functions `pivot_longer()` and `pivot_wider()` from package 'poorman' attempt to replicate the behaviour of the same name functions from package 'tidyr'. In some edge cases, the behaviour differs. Test if the two code chunks above return identical or equal values when `poorman::` is prepended to the names of these two functions. First, ensure that package 'poorman' is installed, then run the code below.

```
poor_long_iris.tb <-
  poorman::pivot_longer(
    iris,
    cols = -Species,
    names_to = "part",
    values_to = "dimension")
identical(long_iris.tb, poor_long_iris.tb)
all.equal(long_iris.tb, poor_long_iris.tb)
class(long_iris.tb)
class(poor_long_iris.tb)
```

What is the difference between the values returned by the two functions? Could switching from package 'tidyr' to package 'poorman' affect code downstream of pivoting?

8.7 Data Manipulation with 'dplyr'

The first advantage a user of the 'dplyr' functions and methods sees is the completeness of the set of operations supported and the symmetry and consistency among the different functions. A second advantage is that almost all the functions are defined not only for objects of class `tibble`, but also for objects of class `data.table` (package 'dtplyr') and for SQL databases (package 'dbplyr'), with consistent syntax (see also section 10.13 on page 412). As discussed above, using a code base that is not yet fully stable has a cost that needs to be balanced against the gain obtained from its use.

8.7.1 Row-wise manipulations

Assuming that the data are stored in long form, row-wise operations are operations combining values from the same observation event—i.e., calculations within a single row of a data frame or tibble (see section 4.4.5 on page 110 for the base R approach). Using functions `mutate()` and `transmute()` we can obtain derived quantities by combining different variables, or variables and constants, or applying a mathematical transformation. We add new variables (columns) retaining existing ones using `mutate()` or we assemble a new tibble containing only the columns we explicitly specify using `transmute()`.

Different from usual R syntax, with `tibble()`, `mutate()` and `transmute()` we can use values passed as arguments, in the statements computing the values passed as later arguments. In many cases, this allows more concise and easier to understand code.

```
tibble(a = 1:5, b = 2 * a)
## # A tibble: 5 x 2
##       a     b
##   <int> <dbl>
## 1     1     2
## 2     2     4
## 3     3     6
## # i 2 more rows
```

Continuing with the example from the previous section, a likely next step would be to split the values in variable `part` into `plant_part` and `part_dim`. We use `mutate()` from 'dplyr' and `str_extract()` from 'stringr' (a package included in the 'tidyverse', aimed at the manipulation of character strings). We use regular expressions (see section 3.4 on page 46) as arguments passed to `pattern`. We do not show it here, but `mutate()` can be used with variables of any `mode`, and calculations can involve values from several columns. It is even possible to operate on values applying a lag or, in other words, using rows displaced relative to the current one.

```
long_iris.tb |>
  mutate(plant_part = str_extract(part, "^[:alpha:]*"),
         part_dimension = str_extract(part, "[:alpha:]*$")) -> long_iris.tb
long_iris.tb
## # A tibble: 600 x 5
##   Species part         dimension plant_part part_dimension
##   <fct>   <chr>            <dbl> <chr>      <chr>
## 1 setosa  Sepal.Length       5.1 Sepal      Length
## 2 setosa  Sepal.Width        3.5 Sepal      Width
## 3 setosa  Petal.Length       1.4 Petal      Length
## # i 597 more rows
```

In the next few chunks, returned values are displayed, while in normal use they would assigned to variables or passed to the next function in a pipe using `|>`. Function `arrange()` is used to sort rows—it makes sorting a data frame or tibble simpler than when using `sort()` or `order()`. Below, `long_iris.tb` rows are sorted based on the values in three of its columns.

```
arrange(long_iris.tb, Species, plant_part, part_dimension)
## # A tibble: 600 x 5
##   Species part         dimension plant_part part_dimension
##   <fct>   <chr>            <dbl> <chr>      <chr>
## 1 setosa  Petal.Length       1.4 Petal      Length
## 2 setosa  Petal.Length       1.4 Petal      Length
## 3 setosa  Petal.Length       1.3 Petal      Length
## # i 597 more rows
```

Function `filter()` can be used to extract a subset of rows—similar to `subset()` but with a syntax consistent with that of other functions in the 'tidyverse'. In this case, 300 out of the original 600 rows are retained.

```
filter(long_iris.tb, plant_part == "Petal")
## # A tibble: 300 x 5
##   Species part         dimension plant_part part_dimension
##   <fct>   <chr>            <dbl> <chr>      <chr>
## 1 setosa  Petal.Length       1.4 Petal      Length
## 2 setosa  Petal.Width        0.2 Petal      Width
## 3 setosa  Petal.Length       1.4 Petal      Length
## # i 297 more rows
```

Function `slice()` can be used to extract a subset of rows based on their positions—an operation that in base R would use positional (numeric) indexes with the `[ , ]` operator: `long_iris.tb[1:5, ]`.

```
slice(long_iris.tb, 1:5)
## # A tibble: 5 x 5
##   Species part         dimension plant_part part_dimension
##   <fct>   <chr>            <dbl> <chr>      <chr>
## 1 setosa  Sepal.Length       5.1 Sepal      Length
## 2 setosa  Sepal.Width        3.5 Sepal      Width
## 3 setosa  Petal.Length       1.4 Petal      Length
## # i 2 more rows
```

Function `select()` can be used to extract a subset of columns—this would be done with positional (numeric) indexes with `[ , ]`, passing them to the second argument as numeric indexes or column names in a vector. It is also possible to use function `subset()` from base R (see section 4.4.1 on page 102). Negative indexes in base R can only be numeric, while `select()` accepts bare column names prepended with a minus for exclusion.

```
select(long_iris.tb, -part)
## # A tibble: 600 x 4
##   Species dimension plant_part part_dimension
##   <fct>       <dbl> <chr>      <chr>
## 1 setosa        5.1 Sepal      Length
## 2 setosa        3.5 Sepal      Width
## 3 setosa        1.4 Petal      Length
## # i 597 more rows
```

In addition, `select()` as other functions in 'dplyr' accepts "selectors" returned by functions `starts_with()`, `ends_with()`, `contains()`, and `matches()` to extract or retain columns. For this example, we use the wide-shaped `iris.tb` instead of `long_iris.tb`.

```
select(iris.tb, -starts_with("Sepal"))
## # A tibble: 150 x 3
##   Petal.Length Petal.Width Species
##          <dbl>       <dbl> <fct>
## 1          1.4         0.2 setosa
## 2          1.4         0.2 setosa
## 3          1.3         0.2 setosa
## # i 147 more rows
```

```
select(iris.tb, Species, matches("pal"))
## # A tibble: 150 x 3
##   Species Sepal.Length Sepal.Width
##   <fct>          <dbl>       <dbl>
## 1 setosa           5.1         3.5
## 2 setosa           4.9         3
## 3 setosa           4.7         3.2
## # i 147 more rows
```

Function `rename()` can be used to rename columns, whereas base R requires the use of both `names()` and `names()<-` and *ad hoc* code to match new and old names. As shown below, the syntax for each column name to be changed is `<new name> = <old name>`. The two names can be given either as bare names as below or as character strings.

```
long_iris.tb |>
select(-part) |>
rename(part = plant_part, size = dimension, dimension = part_dimension)
## # A tibble: 600 x 4
##   Species  size part  dimension
##   <fct>   <dbl> <chr> <chr>
## 1 setosa    5.1 Sepal Length
## 2 setosa    3.5 Sepal Width
## 3 setosa    1.4 Petal Length
## # i 597 more rows
```

Several of the functions described in this section were needed because operator `%>%`) from package 'magrittr' did not support the use of the extraction operators in the rhs using operator syntax. Operator `|>` starting from R version 4.3.0 does not have this limitation (see section 5.5 on page 134), however, the functions from 'dplyr' remain useful as they allow more concise and clear coding of complex conditions.

8.7.2 Group-wise manipulations

Another important operation is to summarise quantities by groups of rows. Contrary to base R, the grammar of data manipulation as implemented in 'dplyr', makes it possible to split this operation into two steps: the setting of the grouping, and the calculation of summaries. This simplifies the code, making it more easily understandable when using pipes compared to the approach of base R `aggregate()` (see section 4.4.2 on page 105).

In early 2023, package 'dplyr' version 1.1.0 added support for per-operation grouping by adding to functions a new parameter (`by` or `.by`). This is still considered an experimental feature that may change. Anyway, it is important to keep in mind that this new approach to grouping is not persistent like that described above. Depending on the circumstances, persistence can simplify the code but also create bugs when not taken into account.

When using persistent grouping, the first step is to use `group_by()` to "tag" a tibble with the grouping. We create a *tibble* and then convert it into a *grouped tibble.* Once we have a grouped tibble, function `summarise()` will recognise the grouping and use it when the summary values are calculated.

```
tibble(numbers = 1:9, Letters = rep(letters[1:3], 3)) |>
  group_by(Letters) |>
  summarise(mean_num = mean(numbers),
            median_num = median(numbers),
            n = n()) |>
  ungroup() # not always needed but safer
## # A tibble: 3 x 4
##   Letters mean_num median_num     n
##   <chr>      <dbl>      <int> <int>
## 1 a              4          4     3
## 2 b              5          5     3
## 3 c              6          6     3
```

In the non-persistent grouping approach, we specify the grouping in the call to `summarise()` (this new feature is labelled as experimental in 'dplyr' version 1.1.3, and may change in future versions).

```
tibble(numbers = 1:9, Letters = rep(letters[1:3], 3)) |>
  summarise(.by = Letters,
            mean_num = mean(numbers),
            median_num = median(numbers),
            n = n())
## # A tibble: 3 x 4
##   Letters mean_num median_num     n
##   <chr>      <dbl>      <int> <int>
## 1 a              4          4     3
## 2 b              5          5     3
## 3 c              6          6     3
```

How is grouping implemented for data frames and tibbles? Best way to find out is to explore how a grouped tibble differs from one that is not grouped.

Tibble `my.tb` is not grouped.

```
my.tb <- tibble(numbers = 1:9, Letters = rep(letters[1:3], 3))
is.grouped_df(my.tb)
## [1] FALSE
class(my.tb)
## [1] "tbl_df"     "tbl"        "data.frame"
names(attributes(my.tb))
## [1] "class"     "row.names" "names"
```

Tibble `my_gr.tb` is grouped by variable, or column, `Letters`. In this case, as our tibble belongs to class `tibble_df`, grouping adds `grouped_df` as the most derived class.

```
my_gr.tb <- group_by(.data = my.tb, Letters)
is.grouped_df(my_gr.tb)
## [1] TRUE
class(my_gr.tb)
## [1] "grouped_df" "tbl_df"     "tbl"        "data.frame"
```

Grouping also adds several attributes with the grouping information in a format suitable for fast selection of group members.

```
names(attributes(my_gr.tb))
## [1] "class"     "row.names" "names"     "groups"
setdiff(attributes(my_gr.tb), attributes(my.tb))
## $class
## [1] "grouped_df" "tbl_df"     "tbl"        "data.frame"
##
## $groups
## # A tibble: 3 x 2
##   Letters       .rows
##   <chr>   <list<int>>
## 1 a               [3]
## 2 b               [3]
## 3 c               [3]
```

A call to `ungroup()` removes the grouping, thereby restoring the original tibble.

```
my_ugr.tb <- ungroup(my_gr.tb)
class(my_ugr.tb)
## [1] "tbl_df"     "tbl"        "data.frame"
names(attributes(my_ugr.tb))
## [1] "class"     "row.names" "names"

all(my.tb == my_gr.tb)
## [1] TRUE
all(my.tb == my_ugr.tb)
## [1] TRUE
identical(my.tb, my_gr.tb)
## [1] FALSE
identical(my.tb, my_ugr.tb)
## [1] TRUE
```

The tests above show that members are in all cases the same as operator == tests for equality at each position in the tibble but not the attributes, while attributes, including `class`, differ between normal tibbles and grouped ones and so they are not *identical* objects.

If we replace `tibble` by `data.frame` in the first statement, and rerun the chunk, the result of the last statement in the chunk is `FALSE` instead of `TRUE`. At the time of writing starting with a `data.frame` object, applying grouping with `group_by()` followed by ungrouping with `ungroup()` has the side effect of converting the data frame into a tibble. This is something to be very much aware of, as there are differences in how the extraction operator `[ , ]` behaves in the two cases. The safe way to write code making use of functions from 'dplyr' and 'tidyr' is to always make sure that subsequent code works correctly with tibbles in addition to with data frames.

8.7.3 Joins

Joins allow us to combine two data sources which share some variables. Variables in common are used to match the corresponding rows before "joining" variables (i.e., columns) from both sources together. There are several *join* functions in 'dplyr'. They differ mainly in how they handle rows that do not have a match between data sources.

We create here some artificial data to demonstrate the use of these functions. We will create two small tibbles, with one column in common and one mismatched row in each.

```
first.tb <- tibble(idx = c(1:4, 5), values1 = "a")
second.tb <- tibble(idx = c(1:4, 6), values2 = "b")
```

Below, we apply the functions exported by 'dplyr': `full_join()`, `left_join()`, `right_join()` and `inner_join()`. These functions always retain all columns, and in case of multiple matches, they keep a row for each matching combination of rows. We repeat each of these examples with the arguments passed to `x` and `y` swapped to show the differences in the behaviour of these functions.

A full join retains all unmatched rows filling missing values with `NA`. By default, the match is done on columns with the same name in `x` and `y`, but this can be changed by passing an argument to parameter by. Using by one can base the match on columns that have different names in x and y, or prevent matching of columns with the same name in x and y (example at end of the section).

```
full_join(x = first.tb, y = second.tb)

## Joining with `by = join_by(idx)`
## # A tibble: 6 x 3
##     idx values1 values2
##   <dbl> <chr>   <chr>
## 1     1 a       b
## 2     2 a       b
## 3     3 a       b
## 4     4 a       b
## 5     5 a       <NA>
## 6     6 <NA>    b
```

```
full_join(x = second.tb, y = first.tb)

## Joining with `by = join_by(idx)`
## # A tibble: 6 x 3
##     idx values2 values1
##   <dbl> <chr>   <chr>
## 1     1 b       a
## 2     2 b       a
## 3     3 b       a
## 4     4 b       a
## 5     6 b       <NA>
## 6     5 <NA>    a
```

Left and right joins retain rows not matched from only one of the two data sources, x and y, respectively.

```
left_join(x = first.tb, y = second.tb)

## Joining with `by = join_by(idx)`
## # A tibble: 5 x 3
##     idx values1 values2
##   <dbl> <chr>   <chr>
## 1     1 a       b
## 2     2 a       b
## 3     3 a       b
## 4     4 a       b
## 5     5 a       <NA>
```

```
left_join(x = second.tb, y = first.tb)

## Joining with `by = join_by(idx)`
## # A tibble: 5 x 3
##     idx values2 values1
##   <dbl> <chr>   <chr>
## 1     1 b       a
## 2     2 b       a
## 3     3 b       a
## 4     4 b       a
## 5     6 b       <NA>
```

```
right_join(x = first.tb, y = second.tb)

## Joining with `by = join_by(idx)`
```

```
## # A tibble: 5 x 3
##       idx values1 values2
##     <dbl> <chr>   <chr>
## 1       1 a       b
## 2       2 a       b
## 3       3 a       b
## 4       4 a       b
## 5       6 <NA>    b
```

```
right_join(x = second.tb, y = first.tb)
```

```
## Joining with `by = join_by(idx)`
## # A tibble: 5 x 3
##       idx values2 values1
##     <dbl> <chr>   <chr>
## 1       1 b       a
## 2       2 b       a
## 3       3 b       a
## 4       4 b       a
## 5       5 <NA>    a
```

An inner join discards rows in x that do not match rows in y and *vice versa*.

```
inner_join(x = first.tb, y = second.tb)
```

```
## Joining with `by = join_by(idx)`
## # A tibble: 4 x 3
##       idx values1 values2
##     <dbl> <chr>   <chr>
## 1       1 a       b
## 2       2 a       b
## 3       3 a       b
## 4       4 a       b
```

```
inner_join(x = second.tb, y = first.tb)
```

```
## Joining with `by = join_by(idx)`
## # A tibble: 4 x 3
##       idx values2 values1
##     <dbl> <chr>   <chr>
## 1       1 b       a
## 2       2 b       a
## 3       3 b       a
## 4       4 b       a
```

Next we apply the *filtering join* functions exported by 'dplyr': `semi_join()` and `anti_join()`. These functions only return a tibble that contains only the columns from x, retaining rows based on their match to rows in y.

A semi join retains rows from x that have a match in y.

```
semi_join(x = first.tb, y = second.tb)
```

```
## Joining with `by = join_by(idx)`
## # A tibble: 4 x 2
##       idx values1
##     <dbl> <chr>
## 1       1 a
## 2       2 a
## 3       3 a
## 4       4 a
```

```
semi_join(x = second.tb, y = first.tb)

## Joining with `by = join_by(idx)`
## # A tibble: 4 x 2
##     idx values2
##   <dbl> <chr>
## 1     1 b
## 2     2 b
## 3     3 b
## 4     4 b
```

A anti-join retains rows from `x` that do not have a match in `y`.

```
anti_join(x = first.tb, y = second.tb)

## Joining with `by = join_by(idx)`
## # A tibble: 1 x 2
##     idx values1
##   <dbl> <chr>
## 1     5 a

anti_join(x = second.tb, y = first.tb)

## Joining with `by = join_by(idx)`
## # A tibble: 1 x 2
##     idx values2
##   <dbl> <chr>
## 1     6 b
```

We here rename column `idx` in `first.tb` to demonstrate the use of `by` to specify which columns should be searched for matches.

```
first2.tb <- rename(first.tb, idx2 = idx)
full_join(x = first2.tb, y = second.tb, by = c("idx2" = "idx"))
## # A tibble: 6 x 3
##    idx2 values1 values2
##   <dbl> <chr>   <chr>
## 1     1 a       b
## 2     2 a       b
## 3     3 a       b
## 4     4 a       b
## 5     5 a       <NA>
## 6     6 <NA>    b
```

8.8 Times and Dates with 'lubridate'

In R and many other computing languages, time values are stored as integer values subject to special interpretation. In R, times are most frequently stored as objects of class `POSIXct` or `POSIXlt`. Package 'lubridate' makes working with dates and times in R much easier.

When dealing with time values, first of all, it is necessary to distinguish universal time coordinates (UTC) and local time coordinates. An instant in time is an absolute value and can be unambiguously described using UTC. Local times are different representations of a given instant in time, using local time coordinates such as CET (central European Time). The relationship between UTC and local times depends on country legislation, national borders, and in some cases, time zones within countries. In addition, many countries make use a seasonal shift in the local time coordinates, the so called "summer time". The dates on which these seasonal shifts are implemented depends on the country or region, and these dates have varied over time. Shifts in local time create gaps and overlaps: some local time values correspond to two different time instants, and the skipped ones do not exist and when encountered should be handled as errors.

Different systems are in use to describe time zones and the corresponding time coordinates. One commonly used is based on three or four letter codes, e.g., EET for Eastern European Time. Another commonly used one is based on the names of continents and cities, e.g., Europe/Helsinki. A third one in common use is simply expressed as an offset in hours, e.g., UTC+3. Most time zones have time shifts of whole hours and few to half hours. To some extent, what names are recognised depends on the operating system under which R is running. See `https://en.wikipedia.org/wiki/List_of_tz_database_time_zones` for a list.

Periodical adjustments introduced by leap years, and even leap seconds need to be taken into account when computing time durations between instants in time, even when using UTC. When carrying out arithmetic operations on dates and times, all these "irregularities" have to be accounted for. The functions and operators from package 'lubridate' implement the necessary corrections for current and historical times.

Times and dates written as text are formatted rather inconsistently depending on the customs of different cultures and languages. Package 'lubridate' also provides functions implementing conversions between character strings and times or dates and back. These `character` to time conversions are based on patterns, and are, in general, reliable if the correct pattern is used. Package 'anytime' defines functions that can decode a broad range of formats, but relying on them can be risky, as not all possible formats are correctly decoded.

Objects of class `POSIXlt`, the class used in R to store dates and times in a partly formatted form, do not necessarily contain time zone information. In many cases, when used in computations `POSIXlt` values are interpreted based on the locale settings under which R is running, e.g., the time zone settings of the computer. Objects of class `Date` do not keep track of the time zone, so do not represent instants in time traceable to UTC.

Whenever possible, it is best to store time data and also dates encoded using UTC as `POSIXct` objects. This eliminates uncertainties that can cause otherwise major difficulties in computations.

`POSIXct` objects are of mode numeric, and thus vectors; because of this, they can be stored as columns in data frames and tibbles. Some statistical functions and even some model fitting functions accept them as input.

Current date can be easily queried, and the returned value is fetched from the computer's clock.

```
this.day <- today()
class(this.day)
## [1] "Date"
as.POSIXct(this.day, tz = "") # local time zone
## [1] "2024-02-17 02:00:00 EET"
```

Similarly, the current instant in time can be retrieved. Resolution is in the order of milliseconds.

```
this.instant <- now()
class(this.instant)
## [1] "POSIXct" "POSIXt"
this.instant
## [1] "2024-02-17 22:35:30 EET"
```

Conversion from character strings to `POSIXct` is straightforward as long as all character strings to be converted have the same or very similarly formatted. A family of functions from 'lubridate' with names like `dmy_hms()` can convert character strings into `POSIXct` objects. These functions are vectorised and can convert a whole character vector in a single operation into a `POSIXct` vector of the same length.

```
dmy_h("04/10/23 15", tz = "EET")
## [1] "2023-10-04 15:00:00 EEST"
dmy_h("04/10/23 3pm", tz = "EET")
## [1] "2023-10-04 15:00:00 EEST"
dmy_h("04/10/23 15 EET") # wrong decoding!
## [1] "2023-10-04 15:00:00 UTC"
```

⚠ Conversion functions with no time components return `Date` objects if no argument is passed to `tz`, while `tz = ""`, as used below, signifies the local time zone.

```
class(ymd("2023-10-04"))
## [1] "Date"
class(ymd("2023-10-04", tz = ""))
## [1] "POSIXct" "POSIXt"
class(today(tzone = ""))
## [1] "Date"
```

Conversions from `Date` into `POSIXct` can give very unexpected results! If you run the statement below, the returned value will be the time difference between the time setting in your computer and UTC!

```
as.POSIXct(ymd("2023-10-04"), tzone = "") - ymd("2023-10-04", tz = "")
## Time difference of 3 hours
```

The computations assume that the value in the `Date` is expressed in UTC, corresponding to 00:00:00 UTC. The time zone difference in UTC at midnight between time zones is not taken into account. Forcing the time zone after conversion in `POSIXct` fixes the problem. Quirks like these make it imperative to do extensive checks when doing conversions involving times and or dates.

```
force_tz(as.POSIXct(ymd("2023-10-04")), tzone = "") - ymd("2023-10-04", tz = "")
## Time difference of 0 secs
```

A difference between to instants in time returns a duration.

```
ymd_hms("2010-05-25 12:05:00") - ymd_hms("1810-05-25 12:00:00")
## Time difference of 73049 days
```

Functions with names in plural, like `years()` ... `seconds()` are constructors of durations, and can be added and subtracted from times.

```
ymd_hms("1810-05-25 12:00:00") + years(200) + minutes(5)
## [1] "2010-05-25 12:05:00 UTC"
ymd_hms("2010-05-25 12:05:00") - ymd_hms("1810-05-25 12:00:00")
## Time difference of 73049 days
ymd("2023-01-01") + seconds(123)
## [1] "2023-01-01 00:02:03 UTC"
```

Functions with names in singular, like `year()` ... `second()` are used to extract and set the implicit components of an instant in time.

```
my.time <- now()
my.time
## [1] "2024-02-17 22:35:30 EET"
year(my.time)
## [1] 2024
hour(my.time)
## [1] 22
second(my.time)
## [1] 30.9487
second(my.time) <- 0
```

Special versions of methods `round()` and `trunc()` are available for times.

```
trunc(my.time, "days")
## [1] "2024-02-17 EET"
round(my.time, "hours")
## [1] "2024-02-17 23:00:00 EET"
```

> 8.5 Working with time data, frequently involves checking that the results of computations are according to our expectations. Sometimes the documentation is not enough and we need to explore with code examples how functions work. For example, take one date in February 2020 and one date in March 2020, and compute the duration between them. Then repeat the computation for year 2022 using the same dates. Which of these years was a leap year?

In the next chapter, I describe data visualisation with package 'ggplot2', frequently also considered part of the 'tidyverse'.

8.9 Further Reading

An in-depth discussion of the 'tidyverse' is outside the scope of this book. Several books describe in detail the use of these packages. As several of them are under active development, recent editions of books such as *R for Data Science* (Wickham et al. 2023) and *R Programming for Data Science* (Peng 2022) are the most useful.

9

R Extensions: Grammar of Graphics

> The commonality between science and art is in trying to see profoundly—to develop strategies of seeing and showing.
>
> Edward Tufte's answer to Charlotte Thralls
> *An Interview with Edward R. Tufte*, 2004

9.1 Aims of This Chapter

Three main data plotting systems are available to R users: base R, package 'lattice' (Sarkar 2008), and package 'ggplot2' (Wickham and Sievert 2016); the last one being the most recent and currently most popular system available in R for plotting data. Even two different sets of graphics primitives (i.e., those used to produce the simplest graphical elements such as lines and symbols) are available in R, those in base R and a newer one in the 'grid' package (Murrell 2019).

In this chapter, you will learn the concepts of the layered grammar of graphics, on which package 'ggplot2' is based. You will also learn how to build several types of data plots with package 'ggplot2'. As a consequence of the popularity and flexibility of 'ggplot2', many contributed packages extending its functionality have been developed and deposited in public repositories. However, I will focus mainly on package 'ggplot2' only briefly describing a few of these extensions.

9.2 Packages Used in This Chapter

If the packages used in this chapter are not yet installed in your computer, you can install them as shown below, as long as package 'learnrbook' is already installed.

```
install.packages(learnrbook::pkgs_ch_ggplot)
```

DOI: 10.1201/9781003404187-9

To run the examples included in this chapter, you need first to load some packages from the library (see section 6.4 on page 179 for details on the use of packages).

```
library(learnrbook)
library(scales)
library(ggplot2)
library(ggrepel)
library(gginnards)
library(broom)
library(ggpmisc)
library(ggbeeswarm)
library(lubridate)
library(tibble)
library(dplyr)
library(patchwork)
```

9.3 The Components of a Plot

I start by briefly presenting concepts central to data visualisation, following the *Data Visualization Handbook* (Koponen and Hildén 2019). Plots are a medium used to convey information, like text. It is worthwhile keeping this in mind. As with text, the design of plots needs to consider what needs to be highlighted to convey the take home message. The style of the plot should match the expectations and the plot-reading abilities of the expected audience. One needs to be careful to avoid ambiguities and most importantly of all not to miss-inform. Data visualisations like text need to be planned, revised, commented upon, and revised again until the best way of expressing our message is found. The flexibility of the grammar of graphics supports very well this approach to designing and producing high quality data visualisations for different audiences.

Of course, when exploring data, fancy details of graphical design are irrelevant, but flexibility remains important as it makes it possible to look at data from many differing angles, highlighting different aspects of them. In the same way as boilerplate text and text templates have specific but limited uses, all-in-one functions for producing plots do not support well the design of original data visualisations. They tend to get the job done, but lack the flexibility needed to do the best job of communicating information. Being this a book about languages, the focus of this chapter is in the layered grammar of graphics.

The plots described in this chapter are classified as *statistical graphics* within the broader field of data visualisation. Plots such as scatter plots include points (geometric objects) that by their position, shape, colour, or some other property directly convey information. The location of these points in the plot "canvas" or "plotting area", given by the values of their x and y coordinates describes properties of the data and any deviation in the mapping of observations to coordinates is misleading, because deviations from the expected mapping conveys wrong/false information to the audience.

A *data label* is connected to an observation but its position can be displaced as long as its link to the corresponding observation can be inferred, e.g., by the direction of an arrow or even simple proximity. Data labels provide ancillary information, such as the name of a gene or place.

Annotations, are additions to a plot that have no connection to individual observations, but rather with all observations taken together, e.g., a text like $n = 200$ indicating the number of observations, usually included in a corner or margin of a plot free of observations.

Axis and tick labels, legends and keys make it possible for the reader to retrieve the original values represented in the plot as graphical elements. Other features of visualisations even when not carrying additional information affect the easy with which a plot can be read and accessibility to readers with visual constraints such as colour blindness. These features include the size of text and symbols, thickness of lines, choice of font face, choice of colour palette, etc.

Because of the different lengths of time available for the audience to interact with visualisations, in general, plots designed to be included in books and journals are unsuitable for oral presentations, and vice versa. It is important to keep in mind the role played by plots in informing the audience, and what information can be expected to be of interest to different audiences and under different situations. The grammar of graphics and its extensions provide enough flexibility to tailor the design of plots to different uses and also to easily create variations of a given plot.

9.4 The Grammar of Graphics

What separates 'ggplot2' from base R and trellis/lattice plotting functions is the use of a layered grammar of graphics (the reason behind 'gg' in the name of package 'ggplot2'). What is meant by grammar in this case is that plots are assembled piece by piece using different "nouns" and "verbs" (Cleveland 1985; Wickham 2010). Instead of using a single function with many arguments, plots are assembled by combining different elements with operators + and %+%. Furthermore, the construction is mostly semantics-based and to a large extent, how plots look when printed, displayed, or exported to a bitmap or vector-graphics file is controlled by themes.

Plotting can be thought as translating or mapping the observations or data into a graphical language. Properties of graphical (or geometrical) objects are used to represent different aspects of the data. An observation can consist of multiple recorded values. Say an observation of air temperature may be defined by a position in 3-dimensional space and a point in time, in addition to the temperature itself. An observation for the size and shape of a plant can consist of height, stem diameter, number of leaves, size of individual leaves, length of roots, fresh mass, dry mass, etc. For example, an effective way of studying and/or communicating the relationship between height and stem diameter in plants, is to plot observations as points using cartesian coordinates, *mapping* stem diameter to the x axis and the height to the y axis.

The grammar of graphics makes it possible to design plots by combining various elements in ways that are nearly orthogonal. In other words, the majority of the possible combinations of "words" yield valid plots as long the rules of the grammar are respected. This flexibility makes 'ggplot2' extremely powerful as types of plots not considered when the 'ggplot2' package was designed can be easily created.

When a ggplot is built, the whole plot and its components are created as R objects that can be saved in the workspace or written to a file as R objects. These objects encode a recipe for constructing the plot, not its final graphical representation. The graphical representation is generated when the object is printed, explicitly or not. Thus, the same "gg" plot object can be rendered into different bitmap and vector graphic formats for display and/or printing.

The transformation of a set of data or observations into a rendered graphic with package 'ggplot2' can be represented as a flow of information, but also as a sequence of actions. However, what avoids that the flexibility from becoming a burden on users is that in most cases adequate defaults are used when the user does not provide explicit "instructions". The recipe to build a plot needs to specify a) the data to use, b) which variable to map to which graphical property (or aesthetic), c) which layers to add and which geometric representation to use, d) the scales that establish the link between data values and aesthetic values, e) a coordinate system (affecting only aesthetics x, y and possibly z), f) a theme to use. The result from constructing a plot object using the grammar of graphics is an R object containing a "recipe for a plot", including the data, which behaves similarly to other R objects.

9.4.1 The words of the grammar

Before building a plot step by step, I introduce the different components of a ggplot recipe, or the words of the grammar of graphics.

Data

The data to be plotted must be available as a `data.frame` or `tibble`, with data stored so that each row represents a single observation event, and the columns are different values observed in that single event. In other words, in long form (so-called "tidy data") as described in chapter 8. The variables to be plotted can be `numeric`, `factor`, `character`, and time or date stored as `POSIXct`. (Some extensions to 'ggplot2' add support for other types of data such as time series).

Mapping

When constructing a plot, data variables have to be mapped to aesthetics (or graphic properties). Most plots will have an x dimension, which is one of the *aesthetics*, and a variable containing numbers (or categories) mapped to it. The position on a 2D plot of, say, a point, will be determined by x and y aesthetics, while in a 3D plot, three aesthetics need to be mapped x, y, and z. Many aesthetics are not related to coordinates, they are properties, like colour, size, shape, line type, or even rotation angle, which add an additional dimension on which to represent the values of variables and/or constants.

Statistics

Statistics are "words" that represent calculation of summaries or some other operation on the values in the data. When *statistics* are used for a computation, the returned value is passed to a *geometry*, and consequently adding a *statistics* also adds a layer to the plot. For example, `stat_smooth()` fits a smoother, and `stat_summary()` applies a summary function such as `mean(()`. Most statistics are applied by group when data have been grouped by mapping additional aesthetics such as colour to a factor.

Geometries

Geometries are "words" that describe the graphics representation of the data: for example, `geom_point()`, plots a point or symbol for each observation or summary value, while `geom_line()`, draws line segments between observations. Some geometries rely by default on statistics, but most "geoms" default to the identity statistics. Each time a *geometry* is used to add a graphical representation of data to a plot, one says that a new *layer* has been added. The grammar of graphics allows plots to contain multiple layers. The name *layer* reflects the fact that each new layer added is plotted on top of the layers already present in the plot, or rather when a plot is printed the layers will be generated in the order they were added to the plot object. For example, one layer in a plot can display the observations, another layer a regression line fitted to them, and a third one may contain annotations such as an equation or a text label.

Positions

Positions are "words" that determine the displacement or not of graphical plot elements relative to their original x and y coordinates. They are one of the arguments accepted by *geometries*. Position `position_identity()` introduces no displacement, and for example, `position_stack()` makes it possible to create stacked bar plots and stacked area plots. Positions will be discussed together with geometries as they are always subordinate to them.

Scales

Scales give the "translation" or mapping between data values and the aesthetic values to be actually plotted. Mapping a variable to the "colour" aesthetic (also recognised when spelled as "color") only tells that different values stored in the mapped variable will be represented by different colours. A scale, such as `scale_colour_continuous()`, will determine which colour in the plot corresponds to which value in the variable. Scales can also define transformations on the data, which are used when mapping data values to aesthetic values. All continuous scales support transformations—e.g., in the case of x and y aesthetics, positions on the plotting region or graphic viewport will be affected by the transformation, while the original values are used for tick labels along the axes or in keys for shapes, colours, etc. Scales are used for all aesthetics, including continuous variables, such as numbers, and categorical ones such as factors. The grammar of graphics allows only one scale per *aesthetic* and plot. This restriction is imposed by design to avoid ambiguity (e.g., it ensures that the red colour will have the same "meaning" in all plot layers where the `colour` *aesthetic* is mapped to data). Scales have limits that are set automatically unless supplied explicitly.

Coordinate systems

The most frequently used coordinate system when plotting data, the cartesian system, is the default for most *geometries*. In the cartesian system, x and y are represented as distances on two orthogonal (at 90°) axes. Additional coordinate systems are available in 'ggplot2' and through extensions. For example, in the polar system of coordinates, the x values are mapped to angles around a central point and y values to the radius. Setting limits to a coordinate system changes the region of the plotting space visible in the plot, but does not discard observations. In other words, when using *statistics*, observations located outside the coordinate limits, i.e., not visible in the rendered plot, will still be included in computations when excluded by coordinate limits but will be ignored when excluded by scale limits.

Themes

How the plots look when displayed or printed can be altered by means of themes. A plot can be saved without adding a theme and then printed or displayed using different themes. Also, individual theme elements can be changed, and whole new themes defined. This adds a lot of flexibility and helps in the separation of the data representation aspects from those related to the graphical design.

Operators

The elements described above are assembled into a ggplot object using operator + and exceptionally using %+%. The choice of these operators makes sense, as ggplot objects are built by sequentially adding members or elements to them.

The functions corresponding to the different elements of the grammar of graphics have distinctive names with the first few letters hinting at their roles: aesthetics mappings (`aes`), geometric elements (`geom_...`), statistics (`stat_...`), scales (`scale_...`), coordinate systems (`coord_...`), and themes (`theme_...`).

9.4.2 The workings of the grammar

A "gg" plot object is an R object of mode `"list"` containing the recipe and data to construct a plot. It is self contained in the sense that the only requirement for rendering it into a graphical representation is the availability of package 'ggplot2'. A "gg" object contains the data in one or more data frames and instructions encoded as functions and parameters, but not yet a rendering of the plot into graphical objects. Both data transformations and rendering of the plot into drawing instructions (encoded as graphical objects or *grobs*) take place at the time of printing or exporting the plot, e.g., when saving a bitmap to a file.

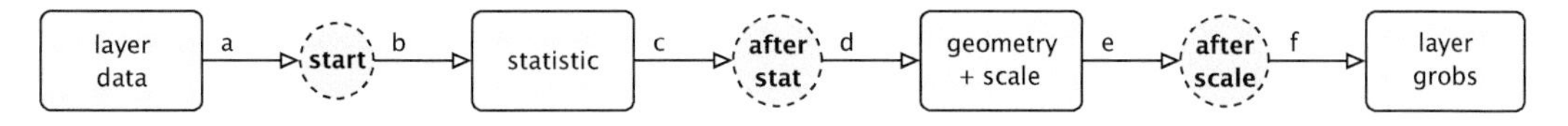

Figure 9.1
Abstract diagram of data transformations in a ggplot layer showing the stages at which mappings between variables and graphic aesthetics take place.

To understand ggplots, one should first think in terms of the graphical organisation of the plot: there is always a background layer onto which other layers composed by different graphical objects are laid. Each layer contains related graphical objects originating from the same data. The last layer added is the topmost and the first one added the lowermost. Graphical objects in upper layers occlude those in the layers below them if their locations overlap. Although frequently layers in a ggplot share the same data and the same mappings to aesthetics, this is not a requirement. It is possible to build ggplots with independent layers, although always with shared scales and plotting area.

A second perspective on ggplots is that of the process of converting the data into a graphical representation that can be printed on paper or viewed on a computer screen. The transformations applied to the data to achieve this can be thought as a data flow process divided into stages. The diagram in Figure 9.1 represents a single self-contained layer in a plot. The data supplied by the user is transformed in stages into instructions to draw a graphical representation. In 'ggplot2' and its documentation, graphical features are called *aesthetics*, with the correspondence between values in the data and values of the aesthetic controlled by *scales*. The values in the data are summarised by *statistics*. However, when no summaries are needed, layers make use of `stat_indentity()`, which copies its input to its output unchanged. *Geometries* provide the "recipe" used to generate graphical objects from the mapped data.

Function `aes()` is used to define mappings to aesthetics. The default for `aes()` is for the mapping to take place at the **start** (leftmost circle in the diagram above), mapping names in the user data to aesthetics such as x, y, colour, and shape. The statistic can alter the mapped data, but in most cases not which aesthetics they are mapped to. Statistics can add default mappings for additional aesthetics. In addition, the default mappings of the data returned by the statistic can be modified by user code at this later stage, **after stat**. Default mappings can be modified again at the **after scale** stage.

Statistics always return a mapping to the same aesthetics that they require as input. However, the values mapped to these aesthetics at the **after stat** stage are in most cases different from those at **start**. Many statistics return additional variables, which are not mapped by default to any aesthetic. These variables facilitate variations on how results from a given type of data summary are added to plots, including the use of a geometry different from the default set by the statistic. In this case, the user has to override default mappings at the **after stat** stage. The additional variables returned by statistics are listed in their documentation. (See section 9.4.5 on page 288 for details.)

As mentioned above, all ggplot layers include a statistic and a geometry. From the perspective of the construction of a plot using the grammar, both `stats` and `geoms` are layer constructor functions. While `stats` take a `geom` as one of their arguments, `geoms` take a `stat` as one of their arguments. Thus, in both cases, a `stat` and a `geom` are added as a layer, and their role and position in the data flow remain the same, i.e., the diagram in Figure 9.1 applies independently of how the layers are added to the plot. The default statistic of many geometries is `stat_identity()` making their behaviour when added to a plot as if the layer they create contained no statistics.

There are some statistics in 'ggplot2' that have companion geometries that can be used (almost) interchangeably. This tends to lead into confusion, and in this book, only geometries that have as default `stat_identity()` are described as geometries in section 9.5. In the case of those that by default use other statistics, like `geom_smooth()` only the companion statistic, `stat_smooth()` for this example, are described in section 9.6.

A ggplot can have a single layer or many layers, but when ggplots have more than one layer, the data flow, computations, and generation of graphical objects takes place independently for each layer. As mentioned above, most ggplots do not have fully independent layers, but the layers share the same data and aesthetic mappings at the **start**. Ahead of this point computations in layers are always independent of those in other layers, except that for a given aesthetic only one scale is allowed per plot.

make it possible

9.4.3 Plot construction

As the use of the grammar is easier to demonstrate by example than to explain with words, I will show how to build plots of increasing complexity, starting from the simplest possible. All elements of a plot have defaults, although in some cases these defaults result in empty plots. Defaults make it possible to create a plot very succinctly. When building a plot step by step, the different viewpoints described in the previous section are relevant: the static structure of the plot's R object, the final graphic output, and the transformations that the data undergo "in transit" from the recipe stored in an object to the graphic output. In this section, I emphasise the syntax of the grammar and how it translates into a plot.

Function `ggplot()` by default constructs an empty plot. This is similar to how `character()`, `numeric()`, etc. construct empty vectors. This empty skeleton of a plot when printed is displayed as an grey rectangle.

```
ggplot()
```

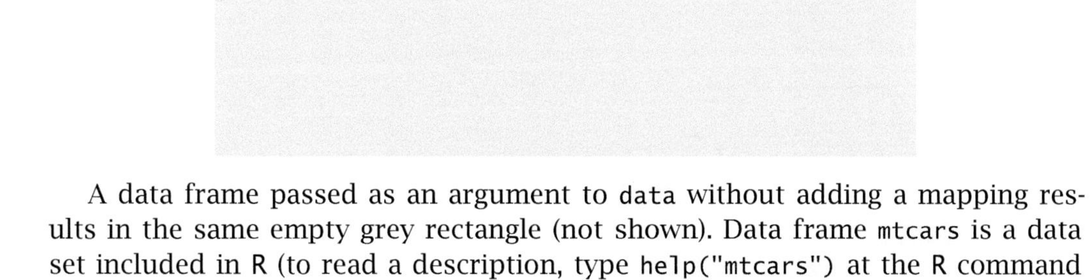

A data frame passed as an argument to `data` without adding a mapping results in the same empty grey rectangle (not shown). Data frame `mtcars` is a data set included in R (to read a description, type `help("mtcars")` at the R command prompt).

```
ggplot(data = mtcars)
```

Once the data are available, a graphical or geometric representation needs to be selected. The geometry used, such as `geom_point()` and `geom_line()`, drawing separate points for the observations or connecting them with lines, respectively, defines the type of plot. A mapping defines which property of the geometric elements will be used to represent the values from a variable in the user's data. Most geometries require mappings to both x and y aesthetics, as they establish the position of the geometrical shapes like points or lines in the plotting area. Additional aesthetics like colour make use of default scales and palettes. These defaults can be overridden with `scale` functions added to the plot (see section 9.10).

Mapping at the **start** stage, `disp` to x and `mpg` to y aesthetics, makes the ranges of the values available. They are used to find default limits for the x and y scales as reflected in the plot axes. The plotting area x and y now match the ranges of the mapped variables, expanded by a small margin. The axis labels also reflect the names of the mapped variables, however, there are no graphical element yet displayed for the individual observations.

```
ggplot(data = mtcars,
       mapping = aes(x = disp, y = mpg))
```

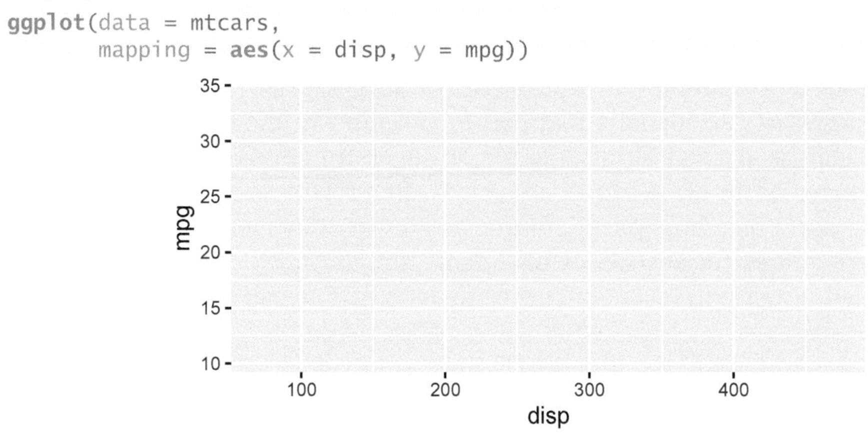

Observations are made visible by the addition of a suitable *geometry* or `geom`

to the plot recipe. Below, adding `geom_point()` makes the observations visible as points or symbols.

```
ggplot(data = mtcars,
       mapping = aes(x = disp, y = mpg)) +
  geom_point()
```

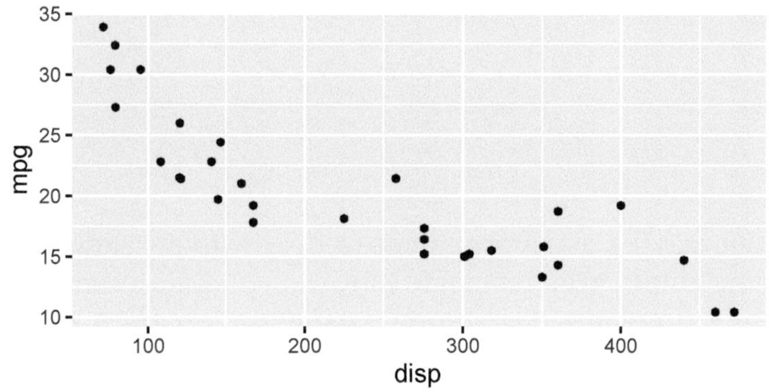

In the examples above, the plots were printed automatically, which is the default at the R console. However, as with other R objects, ggplots can be assigned to a variable.

```
p1 <- ggplot(data = mtcars,
             mapping = aes(x = disp, y = mpg)) +
      geom_point()
```

and printed at a later time, and saved to and read from files on disk.

```
print(p1)
```

Layers and other elements can be also added to a saved ggplot as the saved objects are not the graphical representation of the plots themselves but instead a *recipe* plus data needed to build them.

9.1 As for any R object `str()` displays the structure of `"gg"` objects. In addition, package '`ggplot2`' provides a `summary()` method for `"gg"` plot objects.

As you make progress through the chapter, use these methods to explore the `"gg"` plot objects you construct, paying attention to layers, and global vs. layer-specific data and mappings. You will learn how the plot components are stored as members of `"gg"` plot objects.

Although *aesthetics* are usually mapped to variables in the data, constant aesthetic values can be passed as arguments to layer functions, consistently controlling a property of all elements in a layer. While variables in `data` can be both mapped using `aes()` as whole-plot defaults, as shown above, or within individual layers, constant values for aesthetics have to be set, as shown here, as named arguments passed directly to layer functions, instead of to a call to `aes()`.

```
ggplot(data = mtcars,
       mapping = aes(x = disp, y = mpg)) +
  geom_point(colour = "blue", shape = "square")
```

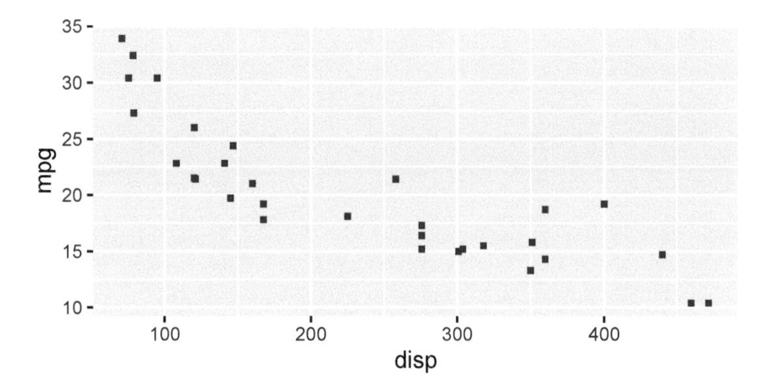

Mapping an aesthetic to a constant value within a call to `aes()` adds a column containing this value to the data frame received as input by the `stat()`. This value is not interpreted as an aesthetic value but instead as a data value. The plot above, but using a call to `aes()`.

```
ggplot(data = mtcars,
       mapping = aes(x = disp, y = mpg)) +
  geom_point(mapping = aes(colour = "blue", shape = "square"))
```

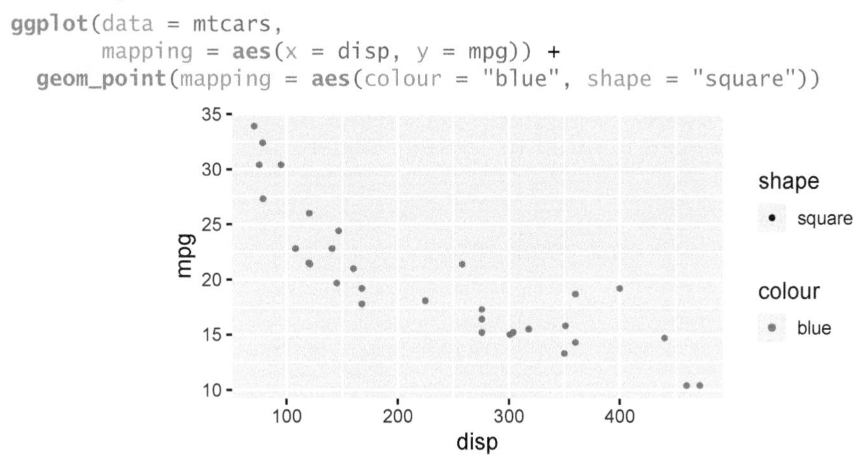

The plot contains red circles instead of blue squares!

In principle, one could correct this plot by adding suitable `scales` but this would be still wasteful by unnecessarily storing many copies of the constant `"blue"` in the `"gg"` plot object.

While a geometry directly constructs during rendering a graphical representation of the observations or summaries in the data it receives as input, a *statistics* or `stat` "sits" in-between the data and a `geom`, applying some computation, usually but not always, to produce a statistical summary of the data. Here `stat_smooth()` fits a linear regression (see section 7.9.1 on page 202) and passes the resulting predicted values to `geom_line()`. Passing `method = "lm"` selects `lm()` as the model fitting function. Passing `formula = y   x` sets the model to be fitted. This plot has two layers, one from geometries `geom_point()` and one from `geom_line()`.

```
ggplot(data = mtcars,
       mapping = aes(x = disp, y = mpg)) +
  geom_point() +
  stat_smooth(geom = "line", method = "lm", formula = y ~ x)
```

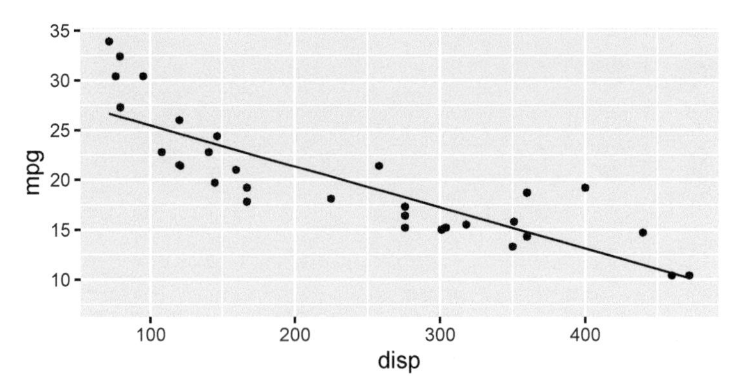

The plots above relied on defaults for *scales*, *coordinates* and *themes*. In the examples below, the defaults are overridden by arguments that produce differently rendered plots. Adding `scale_y_log10()` applies a logarithmic transformation to the values mapped to y. This works like plotting using graph paper with rulings spaced according to a logarithmic scale. Tick marks continue to be expressed in the original units, but statistics are applied to the transformed data. In other words, the transformation specified in the scale affects the values in advance of the **start** stage, before they are mapped to aesthetics and passed to *statistics*. Thus, in this example, the linear regression is fitted to `log10()` transformed y values and the original x values.

```
ggplot(data = mtcars,
       mapping = aes(x = disp, y = mpg)) +
  geom_point() +
  stat_smooth(geom = "line", method = "lm", formula = y ~ x) +
  scale_y_log10()
```

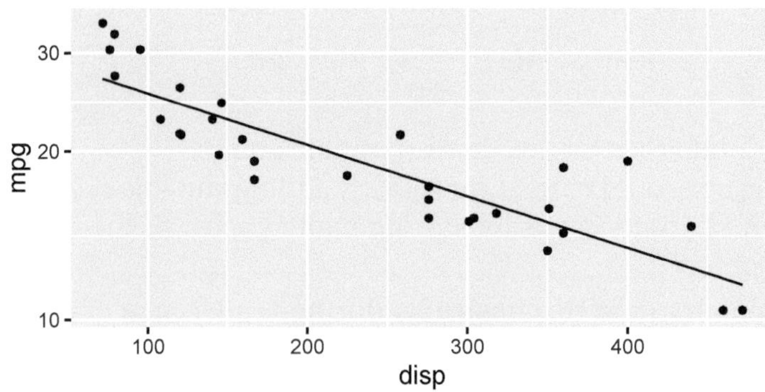

The range limits of a scale can be set manually, instead of automatically as by default. These limits create a virtual *window into the data*: out-of-bounds (oob) observations, those outside the scale limits remain hidden and are not mapped to aesthetics—i.e., these observations are not included in the graphical representation or used in calculations. Crucially, when using *statistics* the computations are only applied to observations that fall within the limits of all scales in use. These limits *indirectly* affect the plotting area when the plotting area is automatically set based on the range of the (within limits) data—even the mapping to values of a different aesthetics may change when a subset of the data is selected by manually setting the limits of a scale.

In contrast to *scale limits*, *coordinates* function as a *zoomed view* into the plotting area, and do not affect which observations are visible to *statistics*. The coordinate system, as expected, is also determined by this grammar element—below, adding cartesian coordinates, which are the default, but setting y limits overrides the default ones.

```
ggplot(data = mtcars,
       mapping = aes(x = disp, y = mpg)) +
  geom_point() +
  stat_smooth(geom = "line", method = "lm", formula = y ~ x) +
  coord_cartesian(ylim = c(15, 25))
```

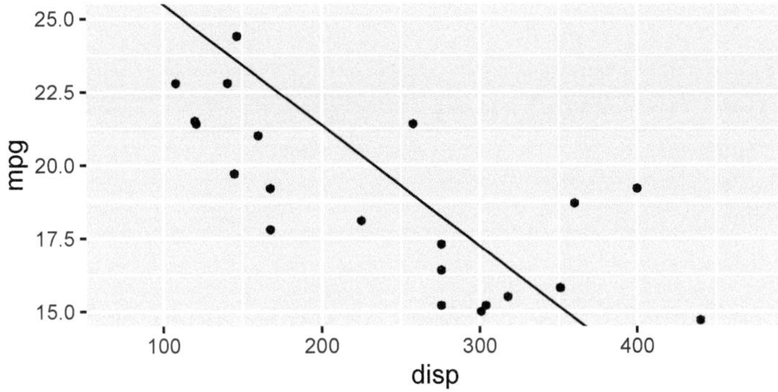

The next example uses a coordinate system transformation. When the transformation is applied to the coordinate system, it affects only the plotting—it sits between the `geom` and the rendering of the plot. The transformation is applied to the values that were returned by *statistics*. The straight line fitted is plotted on the transformed coordinates as a curve, because the model was fitted to the untransformed data obtaining untransformed predicted values. The coordinate transformation is applied to these predicted values and plotted. (Other coordinate systems are described in sections 9.5.6 and 9.12 on pages 306 and 362, respectively.)

```
ggplot(data = mtcars,
       mapping = aes(x = disp, y = mpg)) +
  geom_point() +
  stat_smooth(geom = "line", method = "lm", formula = y ~ x) +
  coord_trans(y = "log10")
```

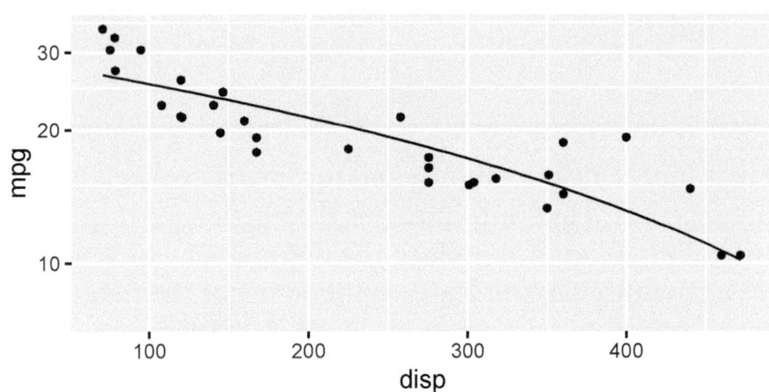

Themes affect the rendering of plots at the time of printing—they can be thought of as style sheets defining the graphic design. A complete theme can override the default gray theme. The plot is the same, the observations are represented in the same way, the limits of the axes are the same and all text is the same. On the other hand, how these elements are rendered by different themes can be drastically different.

```
ggplot(data = mtcars,
       mapping = aes(x = disp, y = mpg)) +
  geom_point() +
  theme_classic()
```

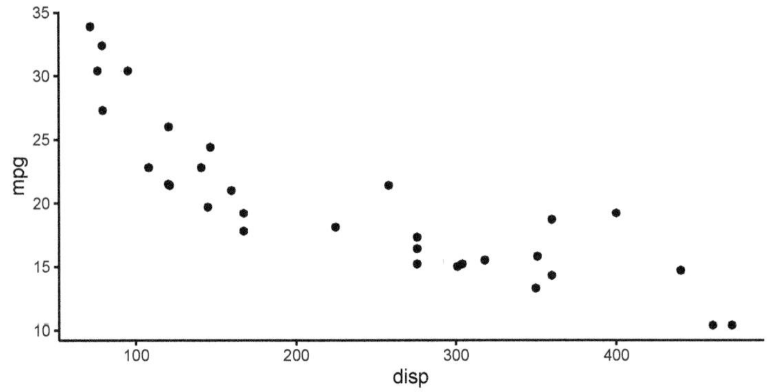

Both the base font size and the base font family can be changed. The base font size controls the size of all text elements, as other sizes are defined relative to the base size. How the plot looks changes when using the same theme as in the previous example, but with a different base point size and font family for text elements. (The use of themes is discussed in section 9.13 on page 364.)

```
ggplot(data = mtcars,
       mapping = aes(x = disp, y = mpg)) +
  geom_point() +
  theme_classic(base_size = 20, base_family = "serif")
```

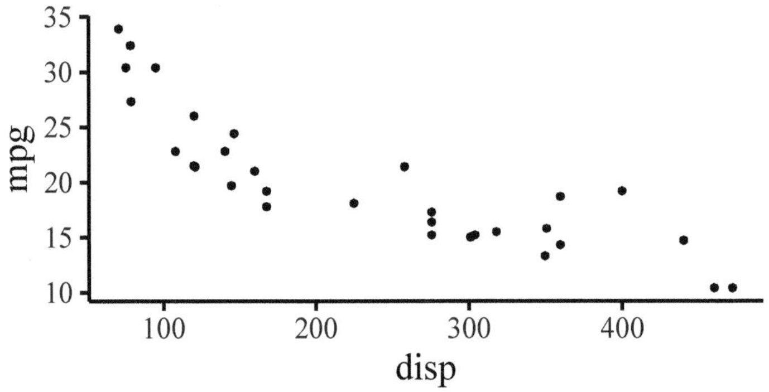

How to set axis labels, tick positions, and tick labels will be discussed in depth in section 9.10 on page 341. Function `labs()` is *a convenience function* used to set the title and subtitle of a plot and to replace the default `name` of scales, here displayed as axis labels. The default `name` of scales is the name of the mapped variable. In the call to `labs()`, the names of aesthetics are used as if they were formal parameters with character strings or R expressions as arguments. Below `x` and `y` are the names of the two *aesthetics* to which two variables in `data` were mapped, `disp` and `mpg`, respectively. Formal parameters `title` and `subtitle` add these plot elements. (The escaped character `\n` stands for new line, see section 3.4 on page 41.)

```
ggplot(data = mtcars,
       mapping = aes(x = disp, y = mpg)) +
  geom_point() +
  labs(x = "Engine displacement (cubic inches)",
       y = "Fuel use efficiency\n(miles per gallon)",
       title = "Motor Trend Car Road Tests",
       subtitle = "Source: 1974 Motor Trend US magazine")
```

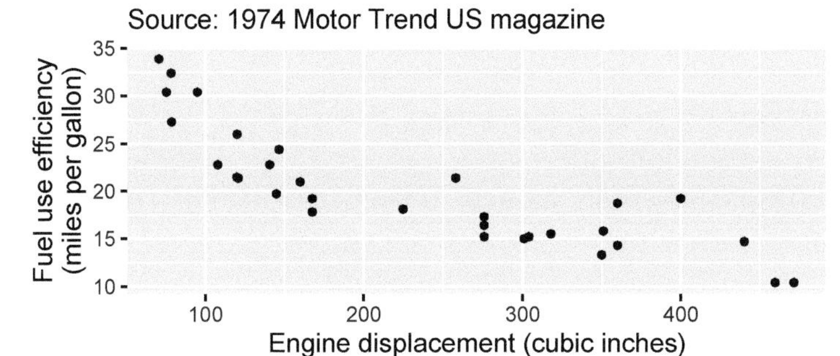

As elsewhere in R, when a value is expected, either a value stored in a variable or a more complex statement returning a suitable value can be passed as an argument to be mapped to an *aesthetic*. In other words, the values to be plotted do not need to be stored as variables (or columns) in the data frame passed as an argument to parameter `data`, they can also be computed from these variables. Below, miles-per-gallon, `mpg` are plotted against the engine displacement per cylinder by dividing `disp` by `cyl` within the call to `aes()`.

```
ggplot(data = mtcars,
       mapping = aes(x = disp / cyl, y = mpg)) +
  geom_point()
```

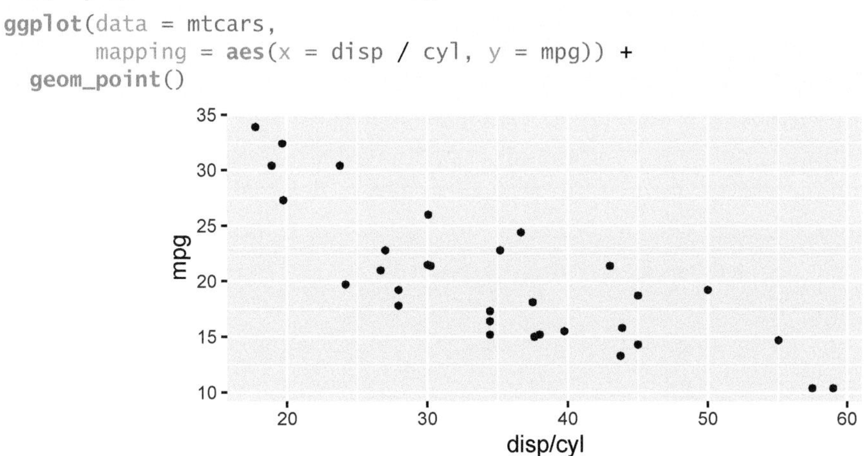

Each of the elements of the grammar exemplified above is implemented in multiple functions, and in addition these functions accept arguments that can be used to modify their behaviour. Multiple data objects as well as multiple mappings can coexist within a single `"gg"` plot object. Packages and user code can define new *geometries*, *statistics*, *scales*, *coordinates*, and even implement new *aesthetics*. Individual elements in a *theme* can be modified and new complete *themes* created, re-used and shared. I describe below how to use the grammar of graphics to construct different types of data visualisations, both simple and complex. Because the different elements interact, I introduce some of them first briefly in sections other than where I describe them in depth.

9.4.4 Plots as R objects

"gg" plot objects and their components behave as other R objects. Operators and methods for the "gg" class are available. As above, a "gg" plot object saved as p1 is used below.

```
p1 <- ggplot(data = mtcars,
             mapping = aes(x = disp, y = mpg)) +
      geom_point()
```

In the previous section, operator + was used to assemble the plots from "anonymous" R objects. Saved or "named" objects can also be combined with +.

```
p1 + stat_smooth(geom = "line", method = "lm", formula = y ~ x)
```

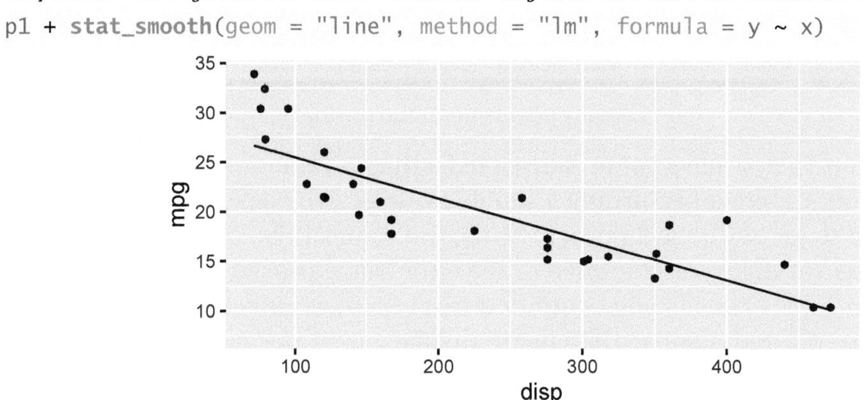

Above, plot elements were added one by one, with operator +. Multiple components can be also added in a single operation. Like individual components, sets of components stored in a list can be saved in a variable and added to multiple plots. This ensures consistency and makes coordinated alterations to a set of plots easier. *Throughout this chapter, I use this approach to achieve conciseness and to highlight what is different and what is not among plots in related examples.*

```
p.ls <- list(
  stat_smooth(geom = "line", method = "lm", formula = y ~ x),
  scale_y_log10())

p1 + p.ls
```

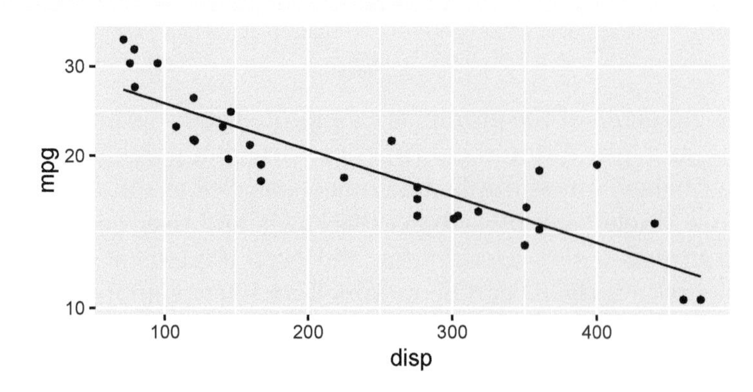

9.2 Reproduce the examples in the previous section, using `p1` defined above as a basis instead of building each plot from scratch.

The separation of plot construction and rendering is possible because `"gg"` objects are self-contained. A copy of the data object passed as an argument to `data` is saved within the plot object, similarly as in model-fit objects. In the example above, `p1` by itself could be saved to a file on disk and loaded into a clean R session, even on another computer, and rendered as long as package 'ggplot2' and its dependencies are available. Another consequence of storing a copy of the data in the plot object, is that later changes to the data object used to create a `"gg"` object are *not* reflected in newly rendered plots from this object: the `"gg"` object needs to be created anew.

The *recipe* for a plot is stored in a `"gg"` plot object. Objects of class `"gg"` are of mode `"list"`. In R, lists can contain heterogeneous members and `"gg"` objects contain data, function definitions, and unevaluated expressions. In other words, the data plus instructions to transform the data, to map them into graphic objects, and various aspects of the rendering from scale limits to type faces to use. (R lists are described in section 4.3 on page 86.)

Top level members of the `"gg"` plot object `p1`, a simple plot, are displayed below with method `summary()`, which shows the components without making explicit the structure of the object.

```
summary(p1)
## data: mpg, cyl, disp, hp, drat, wt, qsec, vs, am, gear, carb [32x11]
## mapping:  x = ~disp, y = ~mpg
## faceting: <ggproto object: Class FacetNull, Facet, gg>
##     compute_layout: function
##     draw_back: function
##     draw_front: function
##     draw_labels: function
##     draw_panels: function
##     finish_data: function
##     init_scales: function
##     map_data: function
##     params: list
##     setup_data: function
##     setup_params: function
##     shrink: TRUE
##     train_scales: function
##     vars: function
##     super:  <ggproto object: Class FacetNull, Facet, gg>
## -----------------------------------
## geom_point: na.rm = FALSE
## stat_identity: na.rm = FALSE
## position_identity
```

Method `str()` shows the structure of objects and can be also used to advantage with ggplots (long output not shown). Alternatively, `names()` extracts the names of the top-level members of `p1`.

```
names(p1)
## [1] "data"          "layers"       "scales"     "guides"     "mapping"
## [6] "theme"         "coordinates"  "facet"      "plot_env"   "layout"
## [11] "labels"
```

9.3 Explore in more detail the different members of object `p1`. For example, the code statement below extracts member `"layers"` from object `p1` and display its structure.

```
str(p1$layers, max.level = 1)
```

How many layers are present in this case?

9.4.5 Scales and mappings

In 'ggplot2', a *mapping* describes which variable in `data` is mapped to which `aesthetic`, or graphic feature of a plot, such as x, y, colour, fill, shape, and linewidth. In 'ggplot2', a *scale* describes the correspondence between *values* in the mapped variable and values of the graphic feature. Below, the numeric variable `cyl` is mapped to the `colour` aesthetic. As the variable is `numeric`, a continuous colour scale is used. Out of the multiple continuous colour scales available, `scale_colour_continuous()` is the default.

```
p2 <-
  ggplot(data = mtcars,
         mapping = aes(x = disp, y = mpg, colour = cyl)) +
  geom_point()
p2
```

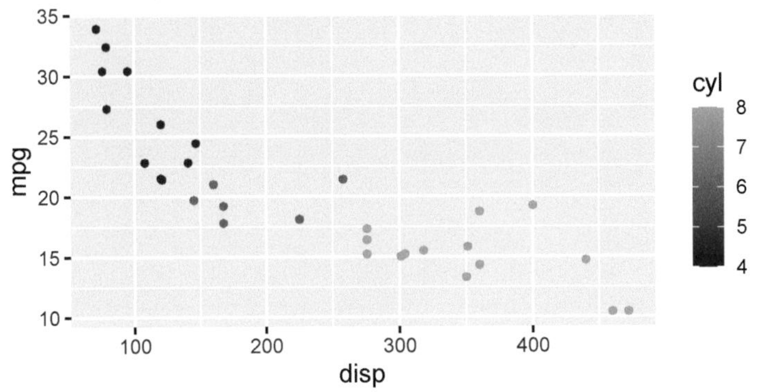

Without changing the `mapping`, a different-looking plot can be created by changing the scale used. Below, in addition, a palette is selected with `option = "magma"` and the range of colours used from this palette adjusted with `end = 0.85`.

```
p2 + scale_colour_viridis_c(option = "magma", end = 0.85)
```

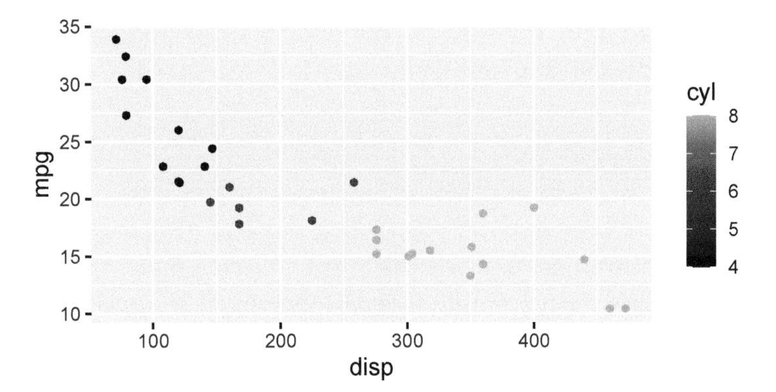

Changing the scale used for the `colour` aesthetic, conceptually does not modify the plot, except for the colours used. There is a separation between the semantic structure of the plot and its graphic design. Still, how the audience interacts and perceives the plot depends on both of these concerns.

Some scales, like those for `colour`, exist in multiple “flavours”, suitable for numeric variables (continuous) or for factors (discrete) values. If `cyl` is converted into a `factor`, a discrete colour scale is used instead of a continuous one. Out of the different discrete scales, `scale_colour_discrete()` is used by default.

```
ggplot(data = mtcars,
       mapping = aes(x = disp, y = mpg, colour = factor(cyl))) +
  geom_point()
```

If `cyl` is converted into an `ordered` factor, an ordinal colour scale is used, by default `scale_colour_ordinal()` (plot not shown).

```
ggplot(data = mtcars,
       mapping = aes(x = disp, y = mpg, colour = ordered(cyl))) +
  geom_point()
```

The scales for other aesthetics work in a similar way as those for colour. Scales are described in detail in section 9.10 on page 344.

In the examples above for simple plots, based on data contained in a single data frame, mappings were established by passing the value returned by the call to `aes()` as the argument to parameter `mapping` of `ggplot()`.

Arguments passed to `data` and/or `mapping` parameters of `ggplot()` work as defaults for all layers in a plot. If arguments are passed to the identically named parameters of a layer function—statistic or geometry—, they are applied to the layer, overriding whole-plot defaults, if they exist. Consequently, the code below creates a plot, `p3`, identical to `p2` above.

```
p3 <-
  ggplot() +
  geom_point(data = mtcars,
             mapping = aes(x = disp, y = mpg, colour = cyl))
p3
```

These examples demonstrate two different approaches that are equally convenient for simple plots with a single layer. However, if a plot has multiple layers based on the same data, the approach used for `p2` makes this clear and is concise. If each layer uses different data and/or different mappings, the second approach is necessary.

In some cases, when flexibility is needed while constructing complex plots with multiple layers other *idioms* can be preferable, e.g., when assembling a plot from "pieces" stored in variables or built programmatically.

The default mapping can also be added directly with the + operator, instead of being passed as an argument to `ggplot()`.

```
ggplot(data = mtcars) +
       aes(x = disp, y = mpg) +
  geom_point()
```

It is also possible to have a default mapping for the whole plot, but no default data.

```
ggplot() +
  aes(x = disp, y = mpg) +
  geom_point(data = mtcars)
```

A mapping saved in a variable (example below), as well as a mapping returned by a function call (shown above for `aes()`), can be passed as an argument to parameter `mapping`

```
my.mapping <- aes(x = disp, y = mpg)
ggplot(data = mtcars,
       mapping = my.mapping) +
  geom_point()
```

In all these examples, the plot remains unchanged (not shown). However, the flexibility of the grammar allows the assembly of plots from separately constructed pieces and reusing these pieces by storing them in variables. These approaches can be very useful in scrips that construct consistently formatted sets of plots, or when the same mapping needs to be used consistently in multiple plots.

The mapping to aesthetics in the call to `aes()` does not have to be to a variable from `data` as in examples above. A a code statement that returns a value computed from one or more variables from `data` is also accepted. Computations during mapping helps avoid the proliferation of variables in the data frames containing observations. In this simple example, `mpg` in miles per gallon is converted into km per litre during mapping.

```
ggplot(data = mtcars,
       mapping =aes(x = disp, y = mpg * 0.43)) +
  geom_point()
```

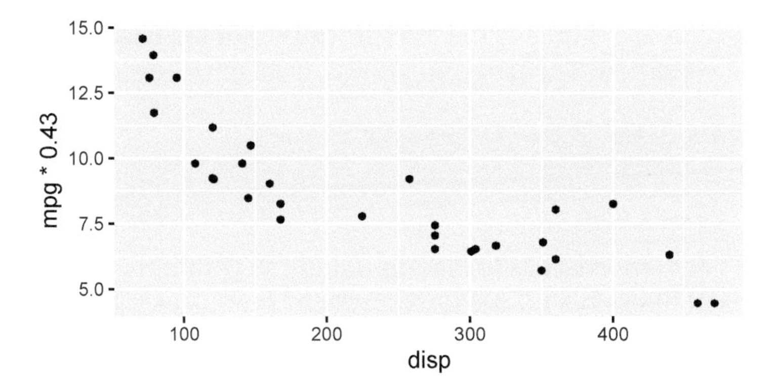

Operations applied to the data before they are plotted are usually implemented in stats. Sometimes it is convenient to directly modify the whole-plot default data before it reaches the layer's stat function. One approach is to pass a function to parameter data of the layer function. This argument must be the definition of a function accepting a data frame as its first argument and returning a data frame. When the argument to data is a function definition instead of the usual data frame, the function is applied to the plot's default data and the data frame returned by the function is used as the data in the layer. In the example below, an anonymous function defined in-line, extracts a subset of the rows. The observations in the extracted rows are highlighted in the plot by overplotting them with smaller yellow shapes.

```
ggplot(data = mtcars,
       mapping = aes(x = disp, y = mpg)) +
  geom_point(size = 4) +
  geom_point(data = function(x){subset(x = x, cyl == 4)},
             colour = "yellow", size = 1.5)
```

The argument passed above to data is a function definition, not a function call. Thus, if a function is passed by name, no parentheses are used. No arguments can be passed to a function, except for the default data passed by position to its first parameter. Consequently, it is not possible to pass function subset directly. The anonymous function above is needed to be able to pass cyl == 4 as argument.

The plot's default data can also be operated upon using the 'magrittr' pipe operator, but not the pipe operator native to R (|>) or the dot-pipe operator from 'wrapr' (see section 8.5 on page 253). In this approach, the dot (.) placeholder at the head of the pipe stands for the plot's default data object. The code statement below uses a pipe as argument for data to call function subset() with cyl == 4 passed as the condition. The plot, not shown, is as in the example above.

```
ggplot(data = mtcars,
       mapping = aes(x = disp, y = mpg)) +
  geom_point(size = 4) +
  geom_point(data = . %>% subset(x = ., cyl == 4), colour = "yellow",
             size = 1.5)
```

A third possible approach is to test the condition within the call to aes(). In this approach, it is not possible to extract a subset of rows. Making some observations invisible by reducing their size seems straightforward. However, setting size = 0 draws a very small point, still visible. Out of various possible approaches, setting size to NA, skips the rows, and na.rm = TRUE silences the expected warning. This is a roundabout approach to subsetting. Notice that scale_size_identity() is also needed. The plot, not shown, when rendered does not differ from the two examples above.

```
ggplot(data = mtcars,
       mapping = aes(x = disp, y = mpg)) +
  geom_point(size = 4) +
  geom_point(colour = "yellow",
             mapping = aes(size = ifelse(cyl == 4, 1.5, NA)),
             na.rm = TRUE) +
  scale_size_identity()
```

As it is usual in R, multiple approaches can be used to the same end.

Late mapping of variables to aesthetics has been possible in 'ggplot2' for a long time using as notation enclosure of the name of a variable returned by a statistic between .., but this notation has been deprecated some time ago and replaced by stat(). In both cases, this imposed a limitation: it was impossible to map a computed variable to the same aesthetic as input to the statistic and to the geometry in the same layer. There were also some other quirks that prevented passing some arguments to the geometry through the dots ... parameter of a statistic.

Since version 3.3.0 of 'ggplot2', the syntax used for mapping variables to aesthetics is based on functions stage(), after_stat() and after_scale(). Function after_stat() replaces both stat() and the .. notation.

The documentation of 'ggplot2' gives several good examples of cases when the new mapping syntax is useful. I give here a different example, a polynomial fitted to data using rlm(). RLM is a procedure that automatically assigns before computing the residual sums of squares, weights to the individual residuals in an attempt to protect the estimated fit from the influence of extreme observations or outliers. When using this and similar methods, it is of interest to plot the residuals together with the weights. One approach is to map weights to a gradient between two colours. The code below constructs a data frame containing artificial data that includes an extreme value or outlier.

```
set.seed(4321)
X <- 0:10
Y <- (X + X^2 + X^3) + rnorm(length(X), mean = 0, sd = mean(X^3) / 4)
df1 <- data.frame(X, Y)
```

```
df2 <- df1
df2[6, "Y"] <-df1[6, "Y"] * 10
```

In the first plot, `after_stat()` is used to map variable `weights` computed by the statistic to the `colour` aesthetic. In `stat_fit_residuals()`, `geom_point()` is used by default. This figure shows the raw residuals with no weights applied (mapped to y by default), and the computed weights (with range 0 to 1) encoded by colours ranging between red and blue.

```
ggplot(data = df2, mapping = aes(x = X, y = Y)) +
  stat_fit_residuals(formula = y ~ poly(x, 3, raw = TRUE), method = "rlm",
                     mapping = aes(colour = after_stat(weights)),
                     show.legend = TRUE) +
  scale_colour_gradient(low = "red", high = "blue", limits = c(0, 1),
                        guide = "colourbar")
```

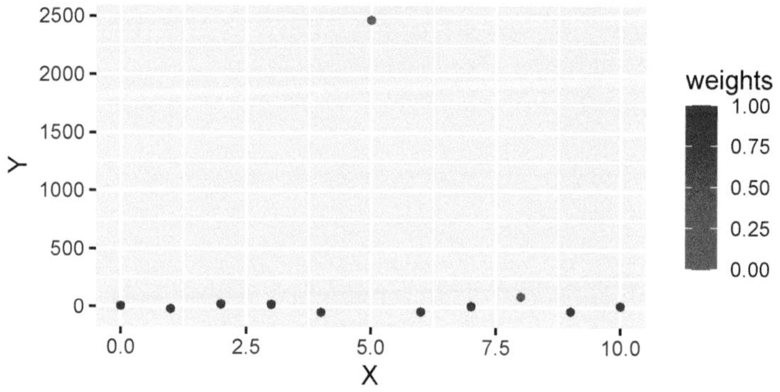

In the second plot, weighted residuals are mapped to the y aesthetic, and weights, as above, to the colour aesthetic. A call to `stage()` can distinguish the mapping ahead of the statistic (`start`) from that after the statistic, i.e., ahead of the geometry. As above, the default geometry, `geom_point()` is used. The mapping in this example can be read as: the variable `X` from the data frame `df2` is mapped to the x aesthetic at all stages. Variable `Y` from the data frame `df2` is mapped to the y aesthetic ahead of the computations in `stat_fit_residuals()`. After the computations, variables `y` and `weights` in the data frame returned by `stat_fit_residuals()` are multiplied and mapped to the y ahead of `geom_point()`.

```
ggplot(df2) +
  stat_fit_residuals(formula = y ~ poly(x, 3, raw = TRUE),
                     method = "rlm",
                     mapping = aes(x = X,
                                   y = stage(start = Y,
                                             after_stat = y * weights),
                                   colour = after_stat(weights)),
                     show.legend = TRUE) +
  scale_colour_gradient(low = "red", high = "blue", limits = c(0, 1),
                        guide = "colourbar")
```

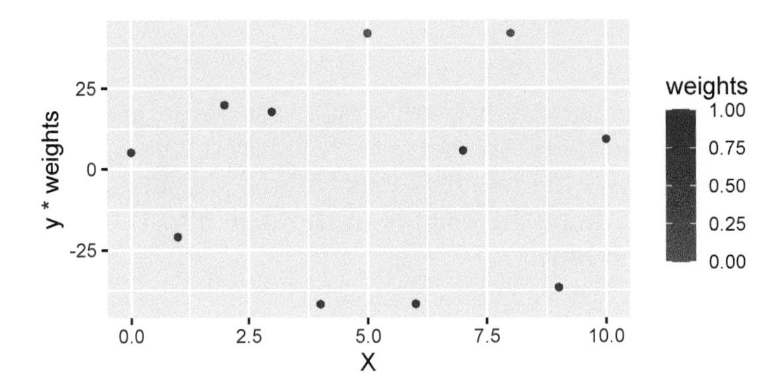

When fitting models to observations with `lm()`, the un-weighted residuals are used to compute the sum of squares unless weights are passed as an argument. In `rlm()`, the weights are computed from the data by the function.

9.5 Geometries

Different geometries support different *aesthetics* (Table 9.1). While `geom_point()` supports `shape`, and `geom_line()` supports `linetype`, both support `x`, `y`, `colour`, and `size`. In this section I describe frequently used `geometries` from package 'ggplot2' and from a few packages that extend 'ggplot2'. The graphic output from some code examples will not be shown, with the expectation that readers will run the code to see the plots.

Mainly for historical reasons, *geometries* accept a *statistic* as an argument, in the same way as *statistics* accept a *geometry* as an argument. In this section I only describe *geometries* which have as a default *statistic* `stat_identity`. In section 9.6.2 (page 317), I describe other *geometries* together with the *statistics* they use by default.

9.5.1 Point

As seen in examples above, `geom_point()`, can be used to add a layer with observations represented by "points" or symbols. In *scatter plots* the variables mapped to x and y aesthetics are both continuous (`numeric`) and in *dot plots* one of them is discrete (`factor` or `ordered`) and the other continuous. The plots in the examples above have been scatter plots.

The first examples of the use of `geom_point()` are for **scatter plots**, as `disp` and `mpg` are `numeric` variables. In the examples above, a third variable, `cyl`, was mapped to `colour`. While the colour aesthetic can be used with all `geoms`, other aesthetics can be used only with some `geoms`, for example the `shape` aesthetic can be used only with `geom_point()` and similar `geoms`, such as `geom_pointrange()`. The values in the `shape` aesthetic are discrete, and consequently only discrete values can be mapped to it.

Table 9.1
'ggplot2' geometries described in section 9.5, packages where they are defined, and the aesthetics supported. The default statistic is in all cases `stat_identity()`.

Geometry	Package	Aesthetics
`geom_point`	'ggplot2'	x, y, shape, size, fill, colour, alpha
`geom_point_s`	'ggpp'	x, y, size, linetype, linewidth, fill, colour, alpha
`geom_pointrange`	'ggplot2'	x, y, ymin, ymax, shape, size, linetype, linewidth, fill, colour, alpha
`geom_errorbar`	'ggplot2'	x, ymin, ymax, linetype, linewidth, colour, alpha
`geom_linerange`	'ggplot2'	x, ymin, ymax, linetype, linewidth, colour, alpha
`geom_line`	'ggplot2'	x, y, linetype, linewidth, colour, alpha
`geom_segment`	'ggplot2'	x, y, xend, yend, linetype, linewidth, colour, alpha
`geom_step`	'ggplot2'	x, y, linetype, linewidth, colour, alpha
`geom_path`	'ggplot2'	x, y, linetype, linewidth, colour, alpha
`geom_curve`	'ggplot2'	x, y, xend or yend, linetype, linewidth, colour, alpha
`geom_area`	'ggplot2'	x, y, (ymin = 0), linetype, linewidth, fill, colour, alpha
`geom_ribbon`	'ggplot2'	x, ymin and ymax, linetype, linewidth, fill, colour, alpha
`geom_align`	'ggplot2'	x or y, xmin or xmax, ymin or ymax, linetype, linewidth, fill, colour, alpha
`geom_rect`	'ggplot2'	xmin, xmax, ymin, ymax, linetype, linewidth, fill, colour, alpha
`geom_tile`	'ggplot2'	x, y, width, height, linetype, linewidth, fill, colour, alpha
`geom_col`	'ggplot2'	x, y, width, linetype, linewidth, fill, colour, alpha
`geom_rug`	'ggplot2'	x or y, linewidth, colour, alpha
`geom_hline`	'ggplot2'	yintercept, linetype, linewidth, colour, alpha
`geom_vline`	'ggplot2'	xintercept, linetype, linewidth, colour, alpha
`geom_abline`	'ggplot2'	intercept, slope, linetype, linewidth, colour, alpha
`geom_text`	'ggplot2'	x, y, label, face, family, angle, size, colour, alpha
`geom_label`	'ggplot2'	x, y, label, face, family, (angle), size, fill, colour, alpha
`geom_text_repel`	'ggrepel'	x, y, label, face, family, angle, size, colour, alpha
`geom_label_repel`	'ggrepel'	x, y, label, face, family, size, fill, colour, alpha
`geom_sf`	'ggplot2'	fill, colour
`geom_table`	'ggpp'	x, y, label, size, colour, angle
`geom_plot`	'ggpp'	x, y, label, vp.width, vp.height, angle
`geom_grob`	'ggpp'	x, y, vp.width, vp.height, label
`geom_blank`	'ggplot2'	—

```
p.base <-
  ggplot(data = mtcars,
       mapping = aes(x = disp, y = mpg, shape = factor(cyl))) +
  geom_point()
p.base
```

⌨ 9.4 Try a different mapping: `disp` → `colour`, `cyl` → `x`, keeping the mapping `mpg` → `y` unchanged. Continue by using `help(mtcars)` and/or `names(mtcars)` to see what other variables are available, and then try the combinations that trigger your curiosity—i.e., explore the data.

Adding `scale_shape_discrete()`, the scale already used by default, but passing `solid = FALSE` in the call creates a version of the same plot based on open shapes, still selected automatically.

```
p.base +
  scale_shape_discrete(solid = FALSE)
```

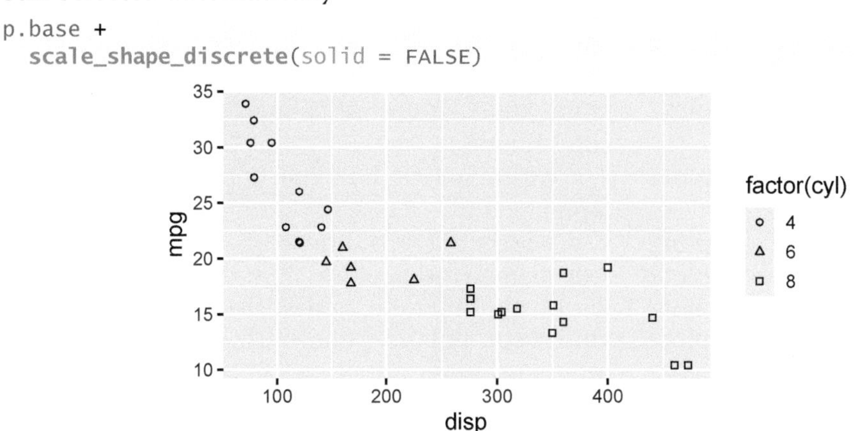

In contrast to "filled" shapes that obey both `colour` and `fill`, "open" shapes obey only `colour`, similarly to "solid" shapes. Function `scale_shape_manual` can be used to set the shape used for each value in the mapped factor. Below, "open" shapes are used, as they reveal partial overlaps better than solid shapes (plot not shown).

```
p.base +
  scale_shape_manual(values = c("circle open",
                                "square open",
                                "diamond open"))
```

It is also possible to use characters as shapes. The character is centred on the position of the observation. As the numbers used as symbols are self-explanatory, the default guide is removed by passing `guide = "none"` (plot not shown).

```
p.base +
 scale_shape_manual(values = c("4", "6", "8"), guide = "none")
```

A variable from `data` can be mapped to more than one aesthetic, allowing redundant aesthetics. This makes possible figures that, even if using colour, are readable when reproduced as black-and-white images and to viewers affected by colour blindness.

```
ggplot(data = mtcars,
       mapping = aes(x = disp, y = mpg,
       shape = factor(cyl), colour = factor(cyl))) +
  geom_point()
```

The next examples of the use of `geom_point()` are for **dot plots**, as `disp` is a `numeric` variable but `factor(cyl)` is discrete. Dot plots are prone to have overlapping observations, and one way of making these points visible is to make them partly transparent by setting a constant value smaller than one for the `alpha` *aesthetic*.

```
ggplot(data = mtcars,
       mapping = aes(x = factor(cyl), y = mpg)) +
  geom_point(alpha = 1/3)
```

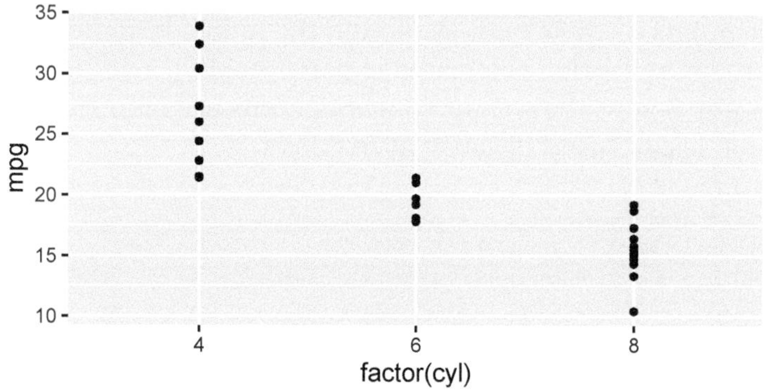

Function `position_identity()`, which is the default, does not alter the coordinates or position of observations, as shown in all examples above. To make overlapping observations visible, instead of making the points semitransparent as above, it is possible randomly displace them along the axis mapped to the discrete variable, x in this case. This is called *jitter*, and can be added using `position_jitter()` as argument to formal parameter `position` of `geoms`. The amount of jitter is set by numeric arguments passed to `width` and/or `height`, given as a fraction of the distance between adjacent factor levels in the plot.

```
ggplot(data = mtcars,
       mapping = aes(x = factor(cyl), y = mpg)) +
  geom_point(position = position_jitter(width = 0.25, heigh = 0))
```

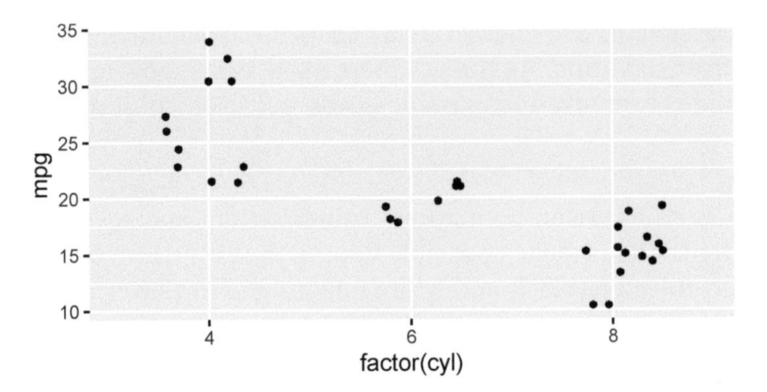

The name as a character string can be also used when no arguments need to be passed to the *position* function, and for some positions by passing numerical arguments to specific parameters of geometries. However, the default width of +0.5 tends to be rarely optimal (plot not shown).

```
ggplot(data = mtcars,
       mapping = aes(x = factor(cyl), y = mpg), colour = factor(cyl)) +
  geom_point(position = "jitter")
```

Bubble plots are scatter- or dot plots in which the size of points or bubbles varies following values of a continuous variable mapped to the `size` *aesthetic*. There are two approaches to this mapping, values in the mapped variable either describe the area of the points or their radii. Although the radius is sometimes used, due to how visual perception works, using area is perceptually closer to a linear mapping compared to radii. Below, the weights of cars in tons are mapped to the area of the points. Open circles are used because of overlaps.

```
ggplot(data = mtcars,
       mapping = aes(x = disp, y = mpg, colour = factor(cyl), size = wt)) +
  scale_size_area() +
  geom_point(shape = "circle open", stroke = 1.5)
```

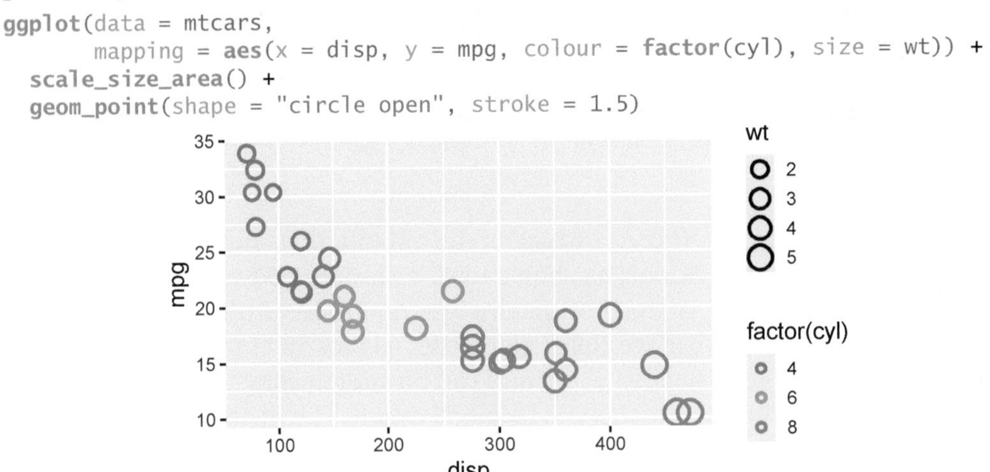

9.5 If a radius-based scale is used instead of an area-based one the perceived size differences are larger, i.e., the "impression" on the viewer is different. In the plot above, replace `scale_size_area()` with `scale_size_radius()`.

Display the plot, look at it carefully. Check the numerical values of some of the weights of the cars, and assess if your perception of the plot matches the numbers behind it.

As a final example summarising the use of `geom_point()`, the scatter plot below combines different *aesthetics* and their *scales*.

```
ggplot(data = mtcars,
       mapping = aes(x = disp, y = mpg, shape = factor(cyl),
                     fill = factor(cyl), size = wt)) +
  geom_point(alpha = 0.33, colour = "black") +
  scale_size_area() +
  scale_shape_manual(values = c("circle filled",
                                "square filled",
                                "diamond filled"))
```

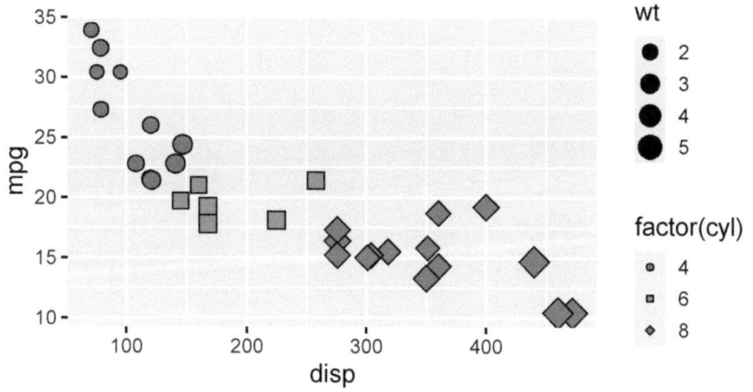

9.6 Play with the code in the chunk above. Remove or change each of the mappings and the scale, display the new plot, and compare it to the one above. Continue playing with the code until you are sure you understand what graphical element in the plot is added or modified by each individual argument or "word" in the code statement.

It is common to draw error bars together with points representing means or medians. These can be added in a single layer with `geom_pointrange()` with values mapped to the `x`, `y`, `ymin` and `ymax` aesthetics, using `y` for the point and `ymin` and `ymax` for the ends of the line segment. Two other *geometries*, `geom_range()` and `geom_errorbar()` draw only a segment or a segment with capped ends. These three `geoms` are frequently used together with `stats` that compute summaries by group. However, summary values calculated before plotting can alternatively be passed as `data`.

9.5.2 Rug

Rarely, rug plots are used by themselves. Instead they are usually an addition to scatter plots. An example of the use of `geom_rug()` follows. They make it easier to see the distribution of observations along the x- and/or y-axes. By default, rugs are drawn on the left and bottom edges of the plotting area. By passing `sides = "btlr"` they are drawn on the bottom, top, left, and right margins. Any combination of the four characters can be used to control the drawing of the rugs.

```
ggplot(data = mtcars,
       mapping = aes(x = disp, y = mpg, colour = factor(cyl))) +
  geom_point() +
  geom_rug(sides = "btlr")
```

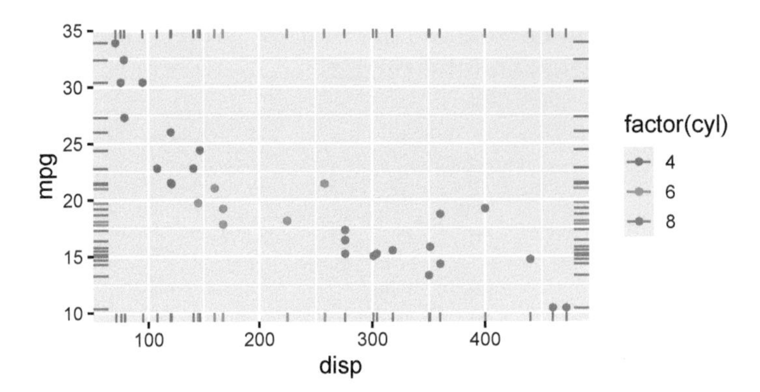

Rug plots are useful when the local density of observations in a continuous variable is not high as, otherwise, rugs become too cluttered and the rug "threads" overlap. When the overlap is moderate, making the segments semitransparent by setting the `alpha` aesthetic to a constant value smaller than one, can make the variation in density easier to appreciate. When the number of observations is large, marginal density plots are preferred.

9.5.3 Line and area

Line plots are normally created using `geom_line()`, and, occasionally using `geom_path()`. These two `geoms` differ in the sequence they follow when connecting values: `geom_line()` connects observations based on the ordering of `x` values while `geom_path()` uses the order in the data. Aesthetic `linewidth` controls the thickness of lines and `linetype` the patterns of dashes and dots.

In a line plot, observations, or the subset of observations in a group, are joined by straight lines. Below, a different data set, `Orange`, with data on the growth of five orange trees (see `help(Orange)`) is used. By mapping `Tree` to `linetype` the observations become grouped, and a line is plotted for each tree.

```
ggplot(data = Orange,
       mapping = aes(x = age, y = circumference, linetype = Tree)) +
  geom_line()
```

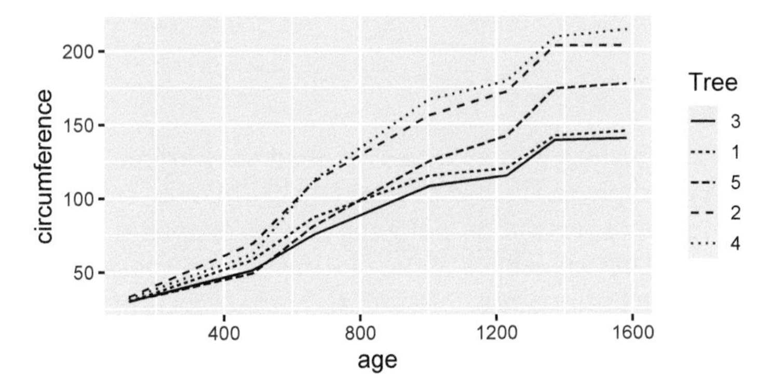

Before 'ggplot2' 3.4.0 the `size` aesthetic controlled the width of lines. Aesthetic `linewidth` was added in 'ggplot2' 3.4.0 and the use of the `size` aesthetic for lines deprecated.

Geometry `geom_step()` plots only vertical and horizontal lines to join the observations, creating a stepped line, or "staircase". Parameter `direction`, with default `"hv"`, controls the ordering of horizontal and vertical lines.

```
ggplot(data = Orange,
       mapping = aes(x = age, y = circumference, linetype = Tree)) +
  geom_step()
```

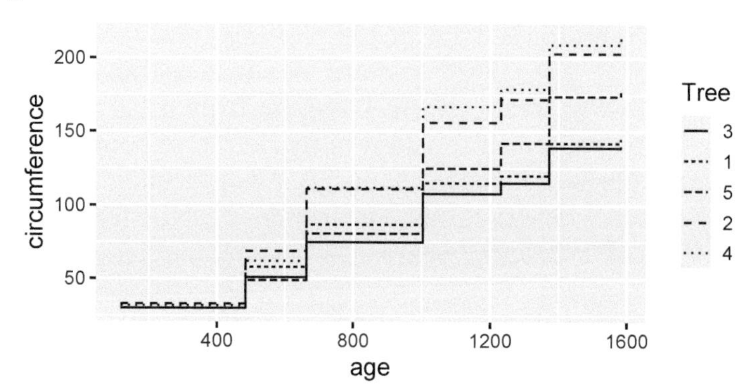

9.7 Using the following toy data, make three plots using `geom_line()`, `geom_path()`, and `geom_step` to add a layer. How do they differ?

```
toy.df <- data.frame(x = c(1,3,2,4), y = c(0,1,0,1))
```

While `geom_line()` draws a line joining observations, `geom_area()` supports, in addition, filling the area below the line according to the `fill` *aesthetic*. In some cases, it is useful to stack the areas, e.g., when the values plotted represent parts of a bigger whole. In the next, contrived, example, the areas representing the growth of the five orange trees are stacked (visually summed) using `position = "stack"` in place of the default `position = "identity"`. The visibility of the lines for individual trees is improved by changing their colour and width from the defaults. (Compare the y axis of the figure below to that drawn using `geom_line()` on page 300.)

```
p1 <- # will be used again later
  ggplot(data = Orange,
         mapping = aes(x = age, y = circumference, fill = Tree)) +
  geom_area(position = "stack", colour = "white", linewidth = 1)
p1
```

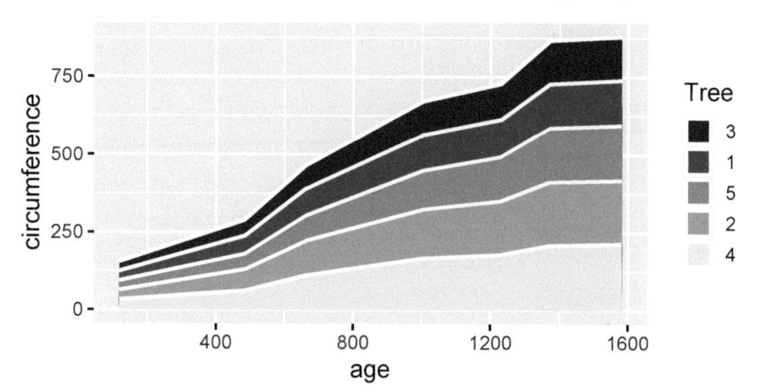

`geom_ribbon()` draws two lines based on the `x`, `ymin` and `ymax` *aesthetics*, with the space between the lines filled according to the `fill` *aesthetic*. `geom_polygon()`

is similar to geom_path() but connects the first and last observations forming a closed polygon that obeys the fill aesthetic.

Finally, three *geometries* for drawing lines across the whole plotting area: geom_hline(), geom_vline() and geom_abline(). The first two draw horizontal and vertical lines, respectively, while the third one draws straight lines according to the *aesthetics* slope and intercept determining the position. The lines drawn with these three geoms extend to the edge of the plotting area.

geom_hline() and geom_vline() require a single parameter (or aesthetic), yintercept and xintercept, respectively. Different from other geoms, the data for these aesthetics can be passed as constant numeric vector containing multiple values. The reason for this is that these geoms are most frequently used to annotate plots rather than plotting observations. Vertical lines can be used to highlight time points, here the ages of 1, 2, and 3 years.

```
p1 +
  geom_vline(xintercept = 365 * 1:3, colour = "gray75") +
  geom_vline(xintercept = 365 * 1:3, linetype = "dashed")
```

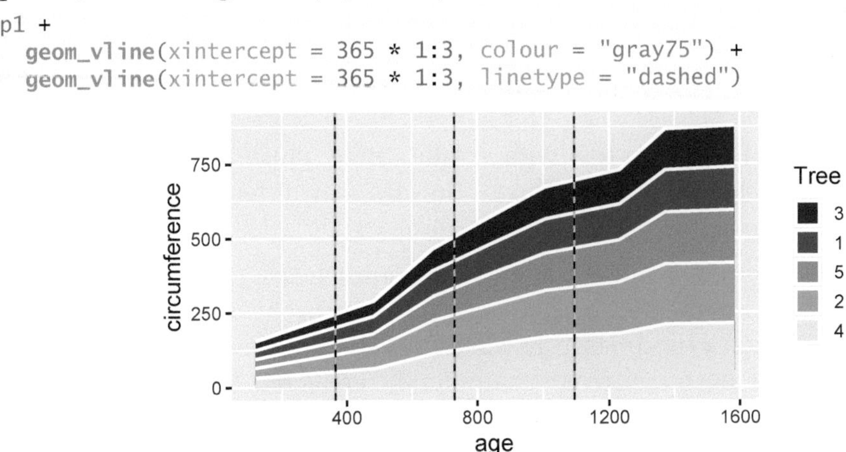

9.8 Change the order of the two layers in the example above. How did the figure change? What order is best? Would the same order be the best for a scatter plot? And would it be necessary to add two geom_vline() layers?

Similarly to geom_hline() and geom_vline(), geom_abline() draws a straight line, accepting as parameters (or as aesthetics) values for the intercept, a, and the slope, b.

Disconnected straight-line segments and arrows, one for each observation or row in the data, can be plotted with geom_segment() which accepts x, xend, y, and yend as mapped aesthetics. geom_spoke(), which uses a polar parametrisation, uses a different set of aesthetics, x, y for origin, and angle and radius for the segment. Similarly, geom_curve() draws curved segments, with the curvature, control points, and angles controlled through parameters. These three *geometries* support arrow heads at the ends of segments or curves, controlled through parameter arrow (not through an aesthetic).

9.5.4 Column

The *geometry* `geom_col()` can be used to create *column plots*, where each bar represents an observation or row in the `data` (frequently means or totals previously computed from the primary observations).

In other contexts, column plots are frequently called bar plots. `R` users not familiar yet with 'ggplot2' are frequently surprised by the default behaviour of `geom_bar()` as it uses `stat_count()` to produce a histogram, rather than plotting values as is (see section 9.6.4 on page 324). `geom_col()` is identical to `geom_bar()` but with `"identity"` as the default statistic.

Using very simple artificial data helps demonstrate how variations of column plots can be obtained. The data are for two groups, hypothetical males and females.

```
set.seed(654321)
my.col.data <-
  data.frame(treatment = factor(rep(c("A", "B", "C"), 2)),
             group = factor(rep(c("male", "female"), c(3, 3))),
             measurement = rnorm(6) + c(5.5, 5, 7))
```

The first plot includes data for `"female"` subjects extracted using a nested call to `subset()`. Except for `x` and `y` default mappings are used for all *aesthetics*.

```
ggplot(subset(my.col.data, group == "female"),
       mapping = aes(x = treatment, y = measurement)) +
  geom_col()
```

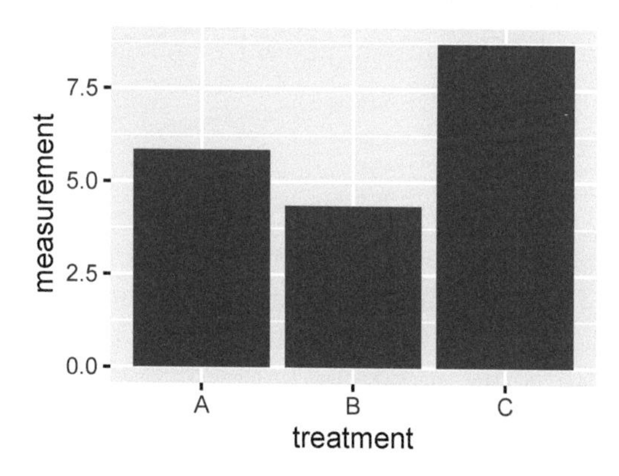

The bars above, are overwhelmingly wide, passing `width = 0.5` makes the bars narrower, using only half the distance between the levels on the x axis. Setting `colour = "white"` overrides the default colour of the lines bordering the bars. Both males and females are included and `group` is mapped to the `fill` aesthetic. The default argument for position in `geom_col()` is `position_stack()`. Function `position_stack()` is similar to `position_stack()` but divides the stacked values by their sum, i.e., the individual stacked "slices" of the column display proportions instead of absolute values.

```
p.base <-
 ggplot(my.col.data,
        mapping = aes(x = treatment, y = measurement, fill = group))

p1 <- p.base + geom_col(width = 0.5) + ggtitle("stack (default)")
```

Using `position = "dodge"` to override the default `position = "stack"` the columns for males and females are plotted side by side.

```
p2 <- p.base + geom_col(position = "dodge") + ggtitle("dodge")
```

The two plots side by side (see section 9.14 on page 369 for details).

```
p1 + p2
```

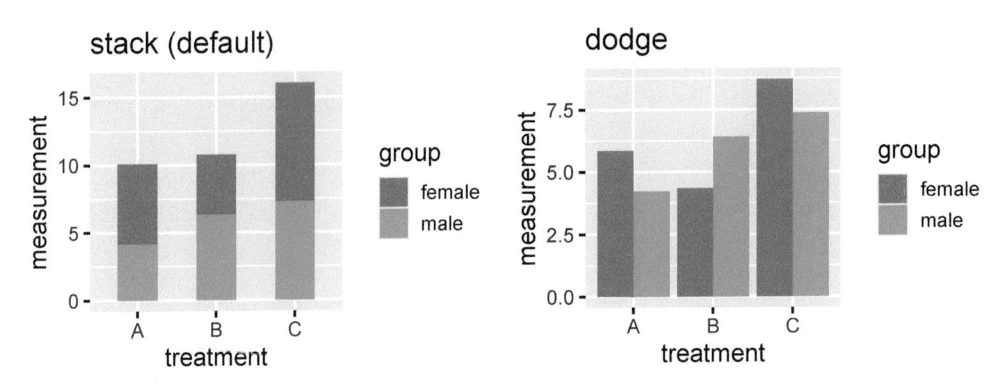

9.9 Change the argument to `position`, or let the default be active, until you understand its effect on the figure. What is the difference between *positions* `"identity"`, `"dodge"`, `"stack"`, and `"fill"`?

9.10 Use constants as arguments for *aesthetics* or map variable `treatment` to one or more of the *aesthetics* recognised by `geom_col()`, such as `colour`, `fill`, `linetype`, `size`, `alpha` and `width`.

9.5.5 Tiles

Tile plots and **heat maps** are useful when observations are available on a regular rectangular 2D grid. The grid can, for example, represent locations in space as well combinations of levels of two discrete classification criteria. The colour or darkness of the tiles informs about the value of the observations. A layer with square or rectangular tiles can be added with `geom_tile()`.

Data from 100 random draws from the *F* distribution with degrees of freedom $\nu_1 = 2, \nu_2 = 20$ are used in the examples.

```
set.seed(1234)
randomf.df <- data.frame(F.value = rf(100, df1 = 2, df2 = 20),
                         x = rep(letters[1:10], 10),
                         y = LETTERS[rep(1:10, rep(10, 10))])
```

`geom_tile()` requires aesthetics x and y, with no defaults, and `width` and `height` with defaults that make all tiles of equal size filling the plotting area. Variable `F.value` is mapped to `fill`.

```
ggplot(data = randomf.df,
       mapping = aes(x, y, fill = F.value)) +
  geom_tile()
```

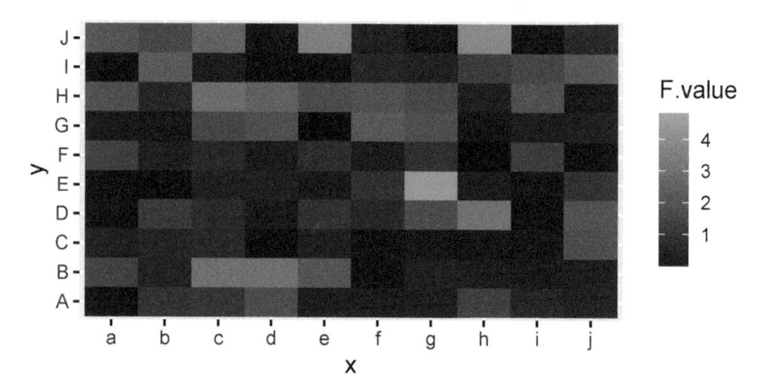

Below, setting `colour = "gray75"` and `linewidth = 1` makes the tile borders visible. Whether highlighting these lines improves or not a tile plot depends on whether the individual tiles correspond to values of a categorical- or continuous variable. For example, when rows of tiles correspond to genes and columns to discrete treatments, visible tile borders are preferable. In contrast, in the case when the tiles are an approximation to a continuous surface like measurements on a regular spatial grid, it is best to suppress tile borders.

```
ggplot(data = randomf.df,
       mapping = aes(x, y, fill = F.value)) +
  geom_tile(colour = "gray75", linewidth = 1)
```

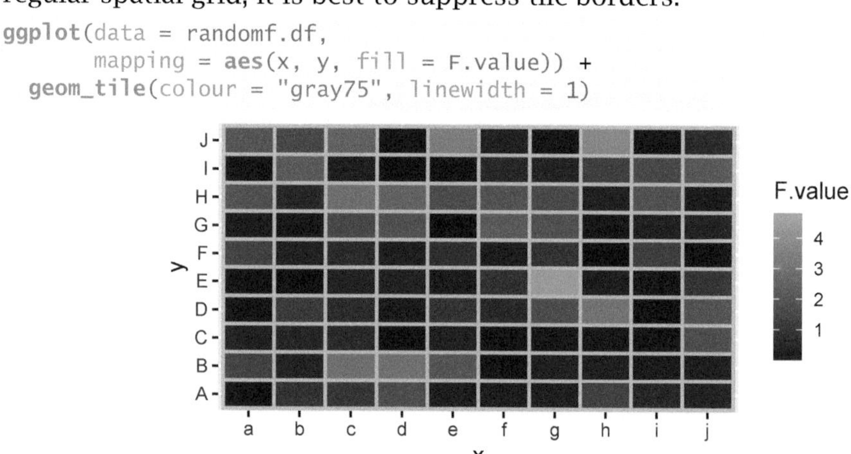

9.11 Play with the arguments passed to parameters `colour` and `size` in the example above, considering what features of the data are most clearly perceived in each of the plots you create.

Continuous fill scales can be used to control the appearance. Below, code for a tile plot based on a gray gradient, with missing values in red, is constructed is shown (plot not shown).

```
ggplot(data = randomf.df,
       mapping = aes(x, y, fill = F.value)) +
  geom_tile(colour = "white") +
  scale_fill_gradient(low = "gray15", high = "gray85", na.value = "red")
```

In contrast to `geom_tile()`, `geom_rect()` draws rectangular tiles based on the position of the corners, mapped to aesthetics `xmin`, `xmax`, `ymin` and `ymax`. In this case, tiles can vary in size and do not need to be contiguous. The filled rectangles can be used, for example, to highlight a rectangular region in a plot (see example on page 313).

9.5.6 Simple features (sf)

'ggplot2' version 3.0.0 or later supports with `geom_sf()`, and its companions, `geom_sf_text()`, `geom_sf_label()`, and `stat_sf()`, the plotting of shape data similarly to geographic information systems (GIS). This makes it possible to display data on maps, for example, using different fill values for different regions. The special *coordinate* `coord_sf()` can be used to select different projections for maps. The *aesthetic* used is called `geometry` and contrary to all the other aesthetics described above, the values to be mapped are of class `sfc` containing *simple features* data with multiple components. Manipulation of simple features data is supported by package '`sf`'. Normal geometries can be use together with `stat_sf_coordinates()` to add other graphical elements to maps. This subject exceeds the scope of this book, so a single and very simple example is shown below.

```
nc <- sf::st_read(system.file("shape/nc.shp", package = "sf"), quiet = TRUE)
ggplot(nc) +
  geom_sf(mapping = aes(fill = AREA), colour = "gray90")
```

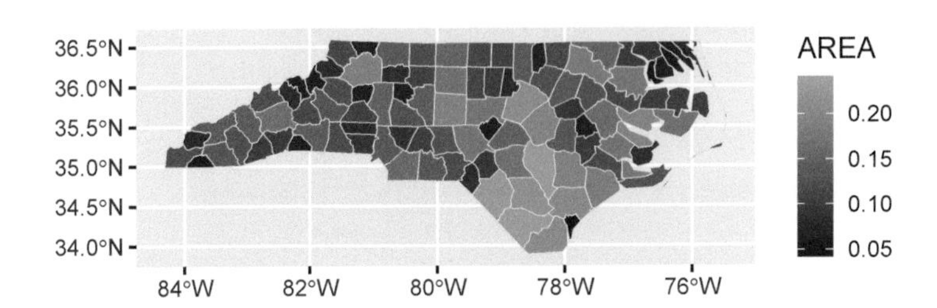

9.5.7 Text

Geometries `geom_text()` or `geom_label()` are used to add textual data labels and annotations to plots.

For `geom_text()` and `geom_label()`, the aesthetic `label` provides the text to be plotted and aesthetics `x` and `y`, the location of the labels. The size of the text is controlled by the `size` aesthetics, while the font is selected by the `family` and `fontface` aesthetics. Below, the whole-plot default mappings for `colour` and `size` aesthetics are overridden within `geom_text()`.

```
ggplot(data = mtcars,
       mapping = aes(x = disp, y = mpg,
                     colour = factor(cyl), size = wt, label = cyl)) +
  geom_point(alpha = 1/3) +
  geom_text(colour = "darkblue", size = 3)
```

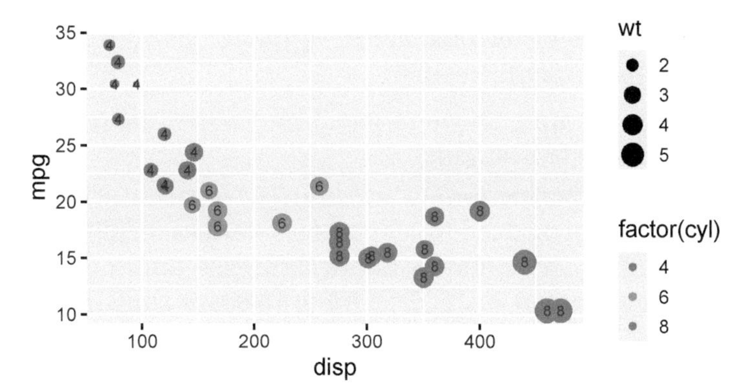

Aesthetics `angle`, expressed in degrees, and `vjust` and `hjust` can be used to rotate the text and adjust its vertical and horizontal justification. The default value of 0.5 for both `hjust` and `vjust` sets the centre of the text at the supplied `x` and `y` coordinates. *"Vertical" and "horizontal" for text justification are relative to the text, not the plot.* This is important when `angle` is different from zero. Values larger than 0.5 shift the label left or down, and values smaller than 0.5, right or up with respect to its `x` and `y` coordinates. A value of 1 or 0 sets the text so that its edge is at the supplied coordinate. Values outside the range 0 ... 1 shift the text even farther away, however, still using units based on the length or height of the text label. Recent versions of 'ggplot2' make possible justification using character constants for alignment: `"left"`, `"middle"`, `"right"`, `"bottom"`, `"center"`, and `"top"`, and two special alignments, `"inward"` and `"outward"`, that automatically vary based on the position in the plotting area.

Below, `geom_text()` or `geom_label()` are used together with `geom_point()` similarly as they are used to add data labels in a plot.

```
my.data <-
  data.frame(x = 1:5,
             y = rep(2, 5),
             label = c("ab", "bc", "cd", "de", "ef"))

ggplot(data = my.data,
       mapping = aes(x, y, label = label)) +
  geom_text(angle = 90, hjust = 1.5, size = 4) +
  geom_point()
```

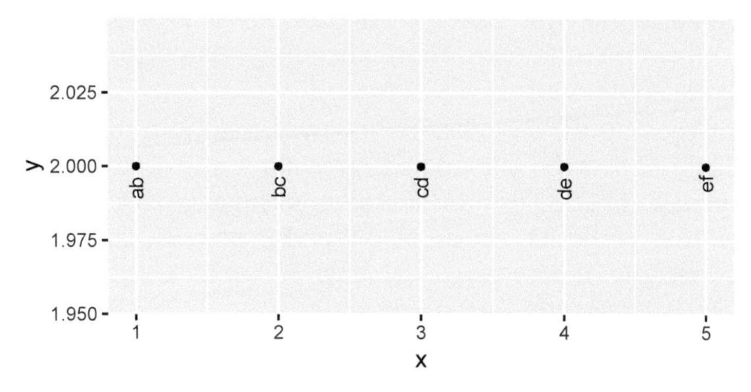

In the case of `geom_label()` the text is enclosed in a box and obeys the `fill` *aesthetic* and additional parameters (described starting at page 309) allowing control of the shape and size of the box. Before 'ggplot2' 3.5.0, `geom_label()` did not support rotation with the `angle` aesthetic.

9.12 Modify the example above to use `geom_label()` instead of `geom_text()` using, in addition, the `fill` aesthetic.

A serif font is set by passing `family = "serif"`. The names `"sans"` (the default), `"serif"` and `"mono"` are recognised by all graphics devices on all operating systems. They do not necessarily correspond to identical fonts in different computers or for different graphic devices, but instead to fonts that are similar. Additional fonts are available for specific graphic devices, such as the 35 "PDF" fonts by the `pdf()` device. In this case, their names can be queried with `names(pdfFonts())`.

```
ggplot(data = my.data,
       mapping = aes(x, y, label = label)) +
  geom_text(angle = 90, hjust = 1.5, size = 4, family = "serif") +
  geom_point()
```

9.13 In the examples above, the character strings were all of the same length, containing a single character. Redo the plots above with longer character strings of various lengths mapped to the `label` *aesthetic*. Do also play with justification of these labels.

R and 'ggplot2' support the use of UNICODE, such as UTF8 character encodings in strings. If your editor or IDE supports their use, then you can type Greek letters and simple maths symbols directly, and they *may* show correctly in labels if a suitable font is loaded and an extended encoding like UTF8 is in use by the operating system. Even if UTF8 is in use, text is not fully portable unless the same font is available, as even if the character positions are standardised for many languages, most UNICODE fonts support at most a small number of languages. In principle, one can use this mechanism to have labels both using other alphabets and languages like Chinese with their numerous symbols mixed in the same figure. Furthermore, the support for fonts and consequently character sets in R is output-device dependent. The font encoding used by R by default depends on the default locale settings of the operating system, which can also lead to garbage printed to the console or wrong characters being plotted running the same code on a different computer from the one where a script was created. Not all is lost, though, as R can be coerced to use system fonts and Google fonts with functions provided by packages 'showtext' and 'extrafont'. Encoding-related problems, especially in MS-Windows, are common.

Plotting (mathematical) expressions involves mapping to the `label` aesthetic character strings that can be parsed as expressions, and setting `parse = TRUE` (see section 9.15 on page 371). Below, the character strings are assembled using `paste()` but, of course, they could have been also typed in as constant values. This use of `paste()` is an example of recycling of shorter vectors, `"alpha["` and `"]"` to match the length of `1:5` (see section 3.10 on page 64).

```
my.data <-
  data.frame(x = 1:5, y = rep(2, 5), label = paste("alpha[", 1:5, "]", sep = ""))
my.data$label
## [1] "alpha[1]" "alpha[2]" "alpha[3]" "alpha[4]" "alpha[5]"
```

Text and labels do not automatically expand the plotting area past their anchoring coordinates. In the example below, `expand_limits(x = 5.2)` ensures that the text is not clipped at the edge of the plotting area.

```
ggplot(data = my.data,
       mapping = aes(x, y, label = label)) +
  geom_text(hjust = -0.2, parse = TRUE, size = 6) +
  geom_point() +
  expand_limits(x = 5.2)
```

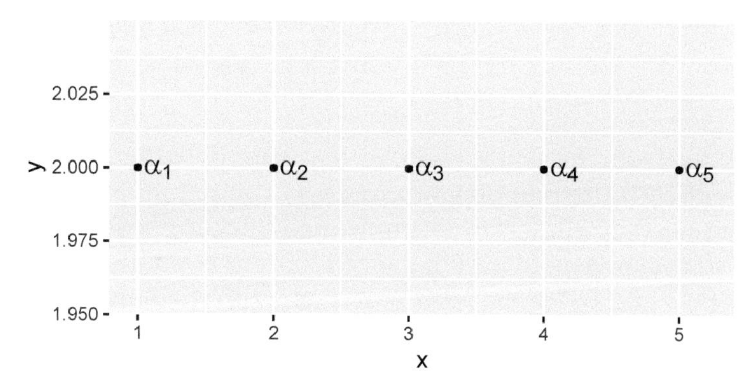

In the example above, the text to be parsed was mapped to the `label` aesthetic using character strings previously added to the data frame `my.data`. It is also possible, and usually preferable, to build suitable character strings with a nested function call, or a code statement, passed as an argument in the call to `aes()` (plot identical to the previous one, not shown).

```
ggplot(data = my.data,
       mapping = aes(x, y, label = paste("alpha[", x, "]", sep = ""))) +
  geom_text(hjust = -0.2, parse = TRUE, size = 6) +
  geom_point()
```

Geometry `geom_label()` obeys the same aesthetics as `geom_text()` (except for `angle` in 'ggplot2' < 3.5.0) and additionally `label.size` for the width of the border line, `label.r` for the roundness of the box corners, `label.padding` for the space between the text boundary and the box boundary, and `fill` for the colour used to fill the boxes' background.

```
my.data <-
  data.frame(x = 1:5, y = rep(2, 5),
             label = c("one", "two", "three", "four", "five"))

ggplot(data = my.data,
       mapping = aes(x, y, label = label)) +
  geom_label(hjust = -0.2, size = 6,
             label.size = 0,
             label.r = unit(0, "lines"),
             label.padding = unit(0.15, "lines"),
             fill = "yellow", alpha = 0.5) +
  geom_point() +
  expand_limits(x = 5.6)
```

9.14 Starting from the example above, play with the arguments to the different parameters and with the mappings to *aesthetics* to get an idea of the variations in the design that they allow. For example, use thicker border lines and increase the padding so that a visually well-balanced margin is retained. You may also try mapping the `fill` and `colour` *aesthetics* to factors in the data.

If the parameter `check_overlap` of `geom_text()` is set to `TRUE`, text overlap will be avoided by suppressing the text that would otherwise overlap other text. *Repulsive* versions of `geom_text()` and `geom_label()`, `geom_text_repel()` and `geom_label_repel()`, are available in package 'ggrepel'. These *geometries* avoid overlaps by automatically repositioning the text or labels. Please read the package documentation for details of how to control the repulsion strength and direction, and the properties of the segments linking the labels to the position of their data coordinates. Nearly all aesthetics supported by `geom_text()` and `geom_label()` are supported by the repulsive versions. However, given that a segment connects the label or text to its anchor point, several properties of these segments can also be controlled with aesthetics or arguments.

```
ggplot(data = mtcars,
       mapping = aes(x = disp, y = mpg,
                     colour = factor(cyl), size = wt, label = cyl)) +
  scale_size() +
  geom_point(alpha = 1/3) +
  geom_text_repel(colour = "black", size = 3,
                  min.segment.length = 0.2, point.padding = 0.1)
```

9.5.8 Plot insets

The support for insets in 'ggplot2' is confined to `annotation_custom()`, which was designed to be used for static annotations expected to be the same in each panel of a plot (the use of annotations is described in section 9.11). Package 'ggpp' provides geoms that mimic `geom_text()` in relation to the *aesthetics* used, but that similarly to `geom_sf()`, expect that the column in `data` mapped to the `label` aesthetics are lists of objects containing multiple pieces of information, rather than atomic vectors. Three geometries are currently available: `geom_table()`, `geom_plot()` and `geom_grob()`.

Given that `geom_table()`, `geom_plot()`, and `geom_grob()` will rarely use a mapping inherited from the whole plot, by default they do not inherit it. Either the mapping should be supplied as an argument to these functions or their parameter `inherit.aes` explicitly set to `TRUE`.

Tables can be added as plot insets with `geom_table()` by mapping a list of data frames (or tibbles) to the `label` *aesthetic*. Positioning, justification, and angle work as for `geom_text()` and are applied to the whole table. The table(s) are constructed as 'grid' `grob` objects and added to the `gg` plot object as a layer.

The code below builds a `tibble` containing summaries from the `mtcars` data set, with the summary values formatted as character strings, adds this tibble as the single member to a list, and stores this list as column named `table.inset` in another `tibble`, named codetable.tb, together with the `x` and `y` coordinates for its location as an inset.

The code uses functions from the 'tidyverse' (see section 8.7.2 on page 262). Data frames and base R functions could have been used instead (see section 4.4.2 on page 105).

```
mtcars |>
  group_by(cyl) |>
  summarize("mean wt" = format(mean(wt), digits = 3),
            "mean disp" = format(mean(disp), digits = 2),
            "mean mpg" = format(mean(mpg), digits = 2)) -> my.table
table.tb <- tibble(x = 500, y = 35, table.inset = list(my.table))
```

As with text labels, justification is interpreted in relation to table-text orientation, however, the default, `"inward"`, rarely needs to be changed if one sets x and y coordinates to the location of the inset corner farthest from the centre of the plot. The inset table is added at its native size, given by the `size` aesthetic, which is applied to the text in it.

```
ggplot(data = mtcars,
       mapping = aes(x = disp, y = mpg, colour = factor(cyl), size = wt)) +
  scale_size() +
  geom_point() +
  geom_table(data = table.tb,
             mapping = aes(x = x, y = y, label = table.inset),
             colour = "black", size = 3)
```

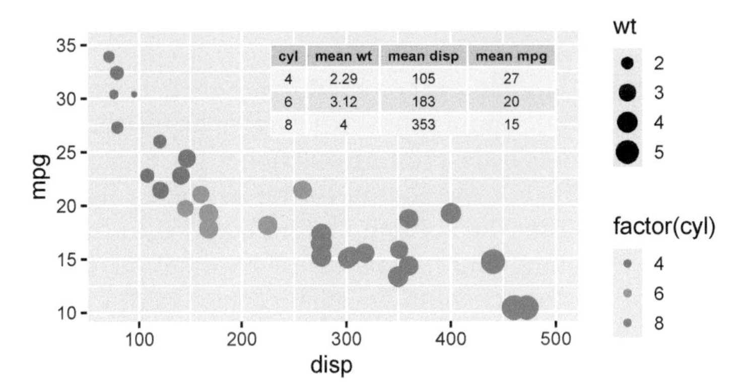

Parsed text, using R's *plotmath* syntax is supported in tables, with fallback to plain text in case of parsing errors, on a cell-by-cell basis.

The *geometry* `geom_table()` uses functions from package 'gridExtra' to build a graphical object for the table. A table theme can be passed as an argument to `geom_table()`.

Geometry `geom_plot()` works similarly to `geom_table()` but insets a ggplot within another ggplot. Thus, instead of expecting a list of data frames or tibbles to be mapped to the `label` aesthetics, it expects a list of ggplots (objects of class `gg`). Inset plots can be very useful for zooming-in on parts of a main plot where observations are crowded and for displaying summaries based on the observations shown in the main plot. The inset plots are nested in viewports which constrain the dimensions of the inset plot. Aesthetics `vp.height` and `vp.width` set the size of the viewports—with defaults of 1/3 of the height and width of the plotting area of the main plot. Themes can be applied separately to the main and inset plots.

In the first example of inset plots, the summaries shown above as numbers in a column in the inset table, are displayed in an inset column plot. We first create a one-row `data.frame` containing the plot to be inset as member of a `list`, and the x and y coordinates in the main plot of the location of the inset. Unlike with a `tibble`, with a `data.frame` we need to use `I()` to protect the `list`.

```
mtcars |>
  group_by(cyl) |>
  summarize(mean.mpg = mean(mpg)) |>
  ggplot(data = _,
         mapping = aes(factor(cyl), mean.mpg, fill = factor(cyl))) +
  scale_fill_discrete(guide = "none") +
  scale_y_continuous(name = NULL) +
    geom_col() +
    theme_bw(8) -> my.plot
plot.tb <- data.frame(x = 500, y = 35, plot.inset = I(list(my.plot)))

ggplot(data = mtcars,
       mapping = aes(x = disp, y = mpg, colour = factor(cyl))) +
  geom_point() +
  geom_plot(data = plot.tb,
            aes(x = x, y = y, label = plot.inset),
            vp.width = 1/2,
            hjust = "inward", vjust = "inward")
```

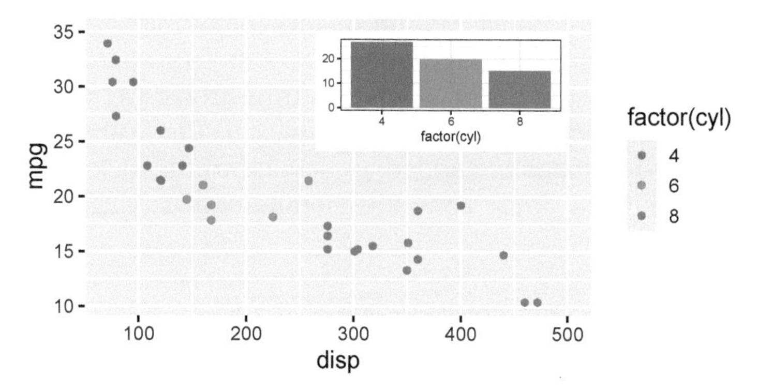

In the second example, the plot inset is a zoom-in into a region of the base plot. The code to build this plot is split into three chunks. `p.main` is the plot to be used as the base for the final plot.

```
p.main <-
  ggplot(data = mtcars,
         mapping = aes(x = disp, y = mpg, colour = factor(cyl))) +
  geom_point()
```

`p.inset`, is the plot to be used as the inset; the call to `coord_cartesian()` zooms-into `p.main`; the call to `labs()` removes the redundant axis labels; the call to `scale_colour_discrete()` removes the redundant guide in the inset; and the calls to `theme_bw()` and `theme()` change the theme and font size for the inset.

```
p.inset <- p.main +
  coord_cartesian(xlim = c(270, 330), ylim = c(14, 19)) +
  labs(x = NULL, y = NULL) +
  scale_colour_discrete(guide = "none") +
  theme_bw(8) + theme(aspect.ratio = 1)
```

As in the previous example, `geom_plot()` adds the inset, in this case with constant values for aesthetics. The call to `annotate()` using `geom_rect()` adds the rectangle highlighting the zoomed-in region in the main plot.

```
p.main +
  geom_plot(x = 480, y = 34, label = list(p.inset), vp.height = 1/2) +
  annotate(geom = "rect", fill = NA, colour = "black",
           xmin = 270, xmax = 330, ymin = 14, ymax = 19,
           linetype = "dotted")
```

Geometry `geom_grob()` differs very little from `geom_plot()` but insets 'grid' graphical objects, called `grob` for short. This approach is very flexible, as grobs can be vector graphics as well as contain rasters (or bitmaps). In most cases, the grobs need to be first created either using functions from package 'grid' to draw them or by converting other types of objects into grobs. Geometry `geom_grob()` is as flexible as `annotation_custom()` with respect to the grobs but behaves as a *geometry*. Below, two bitmaps are added as "labels" to the base plot.

The bitmaps are read from PNG files (contained as examples in package 'gpmisc'.

```
file1.name <-
  system.file("extdata", "Isoquercitin.png",
              package = "ggpp", mustWork = TRUE)
Isoquercitin <- magick::image_read(file1.name)
file2.name <-
  system.file("extdata", "Robinin.png",
              package = "ggpp", mustWork = TRUE)
Robinin <- magick::image_read(file2.name)
```

The two bitmaps are converted into `grobs`, added as two separate members to a list, and the list added as a column to a `data.frame` named, for this example, `grob.tb`. The coordinates for the position of each `grob` as well as the size of each viewport are also added to this `data.frame`.

```
grob.tb <-
  data.frame(x = c(0, 100), y = c(10, 20), height = 1/3, width = c(1/2),
             grobs = I(list(grid::rasterGrob(image = Isoquercitin),
                            grid::rasterGrob(image = Robinin))))
```

The two `grobs` are added as a single plot layer to an empty plot. Insets like these, can be added to any base plot.

```
ggplot() +
  geom_grob(data = grob.tb,
            mapping = aes(x = x, y = y, label = grobs,
                          vp.height = height, vp.width = width),
                          hjust = "inward", vjust = "inward")
```

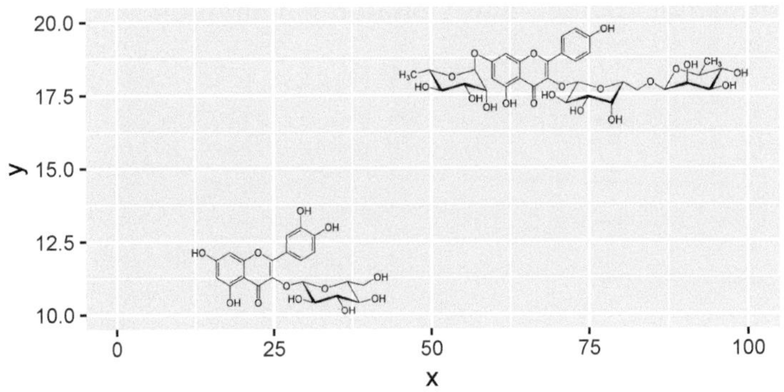

Grid graphics provide the low-level functions that 'ggplot2' uses under the hood. Package 'grid' supports different types of units for expressing the coordinates of positions. In the 'ggplot2' user interface, `"native"` data coordinates are used with only a few exceptions. Package 'grid' supports the use of physical units like `"mm"` as well as relative units like "npc" *normalised parent coordinates.* Positions expressed as npc are numbers in the range 0 to 1, relative to the dimensions of current *viewport*, with origin at the lower left corner. Normalised parent coordinates ("npc") are useful when annotating plots and adding insets at positions relative to the plotting area, as these positions remain always consistent across different plots, or across panels when using facets with free axis limits.

Package 'ggplot2' interprets x and y coordinates in `"native"` data coordinates. Newly, 'ggplot2' >= 3.5.0 interprets "mappings" of variables and constant values enclosed in function `I()` as expressed using "npc" coordinates, skipping the usual mapping based on scales.

```
ggplot(data = mtcars,
       mapping = aes(x = disp, y = mpg, colour = factor(cyl))) +
  geom_point() +
  geom_label(x = I(0.5), y = I(0.9), label = "a label", colour = "black")
```

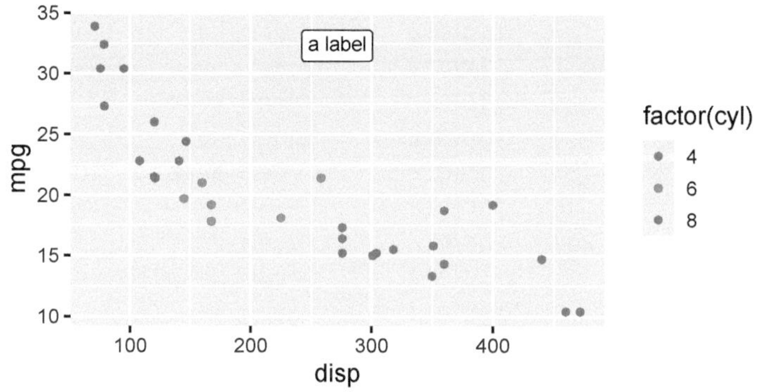

An earlier approach was provided by package 'ggpp' through *pseudo aesthetics* `npcx` and `npcy` and *geometries* that support them can be used with 'ggplot2' <= 3.4.4.

```
ggplot(data = mtcars,
       mapping = aes(x = disp, y = mpg, colour = factor(cyl))) +
  geom_point() +
  geom_label_npc(npcx = 0.5, npcy = 0.9, label = "a label", colour = "black",
            vjust = "center")
```

9.6 Statistics

All statistics, except `stat_identity()`, modify the `data` they receive before passing it to a geometry. Most statistics compute a specific summary from the data, but there are exceptions. More generally, they make it possible to integrate computations on the data into the plotting workflow. This saves effort but more importantly helps ensure that the data and summaries within a given plot are consistent. Table 9.2 list all the statistics used in the chapter.

When a factor is mapped to an aesthetic, each level creates a group. For example, in the first plot example in section 9.5.3 on page 300, the grouping resulted in separate lines. The grouping is not so obvious with other aesthetics but it is not different. Most *statistics* operate separately on the data for each group, returning an independent summary for each group. Mapping a continuous variable to an

Table 9.2
'ggplot2' statistics described in section 9.6, packages where they are defined, their default geometry, and the aesthetics they use as input for computations.

Statistic	Package	Geometry	Aesthetics
stat_function	'ggplot2'	geom_function	x
stat_summary	'ggplot2'	geom_pointrange	x, y
stat_smooth	'ggplot2'	geom_smooth	x, y, weight
stat_poly_line	'ggpmisc'	geom_smooth	x, y, weight
stat_poly_eq	'ggpmisc'	geom_text	x, y, weight
stat_fit_tb	'ggpmisc'	geom_table	x, y, weight
stat_bin	'ggplot2'	geom_bar	x, y
geom_histogram	'ggplot2'	—	x, y
stat_bin2d	'ggplot2'	geom_tile	x, y
stat_bin_hex	'ggplot2'	geom_hex	x, y
stat_density	'ggplot2'	geom_area	x, y
geom_density	'ggplot2'	—	x, y
stat_density_2d	'ggplot2'	geom_density_2d	x, y
stat_boxplot	'ggplot2'	geom_boxplot	x, y
stat_ydensity	'ggplot2'	geom_violin	x, y
geom_violin	'ggplot2'	—	x, y
geom_quasirandom	'ggbeeswarm'	—	x, y
stat_ma_line	'ggpmisc'	geom_smooth	x, y
stat_ma_eq	'ggpmisc'	geom_text	x, y
stat_centroid	'ggpmisc'	geom_point	x, y
stat_quant_line	'ggpmisc'	geom_smooth	x, y
stat_quant_eq	'ggpmisc'	geom_text	x, y
stat_identity	'ggplot2'	geom_point	—

aesthetics does not create groups. All aesthetics, including `x` and `y`, follow this pattern, thus a factor mapped to `x` also creates a group for each level of the factor.

9.6.1 Functions

Statistics `stat_function()` is the simplest to use and understand, even if unusual. It generates y values by applying an R function to a sequence of x values. The range of the `numeric` variable mapped to `x` determines the range of x values used.

Any R function, user defined or not, can be used as long as it is vectorised, with the length of the returned vector equal to the length of the vector passed as an argument to its first parameter. The argument passed to parameter `n` of `geom_function()` determines the length of the generated vector of x values. The data frame returned contains these are the x values and as y values the values returned by the function.

The code to plot the Normal probability distribution function is very simple, relying on the defaults `n = 101` and `geom = "path"`.

```
ggplot(data = data.frame(x = c(-3,3)),
       mapping = aes(x = x)) +
  stat_function(fun = dnorm)
```

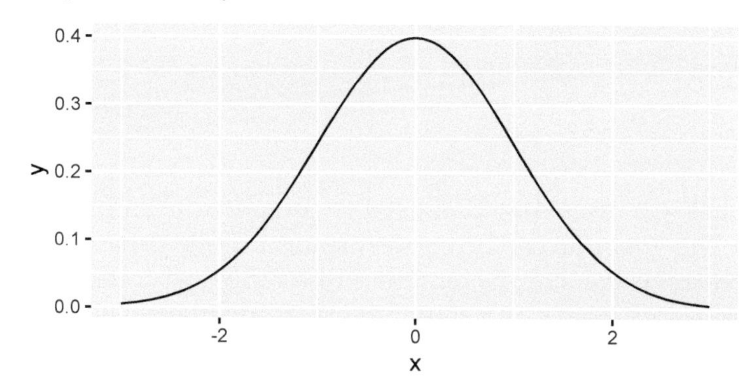

Using a named list, additional arguments can be passed to the function when called to generate the data (plot not shown).

```
ggplot(data = data.frame(x = c(-3,4)),
       mapping = aes(x = x)) +
  stat_function(fun = dnorm, args = list(mean = 1, sd = .5))
```

9.15 Edit the code above so as to plot in the same figure three curves, either for three different values for `mean` or for three different values for `sd`.

Named user-defined functions (not shown), and anonymous functions (below) can also be used.

```
ggplot(data = data.frame(x = 0:1),
       mapping = aes(x = x)) +
  stat_function(fun = function(x, a, b){a + b * x^2},
                args = list(a = 1, b = 1.4))
```

9.16 Edit the code above to use a different function, such as e^{x+k}, adjusting the argument(s) passed through `args` accordingly. Do this by means of an anonymous function, and by means of an equivalent named function defined by your code.

9.6.2 Summaries

The summaries discussed in this section can be superimposed on raw data plots, or plotted on their own. Beware, that if scale limits are manually set, the summaries will be calculated from the subset of observations within these limits. Scale limits can be altered when explicitly defining a scale or by means of functions `xlim()` and `ylim()`. See section 9.12 on page 362 for an explanation of how coordinate limits can be used to zoom into a plot without excluding of x and y values from the data.

It is possible to summarise data on the fly when plotting. The simultaneous calculation of measures of central tendency and of variation in `stat_summary()` allows them to be added together to the same plot layer.

Data frame `fake.data`, constructed below, contains normally distributed artificial values in variable `Y` in two groups, distinguished by the levels of factor `group`.

```
fake.data <- data.frame(
  y = c(rnorm(10, mean = 2, sd = 0.5),
        rnorm(10, mean = 4, sd = 0.7)),
  group = factor(c(rep("A", 10), rep("B", 10))))
```

Below, a base plot is constructed an assigned to `p1.base`.

```
p1.base <-
  ggplot(data = fake.data, mapping = aes(y = y, x = group)) +
  geom_point(shape = "circle open")
```

In `stat_summary()`, the R function used to compute the summaries is passed as an argument. This function can be one returning a single value, like `mean()`, or one returning a central value and the extremes of a range. With the default argument, `stat_summary()` plots means and standard errors, displaying a message.

```
p1.base + stat_summary()

## No summary function supplied, defaulting to `mean_se()`
```

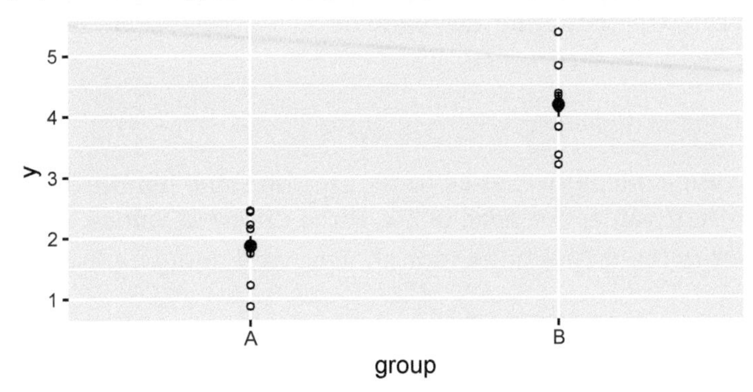

For $\bar{x} \pm$ s.e., the default, `"mean_se"` can be passed as argument to `fun.data` to avoid the message seen above, and for $\bar{x} \pm$ s.d. `"mean_sdl"` should be passed as argument. These functions have to be passed to parameter `fun.data`, while functions that return a single value, like `"mean"`, to `fun`. The `geom` used has to be suitable for the values computed by the `stat`.

Below is code for a similar plot, with means highlighted in red, using `geom_point()`.

```
p1.base +
  stat_summary(fun = "mean", geom = "point",
               colour = "red", shape = "-", size = 15)
```

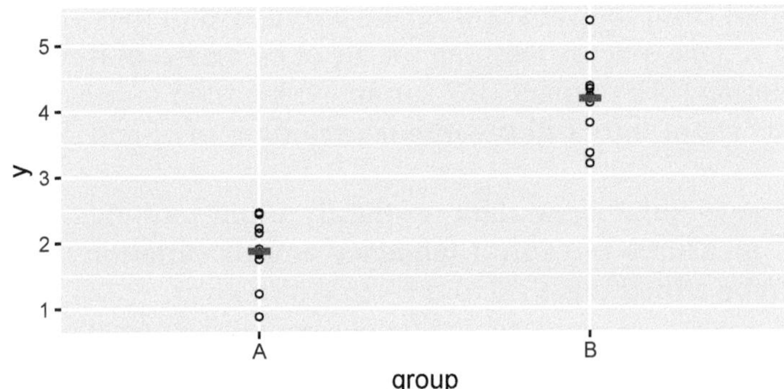

Below, confidence intervals for $P = 0.99$ computed assuming normality are added. Intervals can be also computed without assuming normality, using the empirical distribution estimated from the data by bootstrap using `"mean_cl_boot"` instead of `"mean_cl_normal"`.

```
p1.base +
  stat_summary(fun.data = "mean_cl_normal", fun.args = list(conf.int = 0.99),
               colour = "red", size = 0.7, linewidth = 1, alpha = 0.5)
```

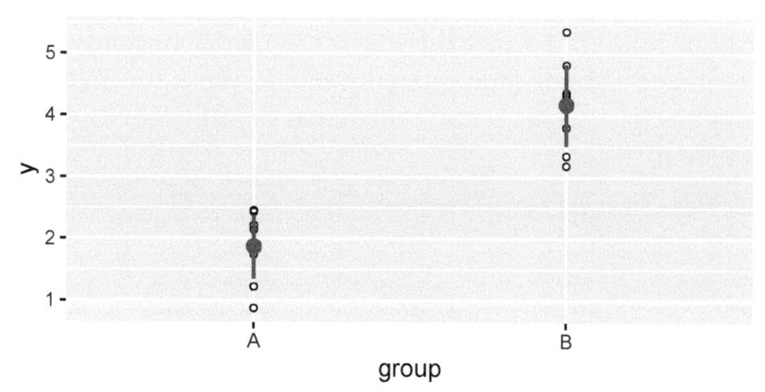

It is possible to use user-defined functions instead of the functions exported by package 'ggplot2' (based on those in package 'Hmisc'). Additional named arguments can be passed to the summary function through parameter `fun.args` of `stat_summary()`.

Means, or other summaries, computed by groups based on the factor mapped to the `x` aesthetic (`class` in this example) can be plotted as columns by passing `"col"` as an argument to parameter `geom`.

```
p2.base <-
  ggplot(data = mpg, mapping = aes(x = class, y = hwy)) +
  stat_summary(geom = "col", fun = mean)
```

Error bars can be added to the column plot. Passing `linewidth = 1` makes the lines of the error bars thicker. The default *geometry* in `stat_summary()` is `geom_pointrange()`, passing `"linerange"` as an argument for `geom` removes the points at the top edge of the bars.

```
p2.base +
  stat_summary(geom = "linerange", fun.data = "mean_cl_normal",
               linewidth = 1, colour = "red")
```

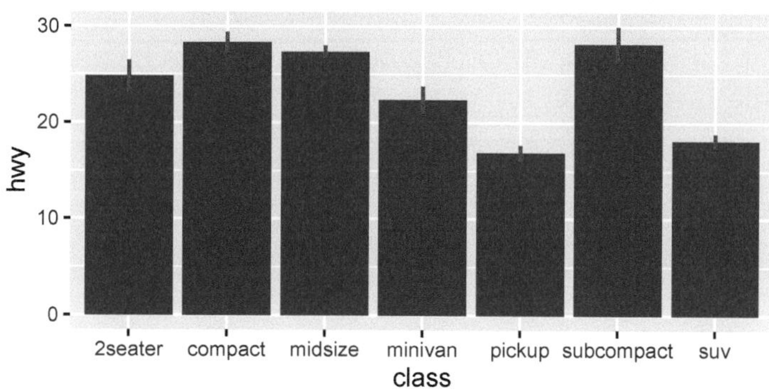

Passing `"errorbar"` instead of `"linerange"` to `geom` results in traditional "capped" error bars. However, this type of error bar has been criticised as adding unnecessary clutter to plots (Tufte 1983). Aesthetic `width` controls the width of the caps at the ends of the tips bars.

When calculated values for the summaries are already available in `data`, equivalent plots can be obtained by mapping the summary values from `data` to the *aesthetics* `x`, `y`, `ymax`, and `ymin` and using the `geoms` `geom_errorbar()` and `geom_linerange()` with their default for `stat`, `stat_identity()`, to add a plot layer.

A layer can be added to a plot directly with a `geom`, possibly passing a `stat` as an argument to it. In this book I have usually avoided this alternative syntax, except when not overriding `stat_identity()`, the usual default. The two code statements below are equivalent.

```
ggplot(data = mpg, mapping = aes(x = class, y = hwy)) +
  geom_col(stat = "summary", fun = mean)

ggplot(data = mpg, mapping = aes(x = class, y = hwy)) +
  stat_summary(geom = "col", fun = mean)
```

9.6.3 Smoothers and models

For describing or highlighting relationships between pairs of continuous variables, using a line, straight or curved, in a plot is very effective. Drawing lines that provide a meaningful and accurate description of the relationship, requires lines based on predictions from models fitted to the observations. Frequently fitted models make possible to assess the reliability of the estimation. See section 7.8 on page 199 for a description of the model fitting procedures underlying the plotting described in the current section.

The statistic `stat_smooth()` fits a smooth curve to observations in the case when the scales for x and y are continuous—the corresponding *geometry* `geom_smooth()` uses this *statistic*, and differs only in how arguments are passed to formal parameters. In the first example, `stat_smooth()` with the default smoother, a spline is used. In `stat_smooth()`, the type of smoother, or `method`, is automatically chosen based on the number of observations, and the choice informed by a message. In statistics, the `formula` must be stated using the names of the x and y aesthetics, rather than the original names of the variables mapped, i.e., in this example, not their name in the `mtcars` data frame. Splines are described in section 7.12 on page 223. When their small enough number makes it possible, observations are usually plotted as points together with the smoother. The observations can be plotted on top of the smoother or the smoother on top of the observations, as done here.

```
p3 <-
  ggplot(data = mtcars, mapping = aes(x = disp, y = mpg)) +
  geom_point()

p3 + stat_smooth(method = "loess", formula = y ~ x)
```

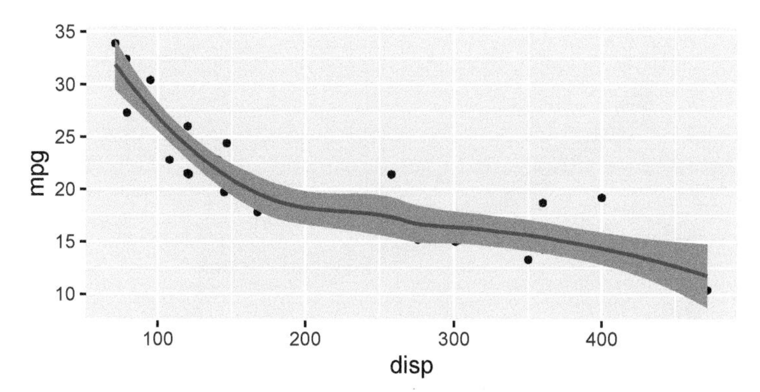

A model different to the default one can be used. Below, a linear regression is fitted with `lm()`. Fitting of linear models is explained in section 7.9 on page 200.

```
p3 + stat_smooth(method = "lm", formula = y ~ x)
```

These data can be grouped, here by mapping `factor(cyl)` to the `colour` *aesthetic*. With three groups, three separate linear regressions are fitted, and displayed as three straight lines. Each one line is delimited by a confidence band for the "true" location of the curve.

```
p3 + aes(colour = factor(cyl)) +
  stat_smooth(method = "lm", formula = y ~ x)
```

To obtain a single fitted smoother, in this case a joint linear regression line for the three groups, the grouping in the layer was disabled by mapping a constant value to the `colour` *aesthetic* in the call to `stat_smooth()`. Values passed to a layer function as argument override the defaults set in `ggplot()`. The use of `"black"` is arbitrary, any other `color` definition known to R could have been used instead.

```
p3 + aes(colour = factor(cyl)) +
  stat_smooth(method = "lm", formula = y ~ x, colour = "black")
```

A different linear model, a second degree polynomial in this example, is fitted below by passing a different argument to `formula` than in the example above for linear regression.

```
p3 + aes(colour = factor(cyl)) +
  stat_smooth(method = "lm", formula = y ~ poly(x, 2), colour = "grey20")
```

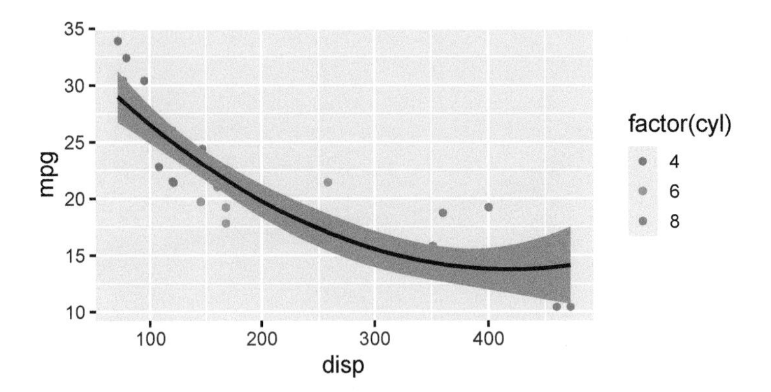

It is possible to use other types of models, including GAM and GLM, as smoothers. I give next two simple examples of the use of `nls()` to fit a model non-linear in its parameters (see section 7.11 on page 220 for details about fitting this same model with `nls()`). In both examples, the model fitted is the Michaelis-Menten equation, describing the rate of a chemical reaction (`rate`) as a function of reactant concentration (`conc`). `Puromycin` is a data set included in the R distribution. Function `SSmicmen()`, used in the first example, is also from R, and is a *self-starting* implementation of the Michaelis-Menten equation. Thanks to this, even though the fit is done with an iterative algorithm, starting values for the parameters to be fitted are not needed. Passing `se = FALSE` suppresses the attempt to compute a confidence band as it is not supported by the `predict()` method for model fits done with function `nls()`.

```
ggplot(data = Puromycin,
       mapping = aes(conc, rate, colour = state)) +
  geom_point() +
  geom_smooth(method = "nls", formula =  y ~ SSmicmen(x, Vm, K), se = FALSE)
```

In the second example, the code describing the equation is passed as an argument to `formula`, with starting values passed as a named list to `start`. The names used for the parameters to be estimated by fitting the model can be chosen at will, within the restrictions of the R language, but of course the names used in `formula` and `start` must match each other. As for other models, `x` and `y` are the names of the aesthetics to which the observations have been mapped (plot not shown).

```
ggplot(data = Puromycin,
       mapping = aes(conc, rate, colour = state)) +
  geom_point() +
  geom_smooth(method = "nls",
              formula =  y ~ (Vmax * x) / (k + x),
              method.args = list(start = list(Vmax = 200, k = 0.05)),
              se = FALSE)
```

In some cases, it is desirable to annotate plots with fitted model equations or fitted parameters. One way of achieving this is by fitting the model and then extracting the parameters to manually construct text strings to use for text or label annotations. However, package '`ggpmisc`' makes it possible to automate such annotations in many cases. This package also provides `stat_poly_line()`, which is similar to `stat_smooth()` but with `method = "lm"` consistently as its default irrespective of the number of observations.

```
my.formula <- y ~ x + I(x^2)
p3 + aes(colour = factor(cyl)) +
  stat_poly_line(formula = my.formula, colour = "black") +
  stat_poly_eq(formula = my.formula, mapping = use_label(c("eq", "F")),
               colour = "black", label.x = "right")
```

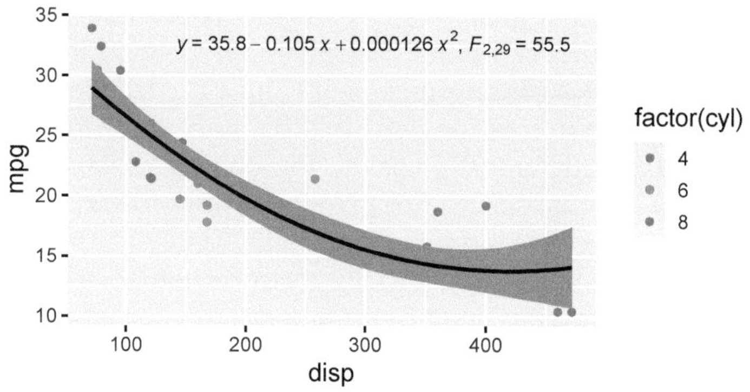

Package '`ggpmisc`' also makes it possible to annotate plots with summary tables from a model fit. The argument passed to `tb.vars` substitutes the names of the columns in the table.

```
ggplot(data = mtcars,
       mapping = aes(x = disp, y = mpg, colour = factor(cyl))) +
  stat_poly_line(formula = my.formula, colour = "black") +
  stat_fit_tb(method.args = list(formula = my.formula),
              colour = "black",
              parse = TRUE,
              tb.vars = c(Parameter = "term",
                          Estimate = "estimate",
                          "s.e." = "std.error",
                          "italic(t)" = "statistic",
                          "italic(P)" = "p.value"),
              label.y = "top", label.x = "right") +
  geom_point() +
  expand_limits(y = 40)
```

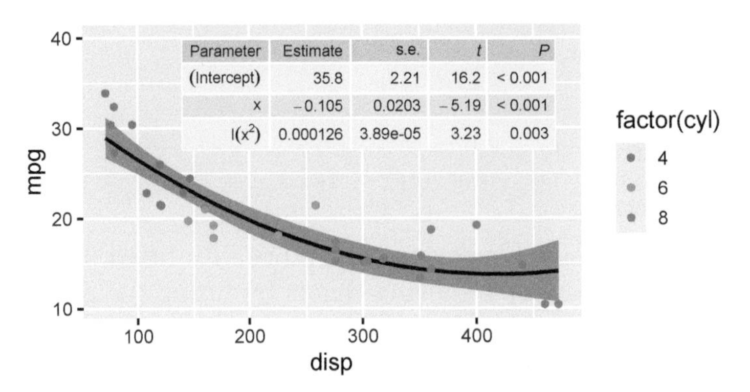

Package 'ggpmisc' provides additional *statistics* for the annotation of plots based on fitted models supported by package 'broom' and its extensions. It also supports lines and equations for quantile regression and major axis regression. Please see the package documentation for details.

9.6.4 Frequencies and counts

When the number of observations is rather small, it is possible rely on the density of graphical elements, such as points, to convey the density of the observations. For example, scatter plots using well-chosen values for transparency, `alpha`, can give a satisfactory impression of the density. Rug plots, described in section 9.5.2 on page 299, can also satisfactorily convey the density of observations along x and/or y axes. Such approaches do not involve computations, while the *statistics* described in this section do. Frequencies by value-range (or bins) and empirical density functions are summaries especially useful when the number of observations is large. These summaries can be computed in one or more dimensions.

Histograms are defined by how the plotted values are calculated. Although histograms are most frequently plotted as bar plots, many bar or "column" plots are not histograms. Although rarely done in practice, a histogram could be plotted using a different *geometry* using `stat_bin()`, the *statistic* used by default by `geom_histogram()`. This *statistic* does binning of observations before computing frequencies, and is suitable for observations on a continuous scales, usually mapped to the `x` aesthetic. When a factor is mapped to `x`, `stat_count()` can be used, the default `stat` of `geom_bar()`. These two *geometries* are described in this section about statistics, because they default to using statistics different from `stat_identity()` and consequently summarise the data.

The code below constructs a data frame containing an artificial data set.

```
set.seed(54321)
my.data <-
  data.frame(X = rnorm(600),
             Y = c(rnorm(300, -1, 1), rnorm(300, 1, 1)),
             group = factor(rep(c("A", "B"), c(300, 300))) )
```

A default and usually suitable number of bins is automatically selected by the `stat_bin()` statistic; however, passing `bins = 15` sets it manually. In a histogram plot the variable mapped onto the `y` *aesthetic* is not from `data` but instead computed in the statistics as the number of observations falling in each *bin*.

```
ggplot(data = my.data, mapping = aes(x = X)) +
  geom_histogram(bins = 15)
```

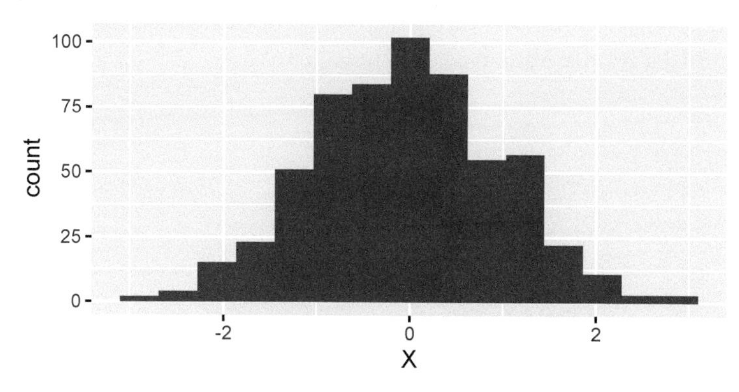

A reason to add layers with `geom_histogram()`, instead of with `stat_bin()` or `stat_count()` is that its name is easier to remember.

```
ggplot(data = my.data,
       mapping = aes(x = Y, fill = group)) +
  stat_bin(bins = 15, position = "dodge")
```

The grouping created by mapping a factor to an additional *aesthetic*, results in two separate histograms. The position of the two groups of bars with respect to each other is controlled with *position* functions (see section 9.9 on page 339 for details). With `position = "dodge"`, bars are plotted side by side; with `position = "stack"`, the default, plotted one above the other; and with `position = "identity"` overlapping. In this last case, adding `alpha = 0.5` makes occluded bars visible. The examples below use `position = "dodge"`.

```
p.base <-
  ggplot(data = my.data,
         mapping = aes(x = Y, fill = group))

p1 <- p.base + geom_histogram(bins = 15, position = "dodge")
```

In addition to `count`, `density`, computed as `count` divided by the number of observations in the group, is returned, and mapped in `p2` using `after_stat()`.

```
p2 <- p.base + geom_histogram(mapping = aes(y = after_stat(density)),
                              bins = 15, position = "dodge")

p1 + p2
```

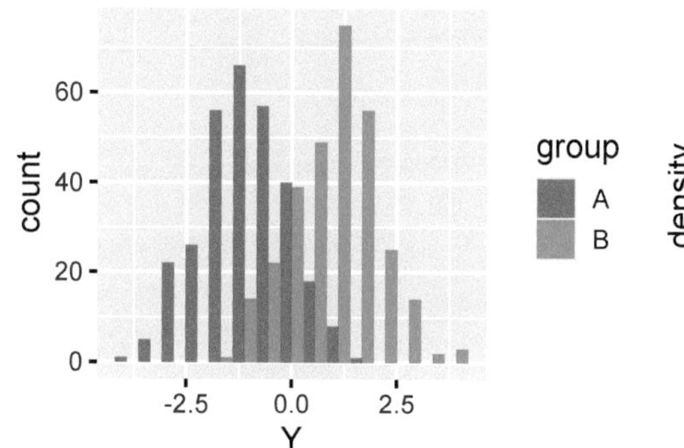

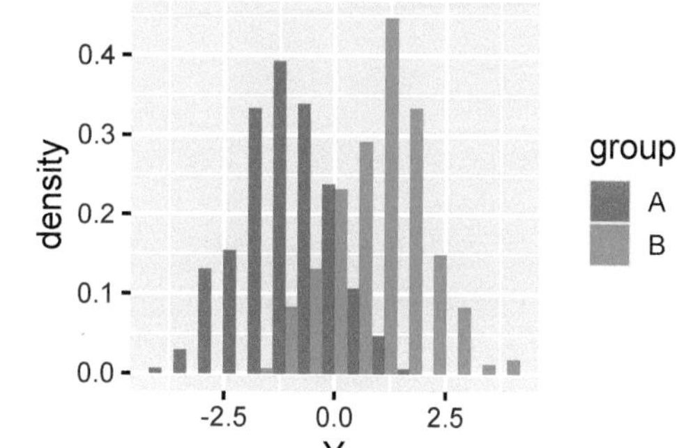

Statistic `stat_bin2d()`, and its matching *geometry* `geom_bin2d()`, by default compute a frequency histogram in two dimensions, along the `x` and `y` *aesthetics*. The `count` for each 2D bin is mapped to the `fill` aesthetic, with a lighter-coloured value being equivalent to a taller bar in a 1D histogram.

```
p.base <-
  ggplot(data = my.data,
         mapping = aes(x = X, y = Y)) +
  facet_wrap(facets = vars(group))
```

```
p.base + stat_bin2d(bins = 8)
```

Statistic `stat_bin_hex()`, and its matching *geometry* `geom_hex()`, differ from `stat_bin2d()` only in their use of hexagonal instead of square bins, and tiles.

```
p.base + stat_bin_hex(bins = 8)
```

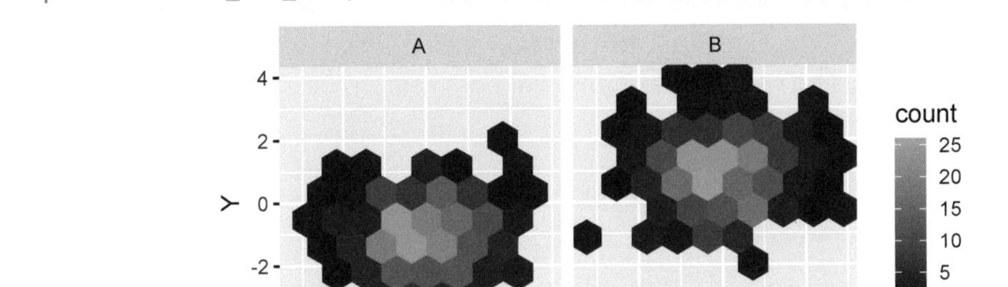

As `stat_bin()`, `stat_bin2d()` and `stat_bin_hex()` compute `density` in addition to `counts` and they can be plotted by mapping them to the `fill` aesthetic.

9.6.5 Density functions

Empirical density functions are the equivalent of a histogram, but are continuous and not calculated using bins, but fitted. They can be estimated in 1 or 2 dimensions (1D or 2D). As with histograms it is possible to use different *geometries* with them. Examples of `geom_density()` used to create 1D density plots follow. A semitransparent fill is used in addition to colour. Density plots for `Y` and `X`, i.e., using as mappings `x = Y` and `x = X`, are shown below side-by-side).

```
p3 <-
  ggplot(data = my.data,
         mapping = aes(x = Y, colour = group, fill = group)) +
  geom_density(alpha = 0.3)

p4 <-
  ggplot(data = my.data,
         mapping = aes(x = X, colour = group, fill = group)) +
  geom_density(alpha = 0.3)
```

Plot composition, as used below, is described in detail in section 9.16 on page 377.

```
p3 + p4 # plot composition
```

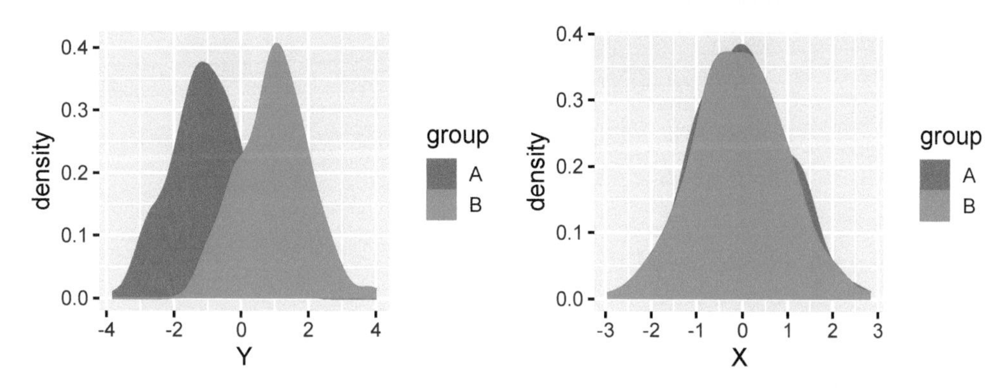

A 2D density plot using the same data as for the 1D plots above. In the first example, `stat_density_2d()` creates two 2D density "maps" shown using isolines, with `group` mapped to the `colour` *aesthetic*. Isolines can be used when the empirical distributions overlap. The 1D plots above show the projections of the 2D density in the plot below onto the two axes.

```
ggplot(data = my.data,
       mapping = aes(x = X, y = Y, colour = group)) +
  stat_density_2d()
```

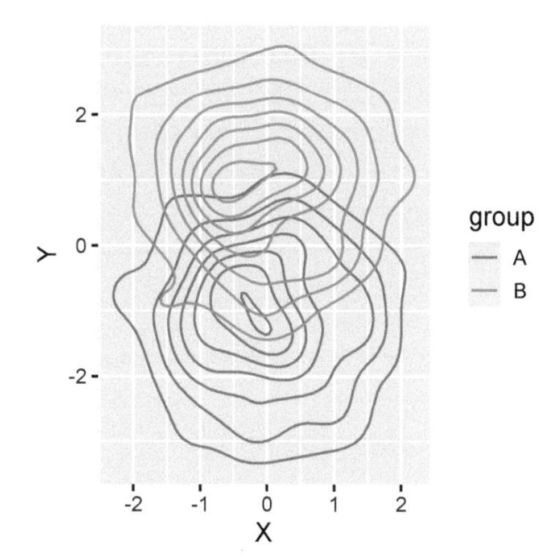

Below, the 2D density for each group is plotted in a separate panel, with `level`, a variable computed by `stat_density_2d()`, mapped to the `fill` *aesthetic*.

```
ggplot(data = my.data,
       mapping = aes(x = X, y = Y)) +
  stat_density_2d(aes(fill = after_stat(level)), geom = "polygon") +
  facet_wrap(facets = vars(group))
```

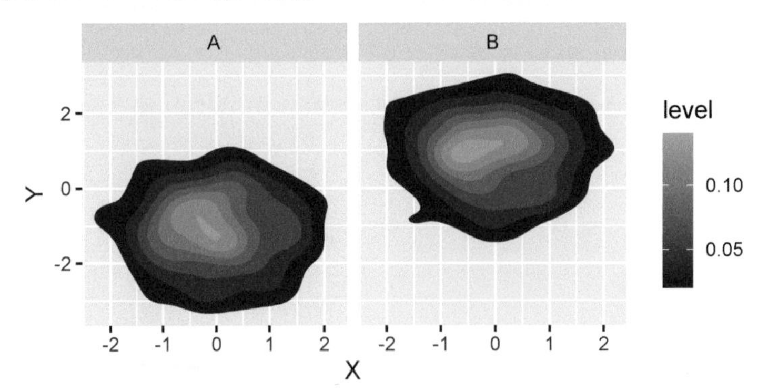

9.6.6 Box and whiskers plots

Box and whiskers plots, or just box plots, are summaries that convey some of the properties of a distribution. They are calculated and plotted with `stat_boxplot()` or the matching `geom_boxplot()`. Although box plots can be plotted based on just a few observations, they are not useful unless each box plot is based on more than 10 to 15 observations. In the next example, a sample of every sixth row from the data frame `my.data` with 600 rows is used.

```
p.base <-
  ggplot(data = my.data[c(TRUE, rep(FALSE, 5)) , ],
         mapping = aes(x = group, y = Y))

p1 <- p.base + stat_boxplot()
```

As with other *statistics*, the appearance obeys both *aesthetics* such as `colour`, and parameters specific to box plots: `outlier.colour`, `outlier.fill`, `outlier.shape`, `outlier.size`, `outlier.stroke`, and `outlier.alpha`, which affect outliers similarly to equivalent `aesthetics`. The shape and width of the "box" can be adjusted with `notch`, `notchwidth` and `varwidth`. Notches in box plots play a similar role as confidence limits play for means.

```
p2 <-
  p.base +
  stat_boxplot(notch = TRUE, width = 0.4,
               outlier.colour = "red", outlier.shape = "*", outlier.size = 5)
```

The two plots have been composed side by side to save space (see section 9.14 on page 369 for details about composing plots).

```
p1 + p2
```

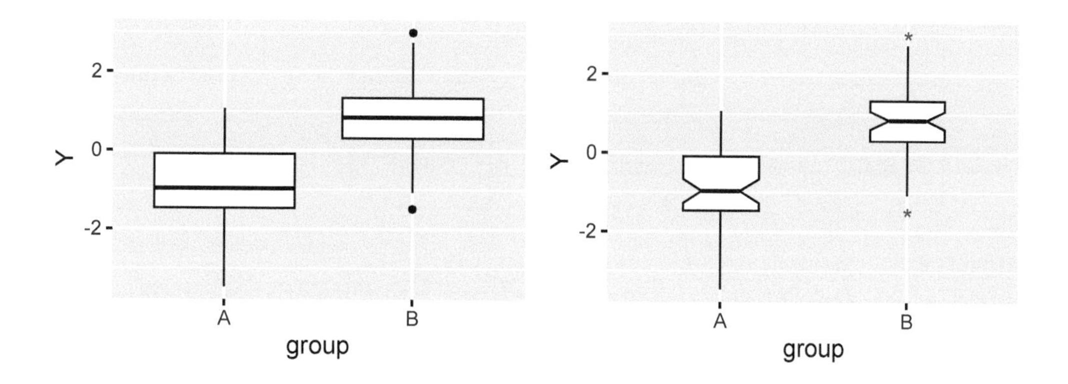

9.6.7 Violin plots

Violin plots are a more recent development than box plots, and usable with relatively large numbers of observations. They could be thought of as being a sort of hybrid between an empirical density function (see section 9.6.5 on page 326) and a box plot (see section 9.6.6 on page 328). As is the case with box plots, they are particularly useful when comparing distributions of related data, side by side. They can be created with `geom_violin()` as shown in the examples below.

```
p3 <- p.base +
  geom_violin(aes(fill = group), alpha = 0.16) +
  geom_point(alpha = 0.33, size = 1.5, colour = "black", shape = 21)
```

As with other *geometries*, their appearance obeys both the usual *aesthetics*, such as colour, and others specific to these types of visual representation.

Other types of displays related to violin plots are *beeswarm* plots and *sina* plots, and can be produced with *geometries* defined in packages 'ggbeeswarm' and 'ggforce', respectively. A minimal example of a beeswarm plot is shown below. See the documentation of the packages for details about the many options in their use.

```
p4 <- p.base + geom_quasirandom()
```

```
p3 + p4
```

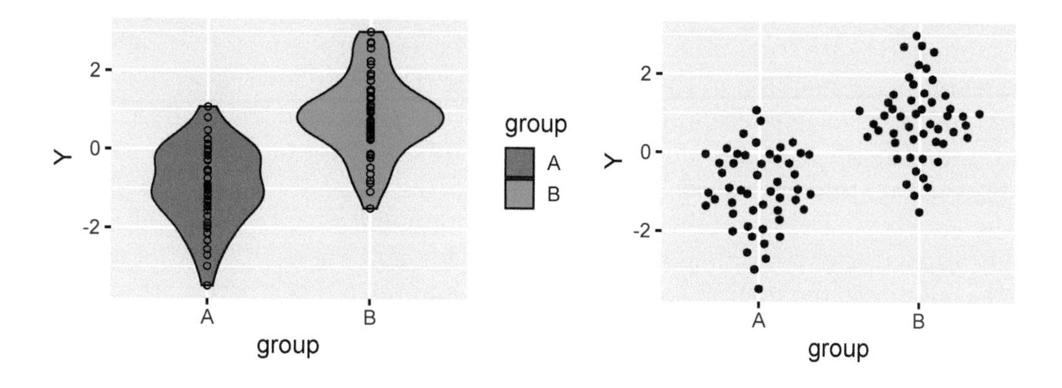

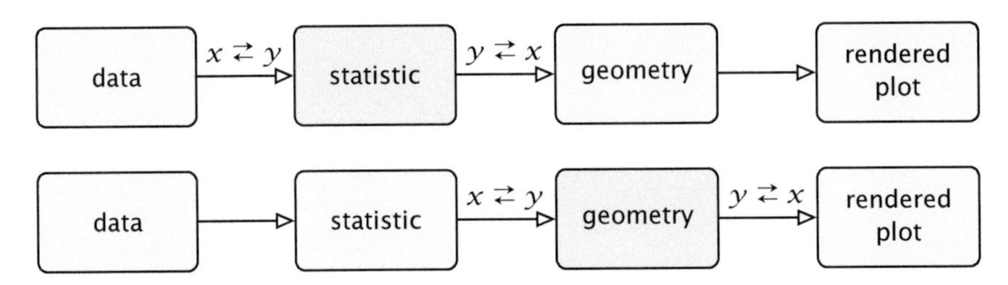

Figure 9.2
Flipped layers. Top diagram, flipped aesthetics in statistic with `orientation = "y"`; bottom diagram, flipped aesthetics in geometry with `orientation = "y"`. During flipping, related aesthetics such as `xmin` and `ymin` are also swapped, but not shown in the diagram.

9.7 Flipped Plot Layers

Although it is the norm to design plots so that the independent variable is on the x axis, i.e., mapped to the `x` aesthetic, there are situations where swapping the roles of x and y is useful. In 'ggplot2', this is described as *flipping the orientation* of a plot or of a plot layer. In the present section, I exemplify both cases where the flipping is automatic and where flipping requires user intervention. Some geometries like `geom_point()` are symmetric on the x and y aesthetics, but others like `geom_line()` operate differently on x and y. This is also the case for most *statistics*.

Starting from 'ggplot2' version 3.3.5, most geometries and statistics where it is meaningful, support flipping using a new syntax. This new approach is different to the flip of the coordinate system (which is expected to be deprecated in the future), and conceptually similar to that implemented by package 'ggstance'. However, instead of defining new horizontal layer functions as in 'ggstance', in 'ggplot2' the orientation of many layer functions can change. This has made package 'ggstance' nearly redundant and the coding of flipped plots easier and more intuitive. Although 'ggplot2' has offered `coord_flip()` for a long time, flipping of plot coordinates affects the whole plot rather than individual layers.

When a factor is mapped to x or y flipping is automatic. A factor creates groups and summaries are computed per group, i.e., per level of the factor irrespective of the factor being mapped to the x or y aesthetic. There are also cases that require user intervention. For example, flipping must be requested manually if both x and y are mapped to continuous variables. This is, for example, the case with `stat_smooth()` and with `geom_line()`.

In *statistics*, passing `orientation = "y"` as argument results in the calculations being applied after swapping the mappings of the x and y aesthetics. After applying the calculations, the mappings of the x and y and related aesthetics are swapped back (Figure 9.2).

In geometries, passing `orientation = "y"` also results in flipping of the aesthetics (Figure 9.2). For example, in `geom_line()`, flipping changes the drawing of the lines. Normally observations are sorted along the x axis before drawing the line segments connecting them. After flipping, as x and y are swapped, observations are sorted along the y axis before drawing the connecting segments. The variables shown on each axis remain the same, as does the position of points drawn with `geom_point()`, but the line connecting them is different: in the example below, only two segments are the same in the flipped plot and in the "normal" one.

```
p.base <-
   ggplot(data = mtcars[1:8, ], mapping = aes(x = hp, y = mpg)) +
    geom_point()
p1 <- p.base + geom_line() + ggtitle("Not flipped")
p2 <- p.base + geom_line(orientation = "y") + ggtitle("Flipped")
p1 + p2
```

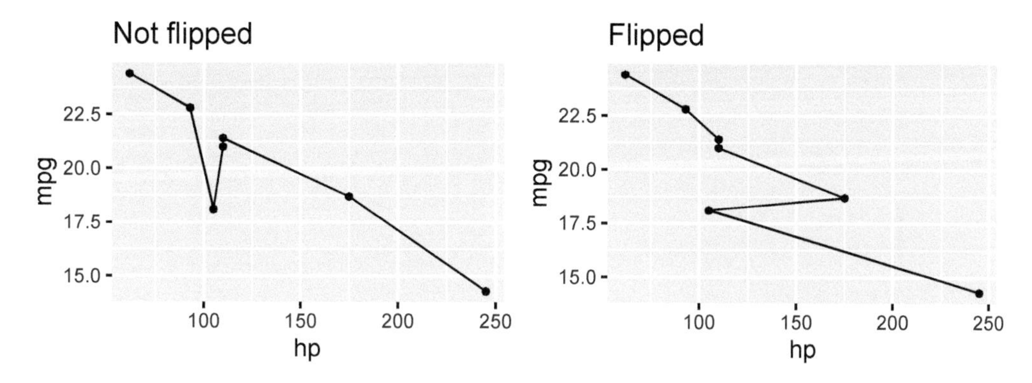

The next pair of examples demonstrates automatic flipping using `stat_boxplot()`. Factor `Species` is mapped first to x and then to y. In both cases, the same boxplots were computed and plotted for each level of the factor. Statistics `stat_boxplot()`, `stat_summary()`, `stat_histogram()` and `stat_density()` behave similarly with respect to automatic flipping.

```
p3 <-
  ggplot(data = iris, mapping = aes(x = Species, y = Sepal.Length)) +
  stat_boxplot()

p4 <-
  ggplot(data = iris, mapping = aes(x = Sepal.Length, y = Species)) +
  stat_boxplot()

p3 + p4
```

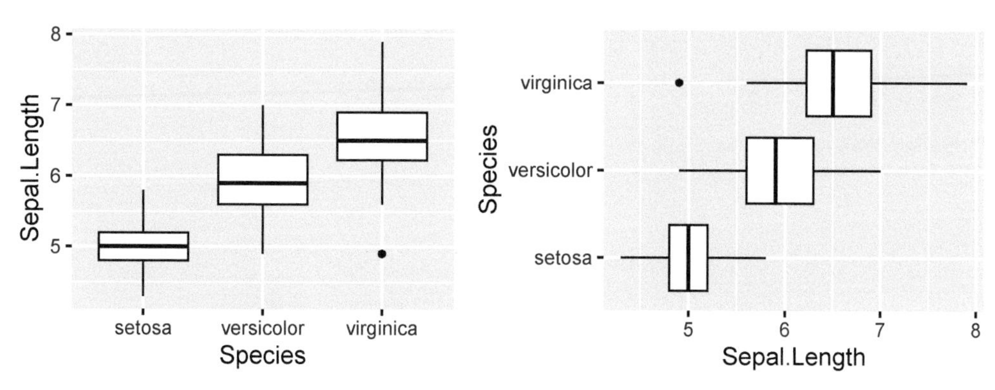

In the case of stats that do computations on a single variable mapped to x or y aesthetics, flipping is also automatic.

```
p5 <-
  ggplot(data = iris,
         mapping = aes(x = Sepal.Length, colour = Species)) +
  stat_density(geom = "line", position = "identity")

p6 <-
  ggplot(data = iris,
         mapping = aes(y = Sepal.Length, colour = Species)) +
  stat_density(geom = "line", position = "identity")

p5 + p6
```

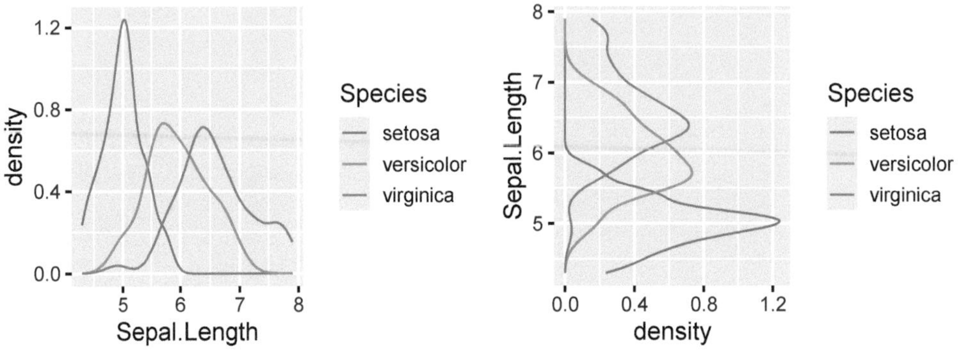

In the case of ordinary least squares (OLS), regressions of y on x and of x on y in most cases yield different fitted lines, even if R^2 is consistent. This is due to the assumption that x values are known, either set or measured without error, i.e., not subject to uncertainty. Under this assumption, all unexplained variation in the data is attributed to y. See section 7.8 on page 199 or consult a Statistics book such as *Modern Statistics for Modern Biology* (Holmes and Huber 2019, pp. 168–170) for additional information.

With two continuous variables mapped, the default is to take x as independent and y as dependent. Passing "x" (the default) or "y" as argument to parameter orientation indicates which of x or y is the independent or explanatory variable.

```
p.base <-
  ggplot(data = iris,
       mapping = aes(x = Sepal.Length, y = Petal.Length)) +
  geom_point() +
  facet_wrap(~Species, scales = "free")

p.base + stat_smooth(method = "lm", formula = y ~ x)
```

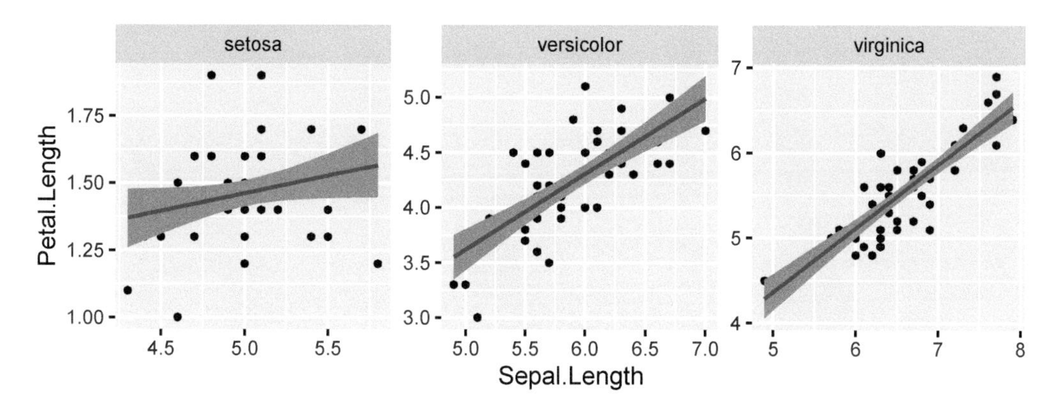

Passing `orientation = "y"` to `geom_smooth()` is equivalent to swapping x and y in the model `formula`. The looser the correlation, the more different are the lines fitted before and after flipping.

```
p.base + stat_smooth(method = "lm", formula = y ~ x, orientation = "y")
```

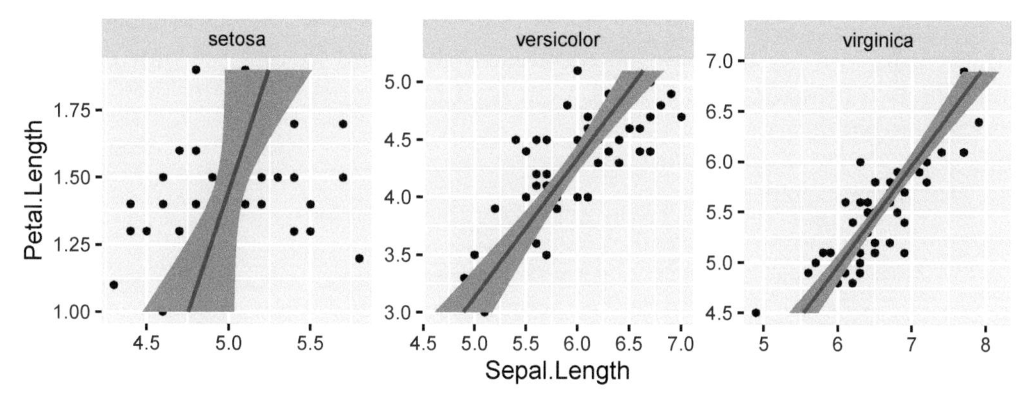

The two variables in the example above, are both response variables, not directly connected by cause and effect, and with measurements subject to similar errors. None, of the two fitted models are close enough to fulfilling the assumptions.

Flipping the orientation of plot layers with `orientation = "y"` is not equivalent to flipping the whole plot with `coord_flip()`. In the first case, which axis is considered independent for computation changes but not the positions of the axes in the plot, while in the second case, the position of the x and y axes in the plot is swapped. So, when coordinates are flipped the x aesthetic is plotted on the vertical axis and the y aesthetic on the horizontal axis, but the role of the variable mapped to the `x` aesthetic remains as explanatory variable. (Use of `coord_flip()` will likely be deprecated in the future.)

```
p.base +
  stat_smooth(method = "lm", formula = y ~ x) +
  coord_flip()
```

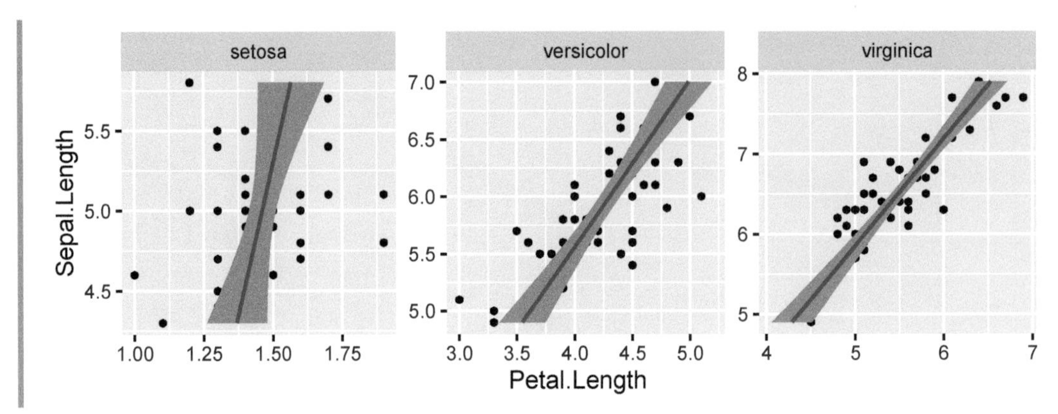

In package 'ggpmisc' (version $\geq$ 0.4.1), statistics related to model fitting have an `orientation` parameter as those from package 'ggplot2' do, but in addition they accept formulas where x is on the lhs and y on the rhs, such as `formula = x ~ y` providing a syntax consistent with R's model fitting functions. With two calls to `stat_poly_line()`, the first using the default `formula = y ~ x`, and the second using `formula = x ~ y` to force the flipping of the fitted model, the plot produced contains two fitted lines per panel, with the flipped ones highlighted as red lines and yellow bands.

```
p.base +
    stat_poly_line() +
    stat_poly_line(formula = x ~ y, colour = "red", fill = "yellow")
```

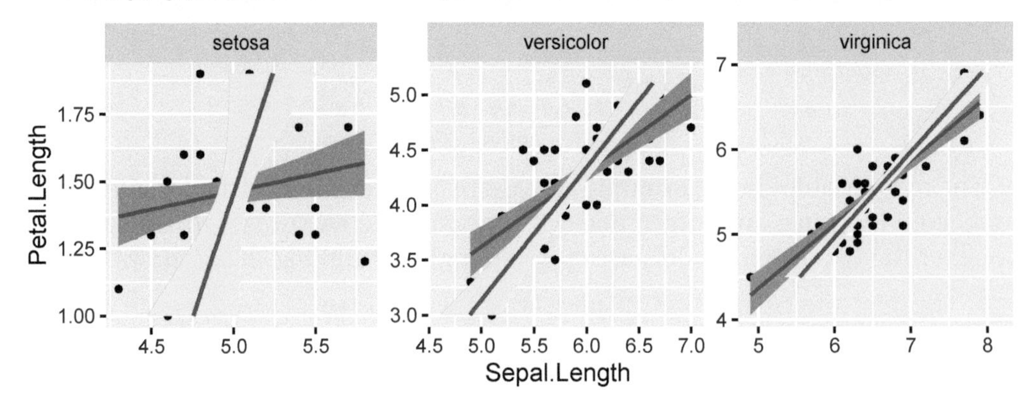

In the case of the `iris` data used for these examples, both approaches used above to linear regression are wrong. In this case, the correct approach is to not assume that there is a variable that can be considered independent and another dependent on it, but instead to use a method like major axis (MA) regression, as below.

```
p.base + stat_ma_line()
```

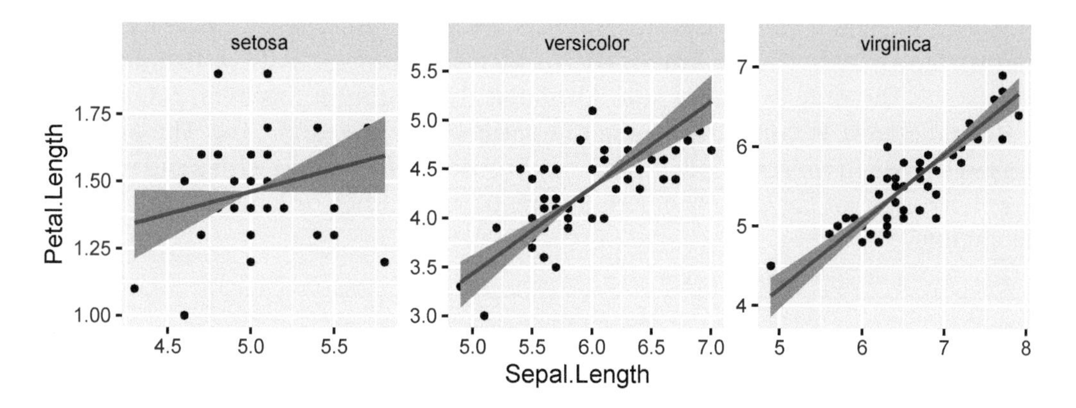

9.8 Facets

Facets are used in a special kind of plots containing multiple panels in which the panels share some properties. These sets of coordinated panels are a useful tool for visualising complex data. These plots became popular through the `trellis` graphs in S, and the 'lattice' package in R. The basic idea is to have rows and/or columns of plots with common scales, all plots showing values for the same response variable. This is useful when there are multiple classification factors in a data set. Similar-looking plots, but with free scales or with the same scale but a 'floating' intercept, are sometimes also useful. In 'ggplot2', there are two possible types of facets: facets organised in a grid and facets along a single 'axis' of variation but, possibly, wrapped into two or more rows. These are produced by adding `facet_grid()` or `facet_wrap()`, respectively. Below, `geom_point()` is used in the examples, but faceting can be used with plots containing layers created with any `geom` or `stat`.

A single-panel plot, saved as `p.base`, will be used through this section to demonstrate how the same plot changes when facets are added.

```
p.base <-
  ggplot(data = mtcars,
         mapping = aes(x = wt, y = mpg)) +
  geom_point()
p.base
```

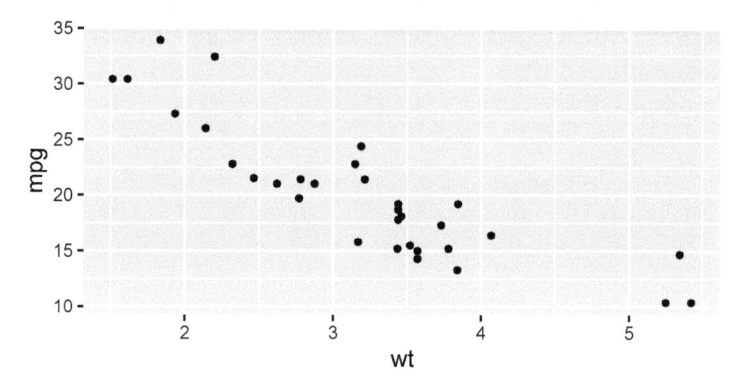

A grid of panels has two dimensions, `rows` and `cols`. These dimensions in the grid of plot panels can be "mapped" to factors. Until recently, a formula-based syntax was the only available one. Although this notation has been retained, the preferred syntax is currently to use the parameters `rows` and `cols`. The argument passed to `cols` in this example is factor `cyl` retrieved from `data` with a call to `vars()`. The "headings" of the panels or *a*re by default the names or labels of the levels of the factor.

```
p.base + facet_grid(cols = vars(cyl))
```

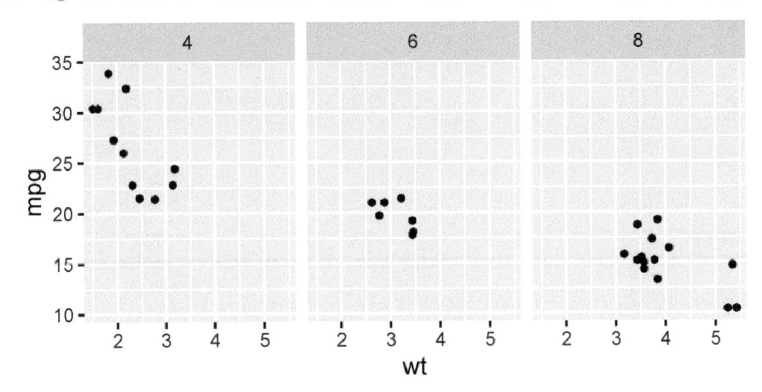

Using `facet_wrap()` the same plot can be coded as follows.

```
p.base + facet_wrap(facets = vars(cyl), nrow = 1)
```

By default, all panels share the same scale limits and share the plotting space evenly, but these defaults can be overridden.

```
p.base + facet_wrap(facets = vars(cyl), nrow = 1, scales = "free_y")
```

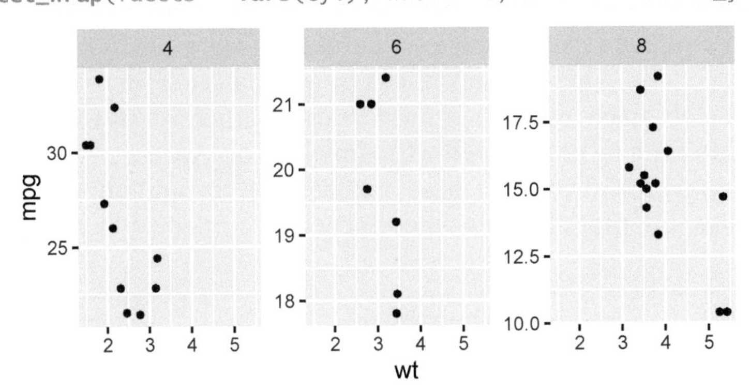

Margins, added with `margins = TRUE`, display an additional column or row of panels with the combined data.

```
p.base + facet_grid(cols = vars(cyl), margins = TRUE)
```

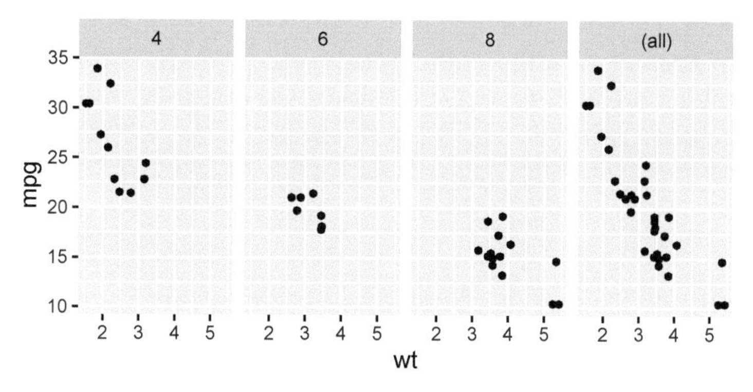

To obtain a 2D grid both `rows` and `cols` have to be passed factors as arguments.

```
p.base + facet_grid(rows = vars(vs), cols = vars(am), labeller = label_both)
```

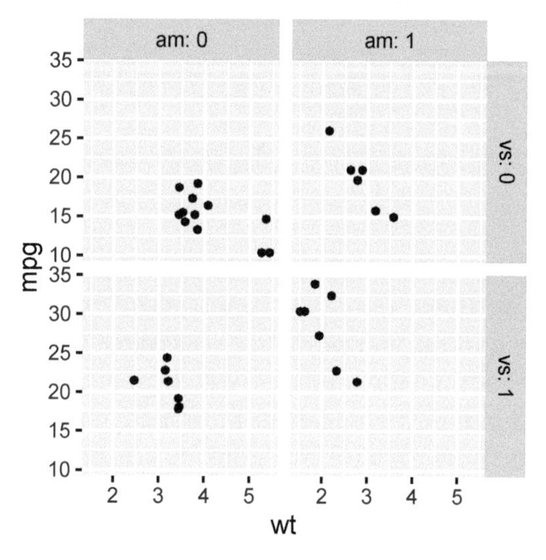

Each faceting dimension can be mapped to more than one factor as below. As the levels are not self-explanatory, `label_both` is passed as argument to `labeller` so that factor names are included in the *strip labels* together with the levels.

```
p.base + facet_grid(cols = vars(vs, am), labeller = label_both)
```

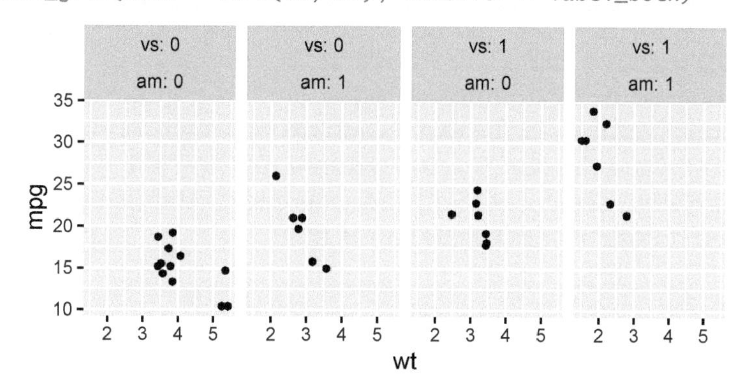

When facetting generates many panels, wrapping them into several rows helps keep the shape of the whole plot manageable. In this example, the number of levels is small, and no wrapping takes place by default. In cases when more panels are present, wrapping into two or more continuation rows is the default. Here, we force wrapping with `nrow = 2`. When using `facet_wrap()` there is only one dimension, and the parameter is called `facets`, instead of `rows` or `cols`.

```
p.base + facet_wrap(facets = vars(cyl), nrow = 2)
```

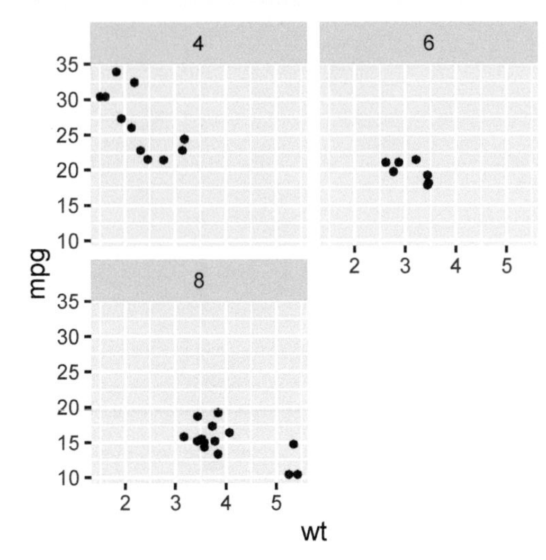

By default, panel headings display the names of the levels of the factor they are based on. Changing these names is one way of changing the labels. This approach can be used to add mathematical expressions or Greek letters in the panel headings. Below, first factor labels in the data frame passed as argument to `data` are set to strings that can be parsed into *plotmath* expressions. Then, in the call to `facet_grid()`, or to `facet_wrap()`, we pass as argument to `labeller` a function definition, `label_parsed`.

```
mtcars$cyl12 <- factor(mtcars$cyl,
                       labels = c("alpha", "beta", "sqrt(x, y)"))
ggplot(data = mtcars,
       mapping = aes(mpg, wt)) +
  geom_point() +
  facet_grid(cols = vars(cyl12), labeller = label_parsed)
```

The labels of the levels of the factor used in faceting can be combined with text, or math, using a "template". Passing as argument to `labeller` function `label_bquote()` and using a plotmath expression as argument for its parameter `cols`, makes this possible. In the expression used below, `.(cyl)` is substituted by the value of `cyl` when the plot is rendered—we use here the name of the variable in the data, `cyl`. See section 9.15 for an example of the use of `bquote()`, the R function based on which `label_bquote()` is built.

```
p.base +
  facet_grid(cols = vars(cyl),
             labeller = label_bquote(cols = .(cyl)~"cylinders"))
```

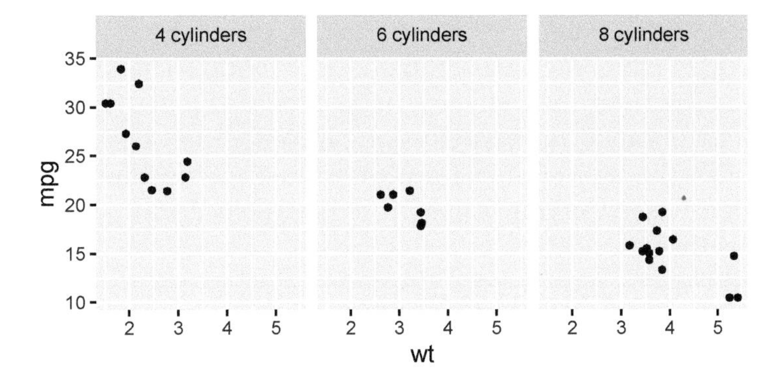

9.9 Positions

Position functions are passed as arguments to the `position` parameter of `geoms`. They displace the positions (the values mapped to `x` and/or `y` aesthetics) away from their original position. Different position functions differ in what displacement is applied. Table 9.3 lists most of the position functions available. Function `position_stack()` and `position_fill()` were already described on page 303, with stacked column and area plots. Function `position_dodge()` was used in plots with side-by-side columns on page 304 and `position_jitter()` was used in dot plot examples on page 297.

The difference between `position_stack()` and `position_fill()` is illustrated by the example below.

```
p.base <-
  ggplot(data = Orange,
         mapping = aes(x = age, y = circumference, fill = Tree))

p1 <- p.base + geom_area(position = "stack", colour = "white", linewidth = 1) +
  ggtitle("stack")
p2 <- p.base + geom_area(position = "fill", colour = "white", linewidth = 1) +
  ggtitle("fill")

p1 + p2
```

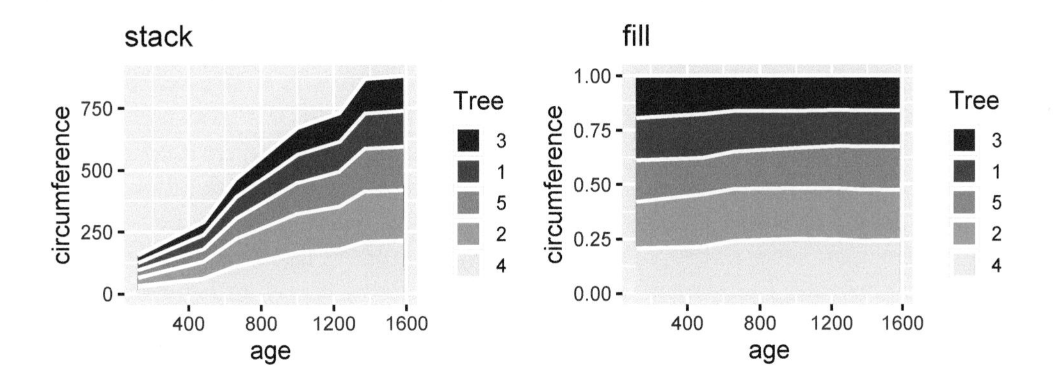

Table 9.3
Position functions from packages 'ggplot2' and 'ggpp'. The table is divided into two sections. A. Positions that only return the modified x and y values. B. Identical positions that additionally return a copy of the unmodified x and y values. The last column describes the type of displacement: fixed uses constant values supplied in the call; random, uses random values for the displacement, within a maximum distance set by the user.

Position	Package	Parameters	Displ.
A. *Origin not kept*			
position_identity	'ggplot2'	—	none
position_stack	'ggplot2'	vjust, reverse	fixed
position_fill	'ggplot2'	vjust, reverse	fixed
position_dodge	'ggplot2'	width, preserve, padding, reverse	fixed
position_dodge2	'ggplot2'	width, preserve, padding, reverse	fixed
position_jitter	'ggplot2'	width, height, seed	rand.
position_nudge	'ggplot2'	x, y	fixed
B. *Origin kept*			
position_stack_keep	'ggpp'	vjust, reverse	fixed
position_fill_keep	'ggpp'	vjust, reverse	fixed
position_dodge_keep	'ggpp'	width, preserve, padding, reverse	fixed
position_dodge2_keep	'ggpp'	width, preserve, padding, reverse	fixed
position_jitter_keep	'ggpp'	width, height, seed	rand.
position_nudge_keep	'ggpp'	x, y	fixed

Position `position_nudge()` is used to consistently displace positions, and is most frequently used with `geom_text()` and `geom_label()` when adding data labels. When position functions are used to add data labels, it is common to add a segment linking the data point to the label. For this to be possible, position functions have to keep the original position. Position functions from package 'ggplot2' discard them while the position functions from packages 'ggpp' and 'ggrepel' keep them in data under a different name. Table 9.3 is divided into sections. The only difference between the position functions in the two sections of the table is in whether the original position is kept or not, i.e., those from package 'ggpp' are backwards compatible with those from package 'ggplot2'.

The displacement introduced by jitter and nudge differ in that jitter is random, and nudge deterministic. In each case, the displacement can be separately adjusted vertically and horizontally. Jitter, as shown above, is useful when we desire to make visible overlapping points. Nudge is most frequently used with data labels to avoid occluding points or other graphical features.

Layer function `geom_point_s()` from package 'ggpp' is used below to make the displacement visible by drawing an arrow connecting original and displaced positions for each observation. We need to use the `_keep` flavour of the position functions for arrows to be drawn.

```
p.base <-
  ggplot(data = mtcars,
         mapping = aes(x = factor(cyl), y = mpg)) +
  geom_point(colour = "blue")
p3 <- p.base +
  geom_point_s(position = position_jitter_keep(width = 0.35, heigh = 0.6),
               colour = "red") +
  ggtitle("jitter")
```

The amount of nudging is set by a distance expressed in data units through parameters `x` and `y`. (Factors have mode `numeric` and each level is represented by an integer, thus distance between levels of a factor is 1.)

```
p4 <- p.base +
  geom_point_s(position = position_nudge_keep(x = 0.25, y = 1),
               colour = "red") +
  ggtitle("nudge")
```

```
p3 + p4
```

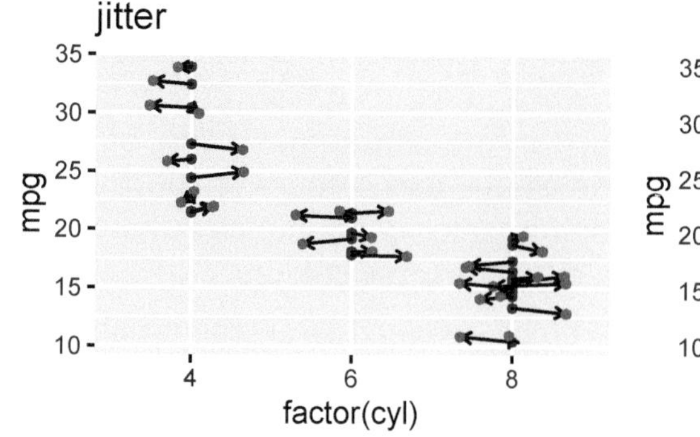

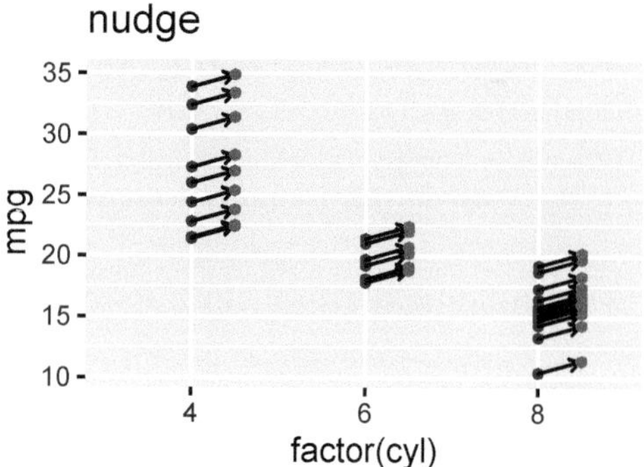

9.10 Scales

In earlier sections of this chapter, most examples have used the default *scales*. In this section, I describe in more detail the use of *scales*. There are *scales* available for all the different *aesthetics* recognised by `geoms`, such as position aesthetics (`x, y, z`), `size`, `shape`, `linewidth`, `linetype`, `colour`, `fill`, `alpha` or transparency, and `angle`. Scales determine how values in `data` are mapped to values of an *aesthetics*, and optionally, also how these values are labelled.

Depending on the characteristics of the variables in `data` being mapped, *scales* can be continuous or discrete, for `numeric` or `factor` variables in `data`, respectively. Some *aesthetics*, like `size` and `colour`, are inherently continuous but others like `linetype` and `shape` are inherently discrete. In the case of inherently continuous aesthetics, both discrete and continuous scales are available, while, obviously for those inherently discrete only discrete scales are available.

The scales used by default have default mappings of data values to aesthetic values (e.g., which colour value corresponds to $cyl = 4$ and which one to $cyl = 8$).

For each *aesthetic*, such as `colour`, there are multiple scales to choose from when creating a plot, both continuous and discrete (e.g., 20 different colour scales in 'ggplot2' 3.4.3). In addition, some scales implement multiple palettes.

As seen in previous sections, *aesthetics* in a plot layer, in addition to being determined by mappings, can also be set to constant values. Aesthetics set to constant values are not mapped to data and are consequently independent of scales.

The most direct mapping to data is `identity`, with the values in the mapped variable directly interpreted as aesthetic values. In a colour scale, say `scale_colour_identity()`, the variable in the data would be encoded with values such as `"red"`, `"blue"`—i.e., valid `R` colours. In a simple mapping using `scale_colour_discrete()` levels of a factor, such as `"treatment"` and `"control"` would be represented as distinct colours with the correspondence between factor levels and individual colours set automatically. In contrast with `scale_colour_manual()` the user explicitly provides the mapping between factor levels and colours by passing arguments to the scale functions' parameters `breaks` and `values`.

The details of the mapping of a continuous variable to an *aesthetic* are controlled with a continuous scale such as `scale_colour_continuous()`. In this case, values in a `numeric` variable will be mapped into a continuous range of colours. How the correspondence between numeric values and colours is controlled can vary among scales. In the case of `colour`, some scales use complex palettes, while others implement simple gradients between two or three colours.

In some scales, missing values, or `NA`, can be assigned an aesthetic value, such as `colour`, while in other cases `NA` values are always skipped instead of plotted. The reverse, mapping values in data to `NA` as aesthetic value is in some cases also possible.

9.10.1 Axis and key labels

First I describe a feature common to all scales, their `name`. The default `name` of all scales is the name of the variable or the expression mapped to it. In the case of the `x`, `y`, and `z` *aesthetics*, the `name` given to the scale is used for the axis labels. For other *aesthetics* the name of the scale becomes the "heading" or *key title* of the guide or key. All scales have a `name` parameter to which a character string or an `R` expression (see section 9.15) can be passed as an argument to override the default. In scales that add a key or guide, passing `guide = "none"` to the scale function removes the key corresponding to the scale.

Convenience functions `xlab()` and `ylab()` can be used to set the axis labels. Convenience function `labs()` can be used to manually set axis labels, key/guide titles, and title and other labels for the plot as a whole. For the names of scales, `labs()` accepts the names of aesthetics as if they were formal parameters and using `title`, `subtitle`, `caption`, `tag`, and `alt` for the labels for the plot as a whole. The text passed to `alt` is not visible in the plot but is expected to be made available to web browsers and used to enhance accessibility. (The size of title and subtitle can seem too big when rendering figures at a small size, see section 9.13 on page 364 on how to replace and modify the theme used.)

```
p.base <-
  ggplot(data = Orange,
         mapping = aes(x = age, y = circumference, colour = Tree)) +
  geom_line() +
  geom_point()
```

```
p.base +
  expand_limits(y = 0) +
  labs(title = "Growth of orange trees",
       subtitle = "Starting from 1968-12-31",
       caption = "see Draper, N. R. and Smith, H. (1998)",
       tag = "A",
       alt = "A data plot",
       x = "Time (d)",
       y = "Circumference (mm)",
       colour = "Tree\nnumber")
```


When passing names directly to scales, the plot title and subtitle can be added with function `ggtitle()` by passing either character strings or R expressions as arguments.

```
p.base +
  expand_limits(y = 0) +
  scale_x_continuous(name = "Time (d)") +
  scale_y_continuous(name = "Circumference (mm)") +
  ggtitle(label = "Growth of orange trees",
          subtitle = "Starting from 1968-12-31")
```

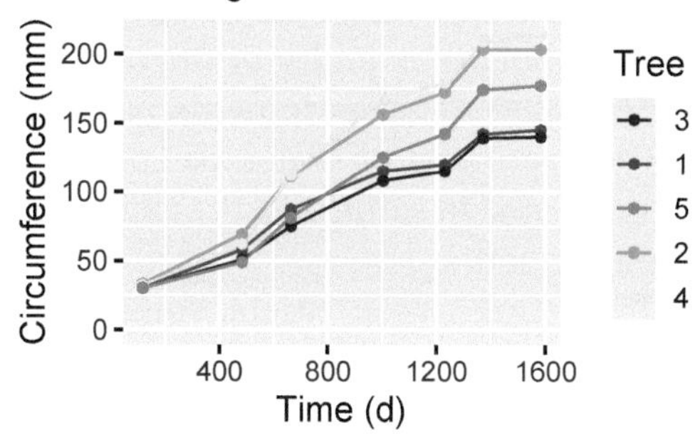

9.17 Make an empty plot (`ggplot()`) and add to it as title an R expression producing $y = b_0 + b_1x + b_2x^2$. (Hint: have a look at the examples for the use of expressions in the `plotmath` demo in R by typing `demo(plotmath)` at the R console.

9.10.2 Continuous scales

I start by listing the most frequently used arguments to the continuous scale functions: `name`, `breaks`, `minor_breaks`, `labels`, `limits`, `expand`, `na.value`, `trans`, `guide`, and `position`. The value of `name` is used for axis labels or the key title (see previous section). The arguments to `breaks` and `minor_breaks` override the default locations of major and minor ticks and grid lines. Setting them to `NULL` suppresses the ticks. By default, the tick labels are generated from the value of `breaks` but an argument to `labels` of the same length as `breaks` will replace these defaults. The values of `limits` determine both the range of values in the data included and the plotting area as described above—by default the out-of-bounds (`oob`) observations are replaced by `NA` but it is possible to instead "squish" these observations towards the edge of the plotting area. The argument to `expand` determines the size of the margins or padding added to the area delimited by `lims` when setting the "visual" plotting area. The value passed to `na.value` is used as a replacement for `NA` valued observations—most useful for `colour` and `fill` aesthetics. The transformation object passed as an argument to `trans` determines the transformation used—the transformation affects the rendering, but breaks and tick labels remain expressed in the original data units. The argument to `guide` determines the type of key or removes the default key. Depending on the scale in question not all these parameters are available. A family of continuous scales, *binned scales*, was added in 'ggplot2' 3.3.0. These scales map a continuous variable from `data` onto a discrete gradient of aesthetic values, but are otherwise very similar.

The code below constructs data frame `fake2.data`, containing artificial data.

```
fake2.data <-
  data.frame(y = c(rnorm(20, mean = 20, sd = 5),
                   rnorm(20, mean = 40, sd = 10)),
             group = factor(c(rep("A", 20), rep("B", 20))),
             z = rnorm(40, mean = 12, sd = 6))
```

Limits

Limits are relevant to all kinds of *scales*. Limits are set through parameter `limits` of the different scale functions. They can also be set with convenience functions `xlim()` and `ylim()` in the case of the `x` and `y` *aesthetics*, and more generally with function `lims()` which like `labs()`, takes arguments named according to the name of the *aesthetics*. The `limits` argument of scales accepts vectors, factors, or a function computing them from `data`. In contrast, the convenience functions do not accept functions as their arguments.

In the next example, by setting "hard" limits, some observations are excluded from the plot, they are not seen by `stats` and `geoms`, i.e., hard limits in scales subset observations in `data` at the `start` stage (see Figure 9.1 on page 277). More precisely, the off-limits observations are converted to NA values before they are passed as data to stats, and subsequently discarded with a warning.

```
p1.base <-
  ggplot(data = fake2.data, mapping = aes(x = z, y = y)) +
  geom_point()

p1 <- p1.base + scale_y_continuous(limits = c(0, 100))
```

To set only one limit leaving the other free, NA is used as a boundary.

```
p2 <-p1.base + scale_y_continuous(limits = c(50, NA))

p1 + p2

## Warning: Removed 37 rows containing missing values or values outside the
scale range
## (`geom_point()`).
```

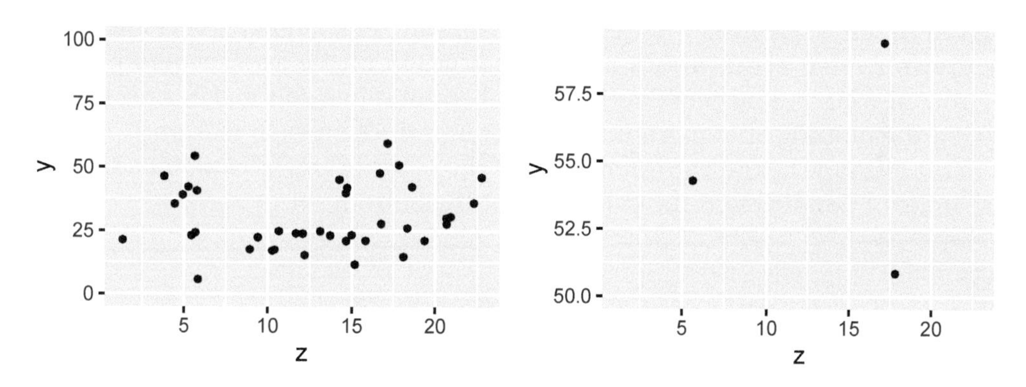

Convenience functions ylim() and xlim() can be used to set the limits to the default x and y scales in use. Below, ylim() is used, but xlim() works identically except for the scale it modifies (plot identical to p2 above, not shown).

```
p1.base +  ylim(50, NA)
```

In general, setting hard limits should be avoided, even though a warning is issued about NA values being omitted, as it is easy to unwillingly subset the data being plotted. It is preferable to use function expand_limits() as it safely *expands* the dynamically computed default limits of a scale—the scale limits will grow past the requested expanded limits when needed to accommodate all observations. The arguments to x and y are numeric vectors of length one or two each, matching how the limits of the x and y continuous scales are defined. Below, the limits are expanded to include the origin.

```
p1.base + expand_limits(y = 0, x = 0)
```

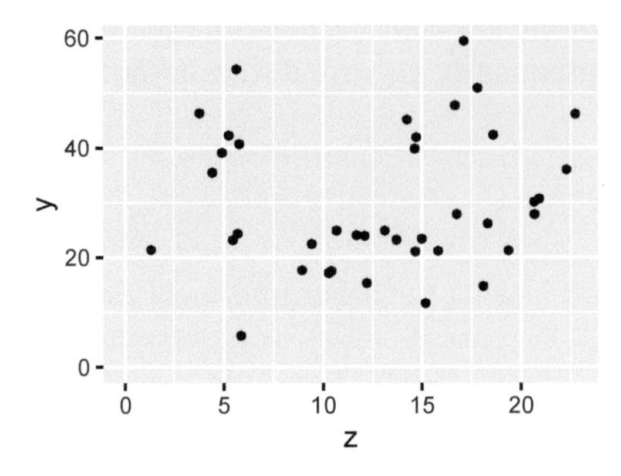

The `expand` parameter of the scales plays a different role than `expand_limits()`. It adds a "margin" or padding around the plotting area. The actual plotting area is given by the scale limits, set either dynamically or manually. Very rarely plots are drawn so that observations are plotted on top of the axes, avoiding this is a key role of `expand`. Rug plots and marginal annotations can make it necessary to expand the plotting area more than the default of 5% on each margin.

In the example below, the upper edge of the plotting area is expanded by adding 0.02 units of padding and the expansion at the bottom set to zero.

```
p2.base <-
  ggplot(data = fake2.data,
         mapping = aes(fill = group, colour = group, x = y)) +
  stat_density(alpha = 0.3, position = "identity")

p1 <-
  p2.base + scale_y_continuous(expand = expansion(add = c(0, 0.01)))
```

Using multipliers has the advantage that the expansion is proportional. A similar effect as above is achieved using multipliers, 10% compared to the range of the `limits` at the top and none at the bottom.

```
p2 <-
  p2.base + scale_y_continuous(expand = expansion(mult = c(0, 0.1)))

p1 + p2
```

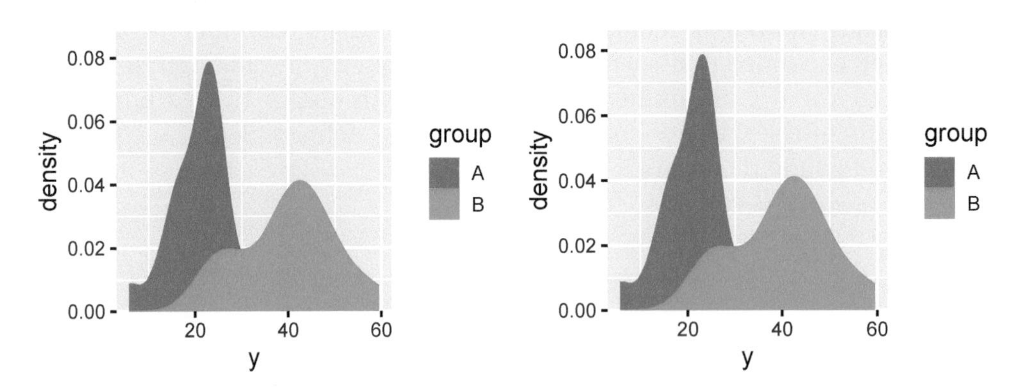

9.18 Compare the rendered plot from `p2.base` to `p1` and `p2` displayed above. What has been the effect of using `expansion()`? Try different values as arguments for `add` and `mult`.

The direction of a scale can be reversed using a transformation (see section 9.10.2 on page 349). Scales `scale_x_reverse()` and `scale_y_reverse()` use by default the necessary transformation. However, inconsistently, `xlim()` and `ylim()` can be used to reverse the scale direction by passing the numeric values for the limits in decreasing order.

9.19 Test what the result is when the first limit is larger than the second one. Is it the same as when setting these same values as limits with `ylim()`? or by replacing `scale_y_continuous()` with `scale_y_reverse()`?

```
p1.base <- scale_y_continuous(limits = c(100, 0))
```

Breaks and their labels

Parameter `breaks` is used not only to set the location of ticks along the axis in scales for the `x` and `y` aesthetics, but also for the keys or guides for other continuous scales such as those for colour. Parameter `labels` is used to set the break labels, including tick labels. The argument passed to each of these parameters can be vector or a function. The default is to compute "good" breaks based on the limits and use to nice numbers suitable for labels. Examples in this section are for continuous scales, see section 9.10.3 on page 351 for break labels in time and date scales.

When manually setting breaks, labels for the `breaks` are automatically computed unless overridden.

```
p3.base <-
  ggplot(data = fake2.data, mapping = aes(x = z, y = y)) +
  geom_point()

p3.base + scale_y_continuous(breaks = c(20, pi * 10, 40, 60))
```

The default breaks are computed by function `pretty_breaks()` from 'scales'. The argument passed to its parameter `n` determines the target number ticks to be generated automatically, but the actual number of ticks computed may be slightly different depending on the range of the data.

```
p3 <-
  p3.base + scale_y_continuous(breaks = pretty_breaks(n = 7))
```

We can set tick labels manually, in parallel to the setting of `breaks` by passing as arguments two vectors of equal length. Below, an expression is used to include a Greek letter in the label.

```
p4 <-
  p3.base +
  scale_y_continuous(breaks = c(20, pi * 10, 40, 60),
                     labels = c("20", expression(10*pi), "40", "60"))
```

```
p3 + p4
```

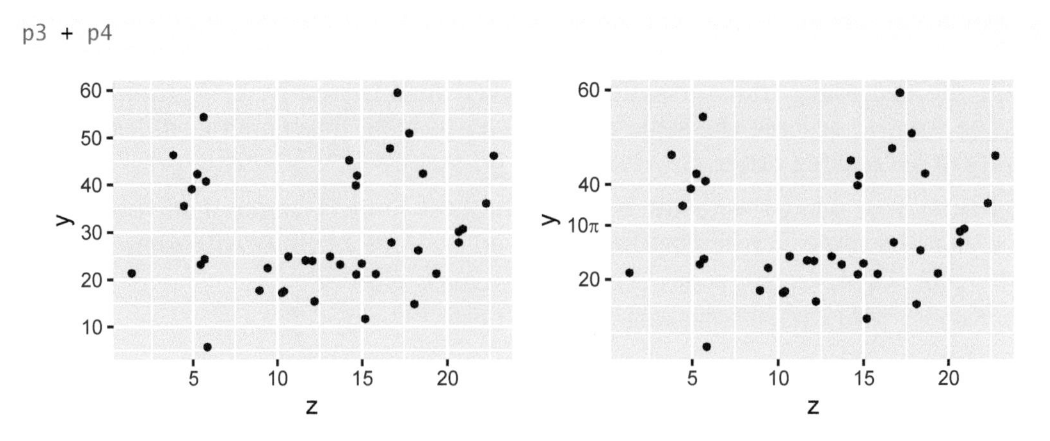

Package 'scales' provides several functions for the automatic generation of tick labels. For example, function `percent()` can be used to display tick labels as percentages when the values mapped from `data` are expressed as decimal fractions. This "transformation" is applied only to the tick labels.

```
p5 <-
  ggplot(data = fake2.data, mapping = aes(x = z, y = y / max(y))) +
  geom_point() +
  scale_y_continuous(labels = percent)
```

For currency, functions `dollar()` and `comma()` can be used to format the numbers in the labels as used for currency. Function `scientific_format()` formats numbers using exponents of 10—useful for logarithmic-transformed scales. Additional functions, `label_number(scale_cut = cut_short_scale())`, `label_log()`, or `label_number(scale_cut = cut_si("g")` provide other options. As shown below, some of these functions can be useful with untransformed continuous scales.

```
p6 <-
  ggplot(data = fake2.data, mapping = aes(x = z, y = y * 1000)) +
  geom_point() +
  scale_y_continuous(name = "Mass",
                     labels = label_number(scale_cut = cut_si("g")))
```

```
p5 + p6
```

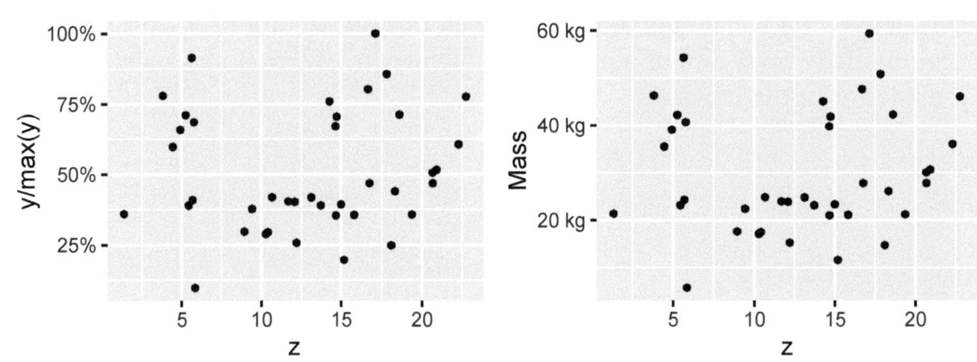

Function `label_number()` and the similar functions listed above, build new functions base on the arguments passed to them, and the values they return are function definitions. Thus, in the example above, even if the statement passed as argument to `labels` is a function call, the value actually "received" by `scale_y_continuous()` is an *ad hoc* function definition created on-the-fly. Some packages define additional functions that work similarly to those from package 'scales'.

Transformed scales

The default scales used by the `x` and `y` aesthetics, `scale_x_continuous()` and `scale_y_continuous()`, accept a user-supplied transformation function as an argument to `trans` with default `trans = "identity"` (no transformation). Package 'scales' defines several transformations that can be used as arguments for `trans`. User-defined transformations can be also implemented and used. In addition, there are predefined convenience scale functions for `log10`, `sqrt` and `reverse`.

Consistently with maths functions in R, the names of the scales are `scale_x_log10()` and `scale_y_log10()`, rather than `scale_y_log()` because in R, function `log()` computes the natural logarithm.

Axis tick-labels display the original values, not transformed ones, and the argument to `breaks` also refers to these. Using `scale_y_log10()` a $\log_{10}$ transformation is applied to the y values.

```
ggplot(data = fake2.data, mapping = aes(x = z, y = y)) +
  geom_point() +
  scale_y_log10(breaks=c(10,20,50,100))
```

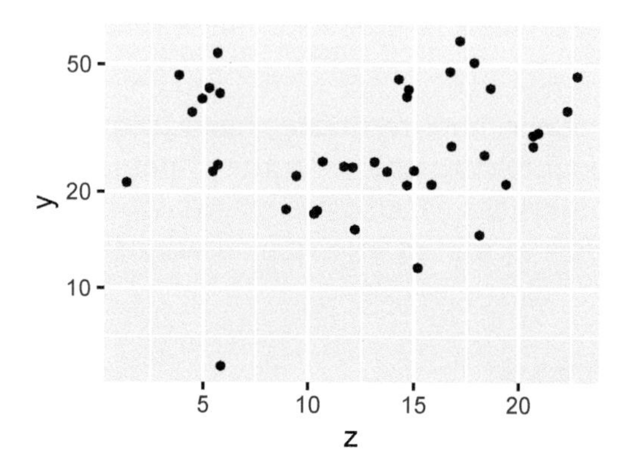

9.20 Using a transformation in a scale is not equivalent to applying the same transformation on the fly when mapping a variable to the x (or y) *aesthetic*. How does the plot produced by the code below differ from the plot using the transformed scale, shown above?

```
ggplot(data = fake2.data, mapping = aes(x = z, y = log10(y))) +
  geom_point()
```

For the most common transformations like `log10()`, scales with those transformations as their default are available. In other cases, as mentioned above, the transformation is set by passing an argument to parameter `trans` of continuous scale functions that by default do not apply a transformation. Below, a predefined transformation, `"reciprocal"` or $1/y$ is used (plot not shown).

```
ggplot(data = fake2.data, mapping = aes(x = z, y = y)) +
  geom_point() +
  scale_y_continuous(trans = "reciprocal")
```

Natural logarithms are important in growth analysis as the slope against time gives the relative growth rate. The growth data for orange trees, from data set `Orange`, are plotted using a `log()` as transformation. Breaks are set using the original values.

```
ggplot(data = Orange,
       mapping = aes(x = age, y = circumference, colour = Tree)) +
  geom_line() +
  geom_point() +
  scale_y_continuous(trans = "log", breaks = c(20, 50, 100, 200))
```

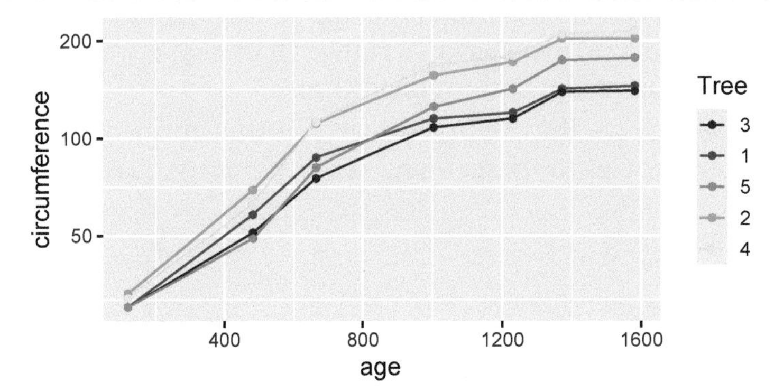

In the examples above, and in practice most frequently, transformations are applied to position aesthetics, `x` and `y`. As the grammar of graphics is consistent, most if not all continuous scales, also accept transformations. In some cases, applying a transformation to a `size` or `colour` scale helps convey the information contained in the data.

Position of x and y axes

The default position of axes can be changed through parameter `position`, using character constants `"bottom"`, `"top"`, `"left"` and `"right"`.

```
ggplot(data = mtcars, mapping = aes(x = wt, y = mpg)) +
  geom_point() +
  scale_x_continuous(position = "top") +
  scale_y_continuous(position = "right")
```

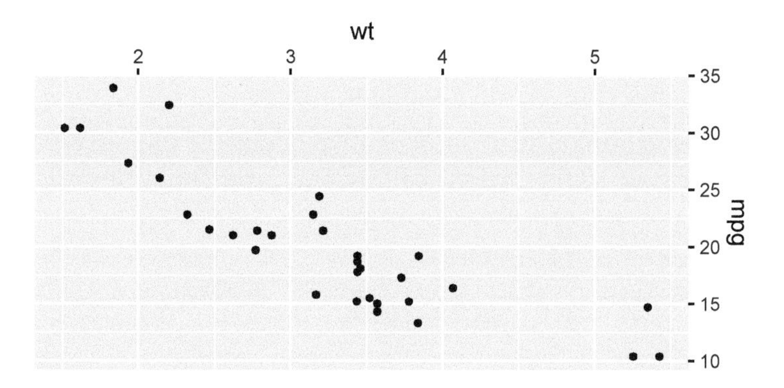

Secondary axes

It is also possible to add secondary axes with ticks displayed in a transformed scale.

```
ggplot(data = mtcars, mapping = aes(x = wt, y = mpg)) +
  geom_point() +
  scale_y_continuous(sec.axis = sec_axis(~ . ^-1, name = "gpm") )
```

It is also possible to use different `breaks` and `labels` than for the main axes, and to provide a different `name` to be used as a secondary axis label.

```
ggplot(data = mtcars, mapping = aes(x = wt, y = mpg)) +
  geom_point() +
  scale_y_continuous(sec.axis = sec_axis(~ . / 2.3521458,
                                         name = expression(km / l),
                                         breaks = c(5, 7.5, 10, 12.5)))
```

9.10.3 Time and date scales for x and y

Time scales are similar to continuous scales for `numeric` values. In R and many other computing languages, time values are stored as integer values subject to special interpretation (see section 8.8 on page 267). Times stored as objects of class `POSIXct` (or `POSIXlt`) can be mapped to continuous *aesthetics* such as `x`, `y`, `colour`, etc. Special scales for different aesthetics are available for time-related data.

Limits and breaks are preferably set using constant values of class `POSIXct`. These are most easily input with the functions in packages ‘lubridate’ or ‘anytime’ that convert dates and times from character strings.

In the next two chunks, scale limits subset a part of the observations present in data. Passing `na.rm = TRUE` when calling the geom functions silences warning messages.

```
ggplot(data = weather_wk_25_2019.tb,
       mapping = aes(x = with_tz(time, tzone = "EET"),
                     y = air_temp_C)) +
  geom_line(na.rm = TRUE) +
  scale_x_datetime(name = NULL,
                   breaks = ymd_h("2019-06-11 12", tz = "EET") + days(0:1),
                   limits = ymd_h("2019-06-11 00", tz = "EET") + days(c(0, 2))) +
  scale_y_continuous(name = "Air temperature (C)") +
  expand_limits(y = 0)
```

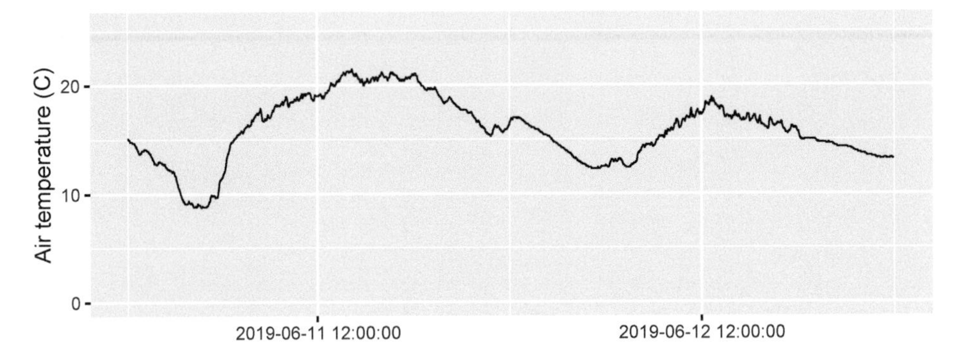

As for numeric scales, breaks and the corresponding labels can be set differently to defaults. For example, if all observations have been collected within a single day, default tick labels will show hours and minutes. With several years, the labels will show only dates. The default labels are frequently good enough. Below, both breaks and the format of the labels are set through parameters passed in the call to `scale_x_datetime()`.

```
ggplot(data = weather_wk_25_2019.tb,
       mapping = aes(x = with_tz(time, tzone = "EET"),
                     y = air_temp_C)) +
  geom_line(na.rm = TRUE) +
  scale_x_datetime(name = NULL,
                   date_breaks = "1 hour",
                   limits = ymd_h("2019-06-16 00", tz = "EET") + hours(c(6, 18)),
                   date_labels = "%H:%M") +
  scale_y_continuous(name = "Air temperature (C)") +
  expand_limits(y = 0)
```

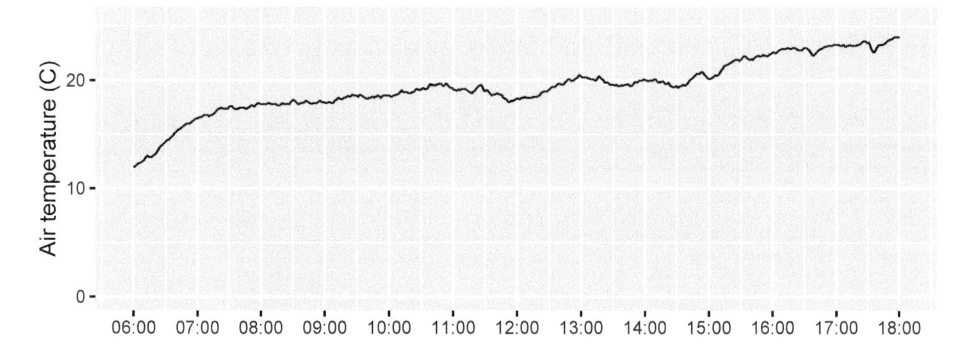

9.21 The formatting strings used are those supported by `strptime()` and `help(strptime)` lists them. Change, in the two examples above, the y-axis labels used and the limits—e.g., include a single hour or a whole week of data, check which tick labels are produced by default and then pass as an argument to `date_labels` different format strings, taking into account that in addition to the *conversion specification* codes, format strings can include additional text.

In *date* scales tick labels are created with functions `label_date()` or `label_date_short()`. In the case of *time* scales, tick labels are created with function `label_time()`. As shown for continuous scales, calls to these functions can passed as argument to the scales.

9.10.4 Discrete scales for x and y

In the case of ordered or unordered factors, the tick labels are by default the names of the factor levels. Consequently, one roundabout way of obtaining the desired tick labels is to set them as factor labels in the data frame. This approach is not recommended as in many cases the text of the desired tick labels may not be a valid R name making more complex by the need to *scape* these names each time they are used. It is best to use simple mnemonic short names for factor levels and variables, and to set suitable labels through scales.

Scales `scale_x_discrete()` and `scale_y_discrete()` can be used to reorder and select the factor levels without altering the data. When using this approach to subset the data, it is necessary to pass `na.rm = TRUE` in the call to layer functions to avoid warnings. Below, arguments passed to `limits` and `labels` in the call `scale_x_discrete` manually convert level names to uppercase (plot not shown, identical plot shown farther down using alternative code).

```
ggplot(data = mpg,
       mapping = aes(x = class, y = hwy)) +
  stat_summary(geom = "col", fun = mean, na.rm = TRUE) +
  scale_x_discrete(limits = c("compact", "subcompact", "midsize"),
                   labels = c("COMPACT", "SUBCOMPACT", "MIDSIZE"))
```

If, as above, replacement is with the same names in upper case, passing function `toupper()` automates the operation. In addition, the code becomes independent of the labels used in the data. This is a more general and less error-prone approach. Any function, user defined or not, that converts the values of `limits` into the desired values can be passed as an argument to `labels`. This example, for completeness, sets scale names and limits, as well as the width of the columns.

```
ggplot(data = mpg,
       mapping = aes(x = class, y = hwy)) +
  stat_summary(geom = "col", fun = mean, na.rm = TRUE, width = 0.6) +
  scale_x_discrete(name = "Vehicle class",
                   limits = c("compact", "subcompact", "midsize"),
                   labels = toupper) +
  scale_y_continuous(name = "Petrol use efficiency (mpg)", limits = c(0, 30))
```

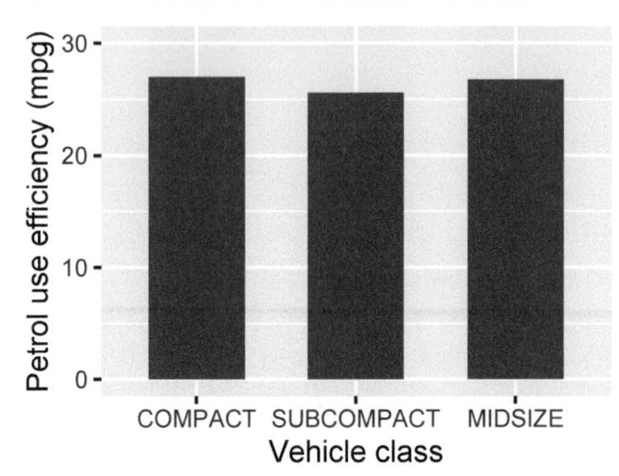

The order of the columns in the plot follows the order of the levels in the factor, thus changing this ordering in factor `mpg$class` works. This approach makes sense when the new ordering needs to be computed based on values in `data`, but can still be applied in the plotting code. Below, the breaks, and together with them the columns, are ordered based on the `mean()` of variable `hwy` by means of a call to `reorder()` within the call to `aes()`.

```
ggplot(data = mpg,
       mapping = aes(x = reorder(x = factor(class), X = hwy, FUN = mean),
                     y = hwy)) +
  stat_summary(geom = "col", fun = mean)
```

9.10.5 Size and line width

The `linewidth` aesthetic was added to package 'ggplot2' in version 3.4.0. Previously, aesthetic `size` described the width of lines as well as the size of text and points or shapes. Below, I describe the scales according to version 3.4.0 and more recent.

For the `size` *aesthetic*, several scales are available, discrete, ordinal, continuous, and binned. They are similar to those already described above. Geometries `geom_point()`, `geom_text()`, and `geom_label()` obey the `size` aesthetic as expected. Size scales can be used with continuous numeric variables, date and times, and with discrete variables. Examples of the use of `scale_size()` and `scale_size_area()` were given in section 9.5.1 on page 294. Scale `scale_size_radius()` is rarely used as it does not match human visual size perception.

A similar set of scales is available for `linewidth` as there is for `size`, discrete, ordinal, continuous, and binned. Geometries `geom_line()`, `geom_hline()`, geom_vline(), geom_abline(), geom_segment(), geom_curve() and related ones, obey the linewidth aesthetic. Geometry geom_pointrange() obeys both aesthetics, as expected, size is used for the size of the point and linewidth for the bar segment. In geometries geom_bar(), geom_col(), geom_area(), geom_ribbon() and all other geometric elements bordered by lines, linewidth controls the width of these lines. Like lines, these borders and segments also obey the linetype aesthetic.

Using linewidth makes code incompatible with versions of 'ggplot2' prior to 3.4.0, while continuing to use size will trigger deprecation messages in newer versions of 'ggplot2'. Eventually, use of size for lines will become an error, so when possible, it is preferable to use the new linewidth aesthetic.

9.10.6 Colour and fill

The colour and fill scales are very similar, but they affect different elements of the plot. All visual elements in a plot obey the colour *aesthetic*, but only elements that have an inner region and a boundary, obey both colour and fill *aesthetics*. The boundary does not need to be rendered as a line when the plot is displayed, but it must exist. This is the case for geom_area() and geom_ribbon() that in recent versions of 'ggplot2' are displayed with lines only on some edges. Only a subset of the shapes supported by geom_point() can be filled. There are separate but equivalent sets of scales available for these two *aesthetics*. I will describe in more detail the colour *aesthetic* and give only some examples for fill. I will, however, start by reviewing how colours are defined and used in R.

Colour definitions in R

Colours can be specified in R not only through character strings with the names of previously defined colours, but also directly as strings describing the RGB (red, green, and blue) components as hexadecimal numbers (on base 16 expressed using 0, 1, 2, 3, 4, 6, 7, 8, 9, A, B, C, D, E, and F as "digits") such as "#FFFFFF" for white or "#000000" for black, or "#FF0000" for the brightest available pure red.

The list of colour names known to R can be obtained be typing colors() at the R console. Differently to package 'ggplot2', base R supports only color as the spelling. Given the number of colours available, subsetting them based on their names is frequently a good first step. Function colors() returns a character vector. Using grep() it is possible to find the names that contain a given character substring, in this example "dark".

```
length(colors())
## [1] 657
grep("dark",colors(), value = TRUE)
##  [1] "darkblue"          "darkcyan"          "darkgoldenrod"     "darkgoldenrod1"
##  [5] "darkgoldenrod2"    "darkgoldenrod3"    "darkgoldenrod4"    "darkgray"
##  [9] "darkgreen"         "darkgrey"          "darkkhaki"         "darkmagenta"
## [13] "darkolivegreen"   "darkolivegreen1"  "darkolivegreen2"  "darkolivegreen3"
## [17] "darkolivegreen4"   "darkorange"        "darkorange1"       "darkorange2"
## [21] "darkorange3"       "darkorange4"       "darkorchid"        "darkorchid1"
## [25] "darkorchid2"       "darkorchid3"       "darkorchid4"       "darkred"
## [29] "darksalmon"        "darkseagreen"      "darkseagreen1"     "darkseagreen2"
```

```
## [33] "darkseagreen3"   "darkseagreen4"   "darkslateblue"   "darkslategray"
## [37] "darkslategray1"  "darkslategray2"  "darkslategray3"  "darkslategray4"
## [41] "darkslategrey"   "darkturquoise"   "darkviolet"
```

The RGB values for an R `color` definition are returned by function `col2rgb()`.

```
col2rgb("purple")
##       [,1]
## red    160
## green   32
## blue   240
col2rgb("#FF0000")
##       [,1]
## red    255
## green    0
## blue     0
```

Colour definitions in R can contain a *transparency* component described by an `alpha` value, which by default is not returned.

```
col2rgb("purple", alpha = TRUE)
##       [,1]
## red    160
## green   32
## blue   240
## alpha  255
```

With function `rgb()` one can define new colours. Enter `help(rgb)` for more details.

```
rgb(1, 1, 0)
## [1] "#FFFF00"
rgb(1, 1, 0, names = "my.color")
##  my.color
## "#FFFF00"
rgb(255, 255, 0, names = "my.color", maxColorValue = 255)
##  my.color
## "#FFFF00"
```

As described above, colours can be defined in the RGB *colour space*; however, other colour models such as HSV (hue, saturation, value) can be also used to define colours.

```
hsv(c(0,0.25,0.5,0.75,1), 0.5, 0.5)
## [1] "#804040" "#608040" "#408080" "#604080" "#804040"
```

Frequently, sets of HSV colours returned by function `hcl()`, using hue, chroma and luminance as inputs, are better for use in scales. While the "value" and "saturation" in HSV are based on physical values, the "chroma" and "luminance" values in HCL are based on human visual perception. Colours with equal luminance will be seen as equally bright by an "average" human. In a scale based on different hues but equal chroma and luminance values, as used by default by package 'ggplot2', all colours are perceived as equally bright. The hues need to be expressed as angles in degrees, with values between zero and 360.

```
hcl(c(0,0.25,0.5,0.75,1) * 360)
## [1] "#FFC5D0" "#D4D8A7" "#99E2D8" "#D5D0FC" "#FFC5D0"
```

It is also important to remember that humans can only distinguish a limited set of colours, and even smaller colour gamuts can be reproduced by screens and printers. Furthermore, variation from individual to individual exists in colour perception, including different types of colour blindness. It is important to take this into account when choosing the colours used in illustrations.

9.10.7 Continuous colour-related scales

Continuous colour scales `scale_colour_continuous()`, `scale_colour_gradient()`, `scale_colour_gradient2()`, `scale_colour_gradientn()`, `scale_colour_date()`, and `scale_colour_datetime()`, give smooth continuous gradients between two or more colours. They are used with `numeric`, `date` and `datetime` data. A matching set of `fill` scales is also available. Other scales like `scale_colour_viridis_c()` and `scale_colour_distiller()` are based on the use of ready-made palettes of sets of colour gradients chosen to work well together under multiple conditions or for human vision including different types of colour blindness.

9.10.8 Discrete colour-related scales

Discrete colour scales, such as `scale_colour_discrete()`, `scale_colour_hue()`, `scale_colour_gray()`, are used with categorical data stored as factors. Some discrete scales, such as `scale_colour_viridis_d()` and `scale_colour_brewer()`, provide multiple discrete sets of colours selectable through palettes. A matching set of discrete `fill` scales is available. Ordinal scales, such as `scale_colour_ordinal()` and `scale_fill_ordinal()`, use palettes that set aesthetic values that ramp in steps between two extreme values. They are used when `ordered` factors are mapped to the aesthetics.

9.10.9 Binned scales

Before version 3.3.0 of '`ggplot2`', only two types of scales were available, continuous and discrete. A third type of scales, called *binned*, (implemented for all the aesthetics where relevant) was added in version 3.3.0. They are used with continuous variables, but they convert the continuous values into discrete ones, using bins corresponding to different ranges of values, and then represent them in the plot using a discrete set of aesthetic values from a gradient. We re-do the figure shown on page 293 but replacing `scale_colour_gradient()` by `scale_colour_binned()`.

```
set.seed(4321)
X <- 0:10
Y <- (X + X^2 + X^3) + rnorm(length(X), mean = 0, sd = mean(X^3) / 4)
df1 <- data.frame(X, Y)
df2 <- df1
df2[6, "Y"] <-df1[6, "Y"] * 10
```

```
ggplot(data = df2) +
  stat_fit_residuals(formula = y ~ poly(x, 3, raw = TRUE),
                     method = "rlm",
                     mapping = aes(x = X,
                                   y = stage(start = Y,
                                             after_stat = y * weights),
                                   colour = after_stat(weights)),
                     show.legend = TRUE) +
  scale_colour_binned(low = "red", high = "blue", limits = c(0, 1),
                      guide = "colourbar", n.breaks = 5)
```

The advantage of binned scales is that they facilitate the fast reading of the plot while their disadvantage is the decreased resolution of the scale. The use of a binned instead of a continuous scale is qualitative. The number of bins can be set by passing an argument to parameter `n.breaks` or alternatively, a `numeric` vector passed as argument to `breaks` can be used to explicitly set bin boundaries. When deciding how many bins to use, one needs to take into account the audience, how the figure will be rendered and displayed, and the length of time available to the viewers to peruse the plot relative to the density of information. Transformations are also allowed in these scales as in others.

9.10.10 Identity scales

In the case of identity scales, the mapping is one to-one to the data. For example, if we map the `colour` or `fill` *aesthetic* to a variable using `scale_colour_identity()` or `scale_fill_identity()`, the mapped variable must already contain valid R `color` definitions. In the case of mapping `alpha`, the variable must contain numeric values in the range 0 to 1.

We use a data frame containing a variable `colours` containing character strings interpretable as the names of `color` definitions known to R. We then use them directly in the plot by passing `scale_colour_identity()`.

```
df3 <- data.frame(X = 1:10, Y = dnorm(10), colours = rep(c("red", "blue"), 5))

ggplot(data = df3, mapping = aes(x = X, y = Y, colour = colours)) +
  geom_point() +
  scale_colour_identity()
```

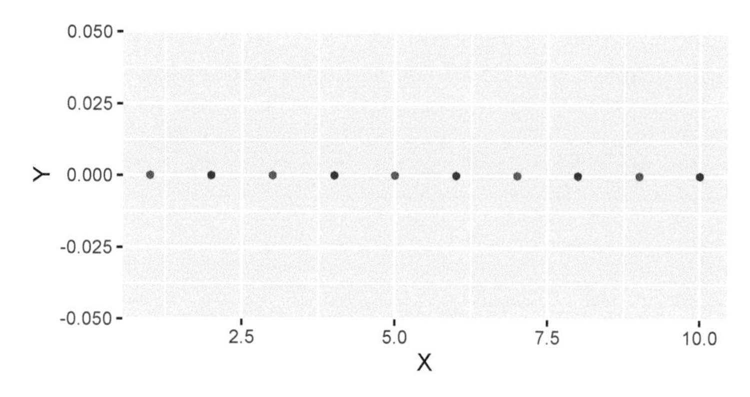

9.22 How does the plot look, if the identity scale is deleted from the example above? Edit and re-run the example code.

While using the identity scale, how would you need to change the code example above, to produce a plot with green and purple points?

The `colour` and `fill` scales used by default by geometries defined in package 'ggplot2' can be changed through R options `"ggplot2.continuous.colour"`, `"ggplot2.discrete.colour"`, `"ggplot2.ordinal.colour"`, `"ggplot2.binned.colour"`, `"ggplot2.continuous.fill"`, `"ggplot2.discrete.fill"`, `"ggplot2.ordinal.fill"` and `"ggplot2.binned.fill"`.

9.11 Adding Annotations

The idea of annotations is that they add plot elements that are not directly connected individual observations in `data`. Some like company logos, could be called "decorations", but others like text indicating the number of observations or even an inset plot or table may convey information about the data set as a whole. They can be drawn referenced to the "native" data coordinates used to plot but the position itself does convey information. Annotations are distinct from data labels. Annotations are added to a ggplot with function `annotate()` as plot layers (each call to `annotate()` creates a new layer). To achieve the behaviour expected of annotations, `annotate()` does not inherit the default `data` or `mapping` of variables to *aesthetics*. Annotations frequently make use of `"text"` or `"label"` *geometries* with character strings as data, possibly to be parsed as expressions. In addition, for example, the `"segment"` geometry can be used to add arrows.

While layers added to a plot using *geometries* and *statistics* respect faceting, layers added with `annotate()` are replicated unchanged in all panels of a faceted plot. The reason is that annotation layers accept *aesthetics* only as constant values which are the same for every panel as no grouping is possible without a `mapping` to `data`. Alternatives, using new geometries, are provided by package 'ggpp'.

Function `annotate()` takes the name of a geometry as its argument, in the example below, `"text"`. Function `aes()` is not used, as only mappings to constant values are accepted. These values can be vectors, thus, layers added with annotate can add multiple graphic objects of the same type to a plot.

```
ggplot(data = fake2.data, mapping = aes(x = z, y = y)) +
  geom_point() +
  annotate(geom = "text",
           label = "origin",
           x = 0, y = 0,
           colour = "blue",
           size=4)
```

9.23 Play with the values of the arguments to `annotate()` to vary the position, size, colour, font family, font face, rotation angle, and justification of the annotation.

It is relatively common to use inset tables, plots, bitmaps, or vector graphics as annotations. In section 9.5.8 on page 311, `geoms` from package '`ggpp`' were used to create insets in plots. An older alternative is to use `annotation_custom()` to add grobs ('`grid`' graphical object) to a ggplot. To add another or the same plot as an inset, it first needs to be converted it into a grob. A ggplot can be converted with function `ggplotGrob()`. In this example, the inset is a zoomed-in window into the main plot. In addition to the grob, coordinates for its location are passed in native data units.

```
p <- ggplot(data = fake2.data, mapping = aes(x = z, y = y)) +
  geom_point()
p + expand_limits(x = 40) +
  annotation_custom(ggplotGrob(p + coord_cartesian(xlim = c(4, 10), ylim = c(20, 30)) +
                                 theme_bw(10)),
                    xmin = 25, xmax = 40, ymin = 30, ymax = 60)
```

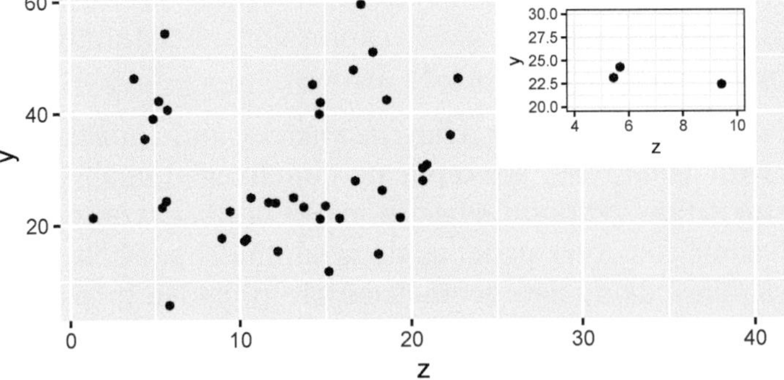

This approach has the limitation, shared with the use of `annotate()`, that if used together with faceting, the inset is added identically to all plot panels.

In the next code example, expressions are used as annotations as well as for tick labels. Do notice that we use recycling and vectorised arithmetic for setting the breaks, as `c(0, 0.5, 1, 1.5, 2) * pi` is equivalent to `c(0, 0.5 * pi, pi, 1.5 * pi, 2 * pi`. Annotations are plotted at their own position, unrelated to any observation in the data, but using the same coordinates and units as for plotting the data.

```
ggplot(data = data.frame(x = c(0, 2 * pi)),
       mapping = aes(x = x)) +
  stat_function(fun = sin) +
  scale_x_continuous(
    breaks = c(0, 0.5, 1, 1.5, 2) * pi,
    labels = c("0", expression(0.5~pi), expression(pi),
             expression(1.5~pi), expression(2~pi))) +
  labs(y = "sin(x)") +
  annotate(geom = "text",
           label = c("+", "-"),
           x = c(0.5, 1.5) * pi, y = c(0.5, -0.5),
           size = 20) +
  annotate(geom = "point",
           colour = "red",
           shape = 21,
           fill = "white",
           x = c(0, 1, 2) * pi, y = 0,
           size = 6)
```

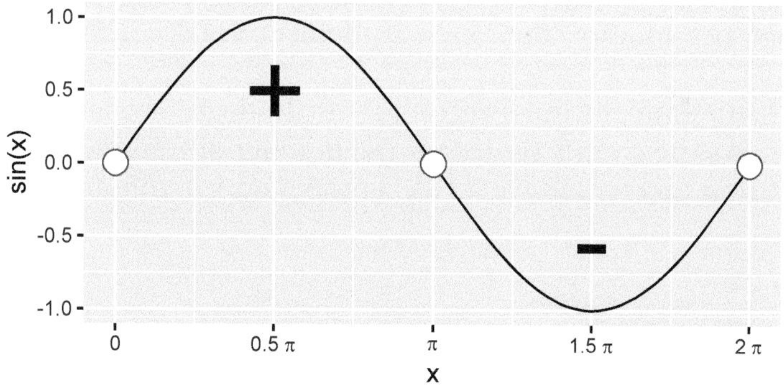

9.24 Modify the plot above to show the cosine instead of the sine function, replacing `sin` with `cos`. This is easy, but the catch is that you will need to relocate the annotations.

Function `annotate()` cannot be used with `geom = "vline"` or `geom = "hline"` as we can use `geom = "line"` or `geom = "segment"`. Instead, `geom_vline()` and/or `geom_hline()` can be used directly passing constant arguments to them. See section 9.5.3 on page 302.

9.12 Coordinates and Circular Plots

The grammar of graphics, as implemented in 'ggplot2', allows many different combinations of its "words", and this is also how circular plots are created. To obtain circular plots, we use the same *geometries*, *statistics*, and *scales* we have been using above, but combined with polar coordinates instead of the default cartesian coordinates. We override the default by adding `coord_polar()` to the plot so that the `x` and `y` *aesthetics* correspond to the angle and radial distance, respectively.

Special systems of coordinates, such as `coord_sf()`, used for maps, support different projections. In contrast, coordinate functions such as `coord_flip()`, `coord_trans()`, and `coord_fixed()` offer variations based on the cartesian system.

9.12.1 Wind-rose plots

Some types of data are more naturally expressed as angles using polar coordinates than on cartesian coordinates. The clearest example is wind direction, from which the name *wind-rose* derives. In some cases of time series data with a strong periodic variation, polar coordinates can be used to highlight phase shifts or changes in frequency. A more mundane application is to plot variation in a response variable through the day with a clock-face-like representation of time of day.

Wind rose plots are frequently histograms drawn on a polar system of coordinates (see section 9.6.4 on page 324). In the examples, we plot wind direction data, measured once per minute during 24 h (dataset `viikki_d29.dat` from package 'learnrbook').

A circular histogram of wind directions with 30-degree-wide bins can be created using `stat_bin()`. The counts represent the number of minutes during 24 h when the wind direction was within each bin, as the data set contains one observation per minute.

```
p.wind <-
  ggplot(data = viikki_d29.dat,
         mapping = aes(x = WindDir_D1_WVT))  +
  stat_bin(colour = "black", fill = "gray50", geom = "bar",
           binwidth = 30, boundary = 0, na.rm = TRUE) +
  coord_polar() +
  scale_x_continuous(breaks = c(0, 90, 180, 270),
                     labels = c("N", "E", "S", "W"),
                     limits = c(0, 360),
                     expand = c(0, 0),
                     name = "Wind direction") +
  scale_y_continuous(name = "Frequency (min/d)")
p.wind
```

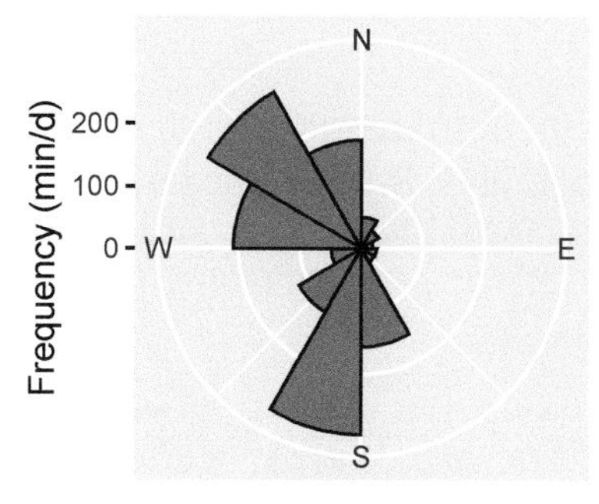

9.25 In the example above, `geom_bar()` was used. Edit the code to use other geometries, e.g., `geom_line()` and `geom_area()`.

A plot created using polar coordinates is not truly circular, but resembles a plot based on cartesian coordinates rolled into a circle. The difference is crucial in the case of some wind-rose plots. In a true circular plot, the data would have to be projected onto a cylinder without any discontinuity. The plot we obtain using `coord_polar()` retains a discontinuity at the North, at the boundary between 0 and 360 degrees. Thus for a histogram computed with `stat_bin()`, one boundary between bins must normally coincide with this divide. In a density plot, the densities on both sides of the North divide are fitted separately, frequently resulting in odd looking plots.

One approach to centring the bins on the cardinal directions would be to pre-compute the frequencies before plotting, pooling the observations for the slices 345–360 and 0–15 degrees into the same bin, and in a separate step, plotting them using `geom_col()` (not shown).

As when using other coordinates we can add facets. In this example, we create a factor based on solar time, to plot separately the observations from before or after local solar noon.

```
p.wind +
  facet_wrap(~factor(ifelse(hour(solar_time) < 12, "AM", "PM")))
```

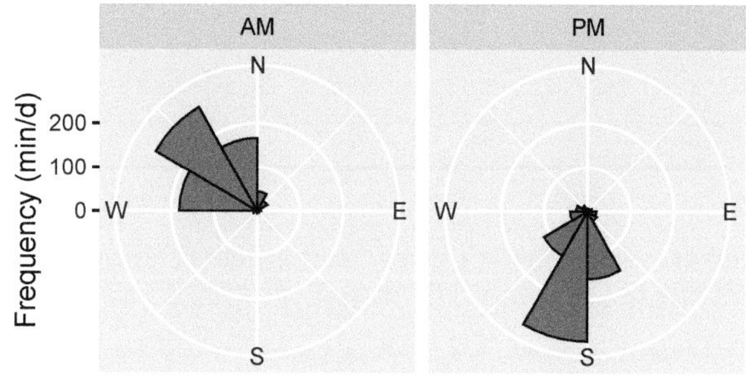

9.12.2 Pie charts

Pie charts are more difficult to read than bar charts because our brain is better at comparing lengths than angles. If used, pie charts should only be used to show composition, or fractional components that add up to a total. In this case, used only if the number of "pie slices" is small (rule of thumb: seven at most), however in general, they are best avoided.

A pie chart of counts is like a bar plot in which instead of heights angles describe the number of counts. `geom_bar()`, which defaults to use `stat_count()`, together with `coord_polar()` creates a pie chart. The brewer gradient scale supplies the palette for the fills, while the colour of the border line is set with `colour = "black")`.

```
ggplot(data = mpg,
       mapping = aes(x = factor(1), fill = factor(class))) +
  geom_bar(width = 1, colour = "black") +
  coord_polar(theta = "y") +
  scale_fill_brewer() +
  scale_x_discrete(breaks = NULL) +
  labs(x = NULL, fill = "Vehicle class")
```

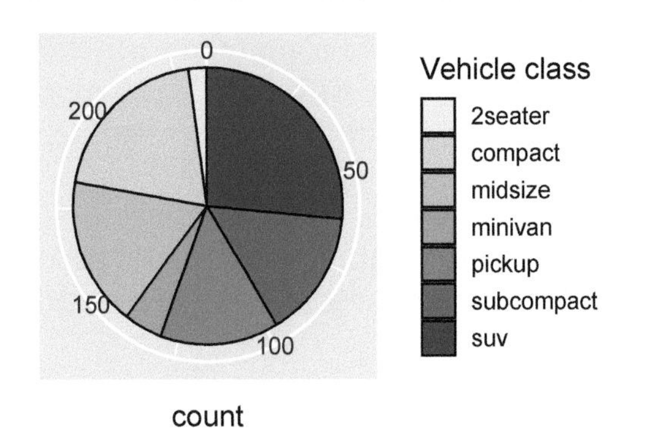

9.26 Edit the code for the pie chart above to obtain a bar chart. Which one of the two plots is easier to read?

9.13 Themes

In 'ggplot2', *themes* are the equivalent of style sheets. They determine how the different elements of a plot are rendered when displayed, printed, or saved to a file. *Themes* do not alter what aesthetics or scales are used to plot the observations or summaries, but instead how text labels, titles, axes, tick marks, plotting-area background, grid lines, etc., are formatted and if displayed or not. Package 'ggplot2' includes several predefined *theme constructors* (usually described as *themes*), and independently developed extension packages define additional ones. These constructors return complete themes, which when added to a plot, replace the theme already present. In addition to choosing among these already available *complete themes*, users can modify the ones already present in a plot by adding *incomplete themes*. When used in this way, *incomplete themes* usually are created on the fly. It is also possible to create new theme constructors that return complete themes, similar to `theme_gray()` from 'ggplot2'.

9.13.1 Complete themes

The theme used by default is `theme_gray()` with default arguments. In 'ggplot2', predefined themes are defined as constructor functions, with parameters. These parameters allow changing some "base" properties. The `base_size` for text elements is given in points, and affects all text elements in the returned theme object because the size of these elements is by default defined relative to the base size. Another parameter, `base_family`, allows the font family to be set. These functions return complete themes.

Themes have no effect on layers produced by *geometries* as themes have no effect on *mappings*, *scales*, or *aesthetics*. In the name `theme_bw()` black-and-white refers to the colour of the background of the plotting area and labels. If the *colour* or fill *aesthetics* are mapped or set to a constant in the figure, these will be respected irrespective of the theme. One cannot convert a colour figure into a black-and-white one by adding a *theme*. For colour gradients an alternative is to use a greyscale gradient by changing the *scale* used to map values to aesthetics. For discrete scales, a different aesthetic can be used, for example, use `shape` or `linetype` instead of `colour`.

Even the default `theme_gray()` can be added to a plot, to replace the default one with a newly constructed one created with arguments different to the defaults ones. Below, a serif font at a larger size than the default is used.

```
ggplot(data = fake2.data,
       mapping = aes(x = z, y = y)) +
  geom_point() +
  theme_gray(base_size = 18,
             base_family = "serif")
```

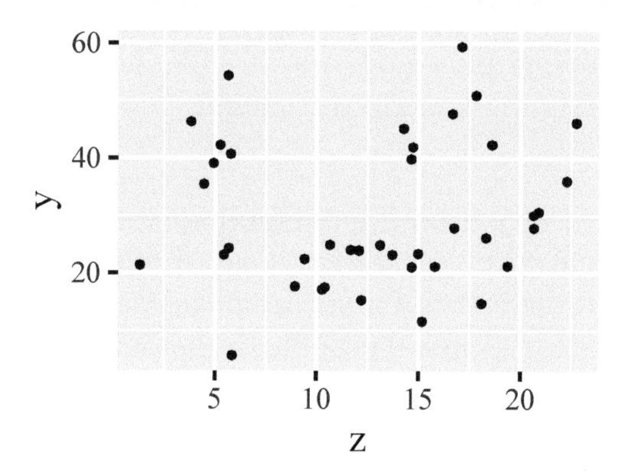

9.27 Change the code in the previous chunk to use, one at a time, each of the predefined themes from 'ggplot2': `theme_bw()`, `theme_classic()`, `theme_minimal()`, `theme_linedraw()`, `theme_light()`, `theme_dark()` and `theme_void()`.

Predefined "themes" like `theme_gray()` are, in reality, not themes but instead are constructors of theme objects. The *themes* they return when called depend on the arguments passed to their parameters. In other words, `theme_gray(base_size = 15)`, creates a different theme than `theme_gray(base_size = 11)`. In this case, as sizes of different text elements are defined relative to the base size, the size of all text elements changes in coordination. Font size changes by *themes* do not affect the size of text or labels in plot layers created with geometries, as their size is controlled by the `size` *aesthetic*.

A frequent idiom is to create a plot without specifying a theme, and then adding the theme when printing or saving it. This can save work, for example, when producing different versions of the same plot for a publication and a talk.

```
p.base <-
  ggplot(data = fake2.data,
         mapping = aes(x = z, y = y)) +
  geom_point()
print(p.base + theme_bw())
```

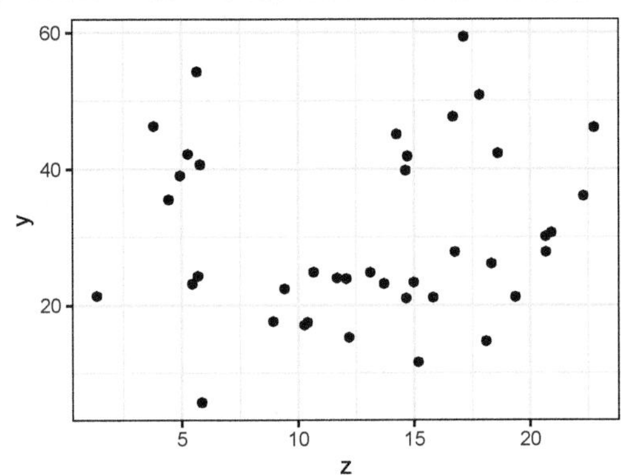

It is also possible to change the theme used by default in the current R session with `theme_set()`.

```
old_theme <- theme_set(theme_bw(15))
```

Similar to other functions used to change options in R, `theme_set()` returns the previous setting. By saving this value to a variable, here `old_theme`, we are able to restore the previous default, or undo the change.

```
theme_set(old_theme)
```

The use of a grey background as default for plots is unusual. This graphic design decision originates in typesetters' goal of maintaining a uniform average luminosity throughout the text and plots in a page. Many scientific journals require or at least prefer a more traditional graphic design. Theme `theme_bw()` is the most versatile of the traditional designs supported as it works well both for individual plots as for plots with facets as it includes a box. Theme `theme_classic()` lacking a box and grid works well for individual plots, but needs to be adjusted when used with facets so as to obtain nice looking plots.

9.13.2 Incomplete themes

To create a significantly different theme, and/or reuse it in multiple plots, it is best to create a new constructor, or a modified complete theme as described in section 9.13.3 on page 368. In other cases, it is enough to tweak individual theme settings for a single plot. Below, overlapping x-axis tick labels are avoided by rotation the axis tick labels. When rotating the labels, it is also necessary to change their justification, as justification is relative to the orientation of the text.

```
ggplot(data = fake2.data,
       mapping = aes(x = z + 1000, y = y)) +
  geom_point() +
  scale_x_continuous(breaks = scales::pretty_breaks(n = 8)) +
  theme(axis.text.x = element_text(angle = 33, hjust = 1, vjust = 1))
```

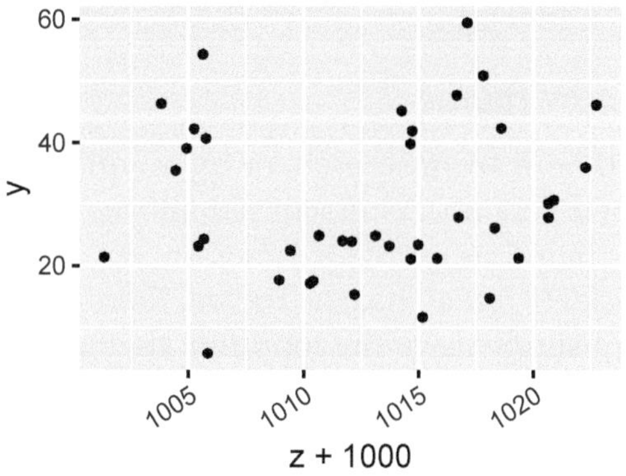

9.28 Play with the code in the last chunk above, modifying the values used for `angle`, `hjust` and `vjust`. (Angles are expressed in degrees, and justification with values between 0 and 1).

A less elegant approach is to use a smaller font size. Within `theme()`, function `rel()` can be used to set size relative to the base size. In this example, we use `axis.text.x` so as to change the size of tick labels only for the x axis.

```
ggplot(fake2.data, aes(z + 100, y)) +
  geom_point() +
  scale_x_continuous(breaks = scales::pretty_breaks(n = 20)) +
  theme(axis.text.x = element_text(size = rel(0.6)))
```

Theme definitions follow a hierarchy, allowing us to modify the formatting of groups of similar elements, as well as of individual elements. In the chunk above, using `axis.text` instead of `axis.text.x`, would have affected the tick labels in both x and y axes.

9.29 Modify the example above, so that the tick labels on the x-axis are blue and those on the y-axis red, and the font size is the same for both axes, but changed from the default. Consult the documentation for `theme()` to find out the names of the elements that need to be given new values. For examples, see *ggplot2: Elegant Graphics for Data Analysis* (Wickham and Sievert 2016) and *R Graphics Cookbook* (Chang 2018).

Formatting of other text elements can be adjusted in a similar way, as well as thickness of axes, length of tick marks, grid lines, etc. However, in most cases, these are graphic design elements that are best kept consistent throughout sets of plots and best handled by creating a new *theme*.

If you both add a *complete theme* and want to modify some of its elements, you should add the whole theme before modifying it with `+ theme(...)`. This may seem obvious once one has a good grasp of the grammar of graphics, but can be at first disconcerting.

It is also possible to modify the default theme used for rendering all subsequent plots.

```
old_theme <- theme_update(text = element_text(colour = "darkred"))
```

Having saved the previous default to `old_theme` it can be restored when needed.

```
theme_set(old_theme)
```

9.13.3 Defining a new theme

Themes can be defined both from scratch, or by modifying existing saved themes, and saving the modified version. As discussed above, it is also possible to define a new, parameterised theme constructor function.

Unless we plan to widely reuse the new theme, there is usually no need to define a new function. We can simply save the modified theme to a variable and add it to different plots as needed. As we will be adding a "ready-build" theme object rather than a function, we do not use parentheses.

```
my_theme <- theme_bw(15) + theme(text = element_text(colour = "darkred"))
p.base + my_theme
```

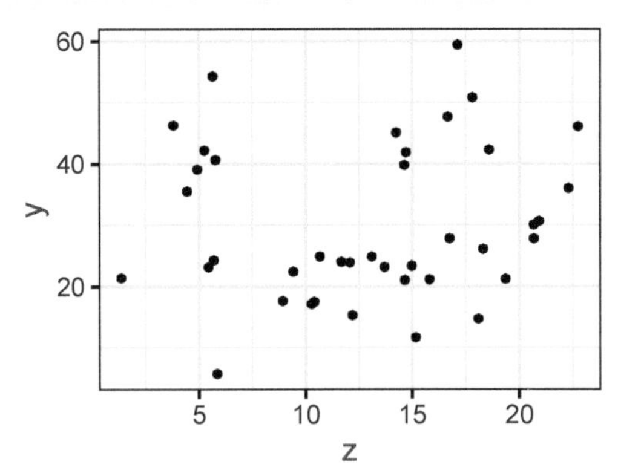

Creating a new theme constructor similar to those from package 'ggplot2' can be fairly simple if the changes are few. As the implementation details of theme objects may change in future versions of 'ggplot2', the safest approach is to rely only on the public interface of the package. The functions exported by package 'ggplot2' can be wrapped inside a new function that modifies the theme before returning it. The interface, parameters, of the wrapped function can be included in the new one and the arguments passed along to the wrapped function, as is or modified. If needed, additional parameters can be handled by code in the wrapper function. Below, a wrapper on `theme_gray()` is constructed retaining a compatible interface, but adding a new base parameter, `base_colour`. A different default is used for `base_family`. The key detail is passing `complete = TRUE` to `theme()`, as this tags the returned theme as being usable by itself, resulting in replacement of any theme already in a plot when it is added.

```
my_theme_gray <-
  function (base_size = 11,
            base_family = "serif",
            base_line_size = base_size/22,
            base_rect_size = base_size/22,
            base_colour = "darkblue") {

    theme_gray(base_size = base_size,
               base_family = base_family,
               base_line_size = base_line_size,
               base_rect_size = base_rect_size) +

    theme(line = element_line(colour = base_colour),
          rect = element_rect(colour = base_colour),
          text = element_text(colour = base_colour),
          title = element_text(colour = base_colour),
          axis.text = element_text(colour = base_colour),
          complete = TRUE)
  }
```

Our own theme constructor, created without too much effort, is ready to be used. To avoid surprising users, it is good to make `my_theme_grey()` a synonym of `my_theme_gray()` following ‘ggplot2’ practice.

```
my_theme_grey <- my_theme_gray
```

A plot created using `my_theme_gray()` with text colour set to dark red.

```
p.base + my_theme_gray(15, base_colour = "darkred")
```

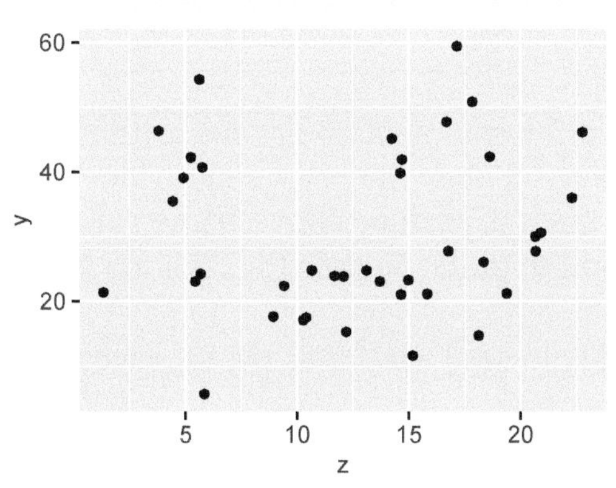

9.14 Composing Plots

While facets make it possible to create plots with panels that share the same mappings and data (see section 9.8 on page 335), plot composition makes it possible to combine separately created "gg" plot objects into a single plot. Composition before rendering makes it possible to automate the correct alignments, ensure consistency of text size and even merge duplicate guide or keys. Composite plots can save space on the screen or page, but more importantly can bring together data visualisations that need to be compared or read as a whole.

Package 'patchwork' defines a simple grammar for composing plots created with 'ggplot2', that I have used earlier in the chapter to display pairs of plots side by side. Composition with 'patchwork' can also include grid graphical objects. The plot composition grammar uses operators +, | and /, although 'patchwork' provides additional tools for defining complex layouts of panels. While + allows different layouts, | composes panels side by side, and / composes panels on top of each other. The plots to be used as panels can be grouped using parentheses. The operands must be whole plots, below, this ensured by saving each plot to a variable. When composing anonymous plots they must be enclosed in parentheses, to ensure that the correct operators are dispatched.

Three simple plots, p1, p2 and p3 will be used below.

```
p1 <- ggplot(mpg, aes(displ, cty, colour = factor(cyl))) +
        geom_point() +
        theme(legend.position = "top")
p2 <- ggplot(mpg, aes(displ, cty, colour = factor(year))) +
        geom_point() +
        theme(legend.position = "top")
p3 <- ggplot(mpg, aes(factor(model), cty)) +
        geom_point() +
        theme(axis.text.x =
                element_text(angle = 90, hjust = 1, vjust = 0.5))
```

9.30 A combined plot can be simply assembled using the operators (plot not shown).

```
p1 | (p2 / p3)
```

```
(p1 | p2) / p3
```

The operators used for composition are the arithmetic ones, and even if used for a different purpose still obey the precedence rules of mathematics. The order of precedence can be altered, as done above, using parentheses. Run the examples above after creating three plots. Modify the code trying different ways of organising the three panels.

A title for the whole plot and a letter as tag for each panel are added as a whole-plot annotation.

```
((p1 | p2) / p3) +
  plot_annotation(title = "Fuel use in city traffic:", tag_levels = 'a')
```

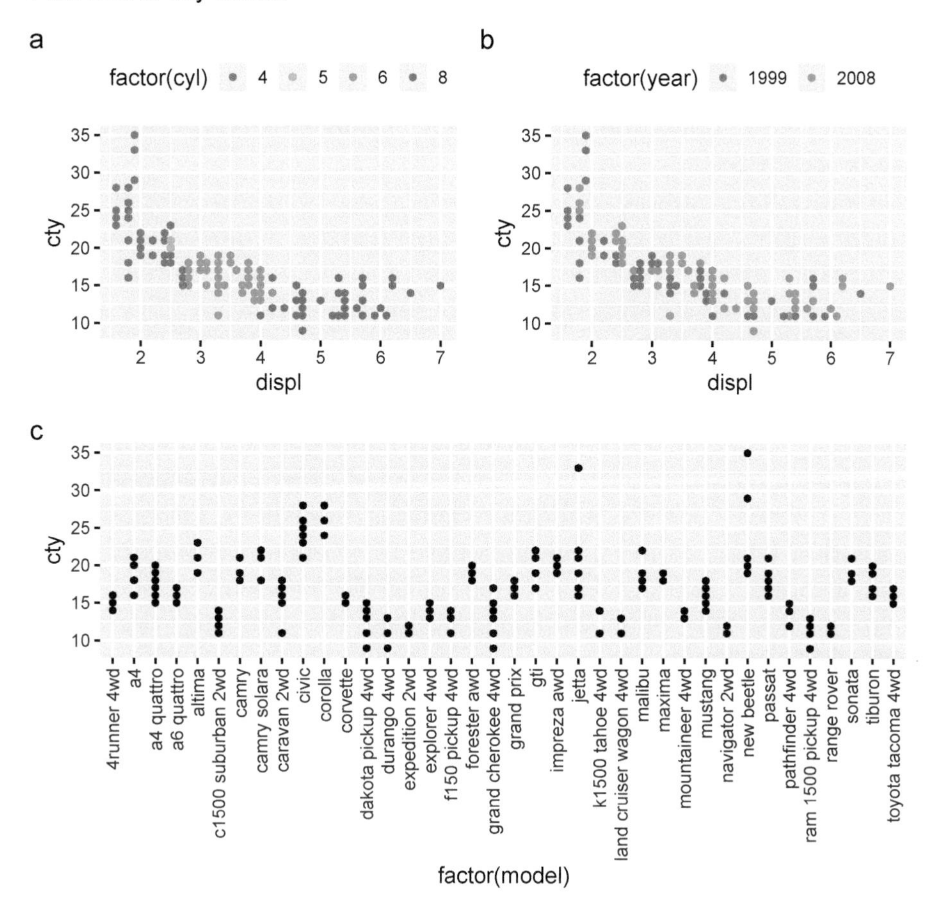

Package 'patchwork' has in recent versions tools for the creation of complex layouts, addition of insets and combining in the same layout plots and other graphic objects such as bitmaps, photographs, and even tables.

9.31 Package 'patchwork' can be very useful. Study the documentation and its examples, and try to think how it could be useful to you. Then try to compose plots like those you could use in your work or studies.

9.15 Using `plotmath` Expressions

Plotmath expression are similar to R expressions, but they are targeted at the creation of mathematical annotations. In some respects, they are similar to the math mode in LaTeX. They are used in graphical output like plots. The syntax sometimes feels awkward and takes some time to be learnt, but it gets the job done.

The main limitation to producing rich text annotations in R similar to those possible using LaTeX or using HTML is at the core of the R program. There is work in progress and improvements can be expected in coming years. Meanwhile, the already implemented enhancements gradually appear as enhanced features in 'ggplot2' and its extensions.

Package 'ggtext' provides rich-text (basic HTML and Markdown) support for 'ggplot2', both for annotations and for data visualisation. This is an alternative to the use of R expressions.

In sections 9.6.1 and 9.5.7, simple examples of the use of R expressions for labelling plots were given. The `demo(plotmath)` demo and the help page `help(plotmath)` provide enough information to start using expressions in plots. Although expressions are shown here in the context of plotting, they are also used in other contexts in R code.

In general, it is possible to create *expressions* explicitly with function `expression()` or by parsing a character string. In the case of 'ggplot2' for some plot elements, layers created with `geom_text()` and `geom_label()`, and the strip labels of facets the parsing is delayed and applied to mapped character variables in `data`. In contrast, for titles, subtitles, captions, axis-labels, etc. (anything that is defined within `labs()`), the expressions have to be entered explicitly, or saved as such into a variable, and the variable passed as an argument.

When plotting expressions using `geom_text()`, the parsing of character strings is signalled by passing `parse = TRUE` in the call to the layer function. In the case of facets' strip labels, parsing or not depends on the *labeller* function used. An additional twist is the possibility of combining static character strings with values taken from `data` (see section 9.8 on page 335).

The most difficult thing to remember when writing expressions is how to connect the different parts. A tilde (~) adds space in between symbols. Asterisk (*) can be also used as a connector. The * is usually needed when dealing with numbers next to symbols. Using whitespace is allowed in some situations, but not in others. To include within an expression text that should not be parsed, it must be enclosed in quotation marks, which may need themselves to be quoted. For a long list of examples, have a look at the output and code displayed by `demo(plotmath)` at the R command prompt.

Expressions are frequently used for axis labels, e.g., when the units or symbols require the use of superscripts or Greek letters. In this case, they are usually entered as expressions.

```
p1 + labs(y = expression("Fuel use"~~(m~g^{-1})),
          x = "Engine displacement (L)",
          colour = "Engine\ncylinders") +
     theme(legend.position = "right")
```

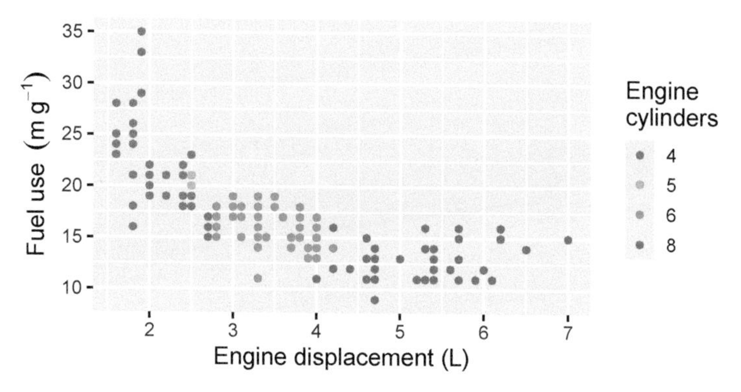


```
set.seed(54321) # make sure we always generate the same data
my.data <-
  data.frame(x = 1:5,
             y = rnorm(5),
             greek.label = paste("alpha[", 1:5, "]", sep = ""))
```

In the example below, the x-axis label is a Greek α character with i as subscript, and the y-axis label includes a superscript in the units. The title we use is a character string, while the subtitle is a rather complex expression.

Each observation has as data label a subscripted $alpha$. When using a *geometry*, instead of directly using an expression, we map to the `label` *aesthetic* character strings to be parsed into expressions. In other words, character strings, that are written using the syntax of expressions. We need to set `parse = TRUE` in the call to the *geometry* so that the strings instead of being plotted as is, are parsed into expressions before the plot is rendered.

```
ggplot(my.data, aes(x, y, label = greek.label)) +
   geom_point() +
   geom_text(angle = 45, hjust = 1.2, parse = TRUE) +
   labs(x = expression(alpha[i]),
        y = expression(Speed~~(m~s^{-1})),
        title = "Using expressions",
        subtitle = expression(sqrt(alpha[1] + frac(beta, gamma))))
```

As parsing character strings is an alternative way of creating expressions, this approach can be also used in other situations. For example, a character string stored in a variable can be parsed with `parse()` as done below for `subtitle`. Tick labels are also set to expressions, taking advantage that `expression()` accepts multiple arguments separated by commas returning a vector of expressions.

```
my_eq.char <- "alpha[i]"
ggplot(my.data, aes(x, y)) +
   geom_point() +
   labs(title = parse(text = my_eq.char)) +
   scale_x_continuous(name = expression(alpha[i]),
                      breaks = c(1,3,5),
                      labels = expression(alpha[1], alpha[3], alpha[5]))
```

α_i

-0.4
-0.8
-1.2
-1.6
y
α_1 α_3 α_5
α_i

A different approach (no example shown) would be to call `parse()` explicitly for each individual label, something that might be needed if the tick labels need to be "assembled" programmatically instead of set as constants.

Differences between `parse()` and `expression()`. Function `parse()` takes as an argument a character string. This is very useful as the character string can be created programmatically. When using `expression()` this is not possible, except for substitution at execution time of the value of variables into the expression. See the help pages for both functions.

Function `expression()` accepts its arguments without any delimiters. Function `parse()` takes a single character string as an argument to be parsed, in which case quotation marks within the string need to be *escaped* (using \" where a literal " is desired). In both cases, a character string can be embedded using one of the functions `plain()`, `italic()`, `bold()` or `bolditalic()` which also affect the font used. The argument to these functions needs to be a character string delimited by quotation marks if it is not to be parsed.

When using `expression()`, bare quotation marks can be embedded,

```
ggplot(cars, aes(speed, dist)) +
  geom_point() +
  xlab(expression(x[1]*"  test"))
```

while in the case of `parse()` they need to be *escaped*,

```
ggplot(cars, aes(speed, dist)) +
  geom_point() +
  xlab(parse(text = "x[1]*\"  test\""))
```

and in some cases will be enclosed within a format function.

```
ggplot(cars, aes(speed, dist)) +
  geom_point() +
  xlab(parse(text = "x[1]*italic(\"  test\")"))
```

Some additional remarks. If `expression()` is passed multiple arguments, it returns a vector of expressions. Where `ggplot()` expects a single value as an argument, as in the case of axis labels, only the first member of the vector will be used.

```
ggplot(cars, aes(speed, dist)) +
  geom_point() +
  xlab(expression(x[1], "  test"))
```

Depending on the location within a expression, spaces maybe ignored, or illegal. To juxtapose elements without adding space, use *, and to explicitly insert whitespace, use ~. As shown above, spaces are accepted within quoted text. Consequently, the following alternatives can also be used.

```
xlab(parse(text = "x[1]~~~~\"test\""))
```

```
xlab(parse(text = "x[1]~~~~plain(test)"))
```

However, unquoted whitespace is discarded.

```
xlab(parse(text = "x[1]*plain(   test)"))
```

Finally, it can be surprising that trailing zeros in numeric values appearing within an expression are dropped.

Function `paste()` was used above to insert values stored in a variable; functions `format()`, `sprintf()` and `strftime()` allow the conversion into character strings of other values. These functions can be used when creating plots to generate suitable character strings for the `label` *aesthetic* out of numeric, logical, date, time, and even character values. They can be, for example, used to create labels within a call to `aes()`.

```
sprintf("log(%.3f) = %.3f", 5, log(5))
## [1] "log(5.000) = 1.609"
sprintf("log(%.3g) = %.3g", 5, log(5))
## [1] "log(5) = 1.61"
```

9.32 Study the chunk above. If you are familiar with C or C++ function `sprintf()` will already be familiar to you, otherwise study its help page.

Play with functions `format()`, `sprintf()` and `strftime()`, using them to convert and format different types of data into character strings with different numbers of significant digits, scientific notation, decimal format, different field width, justification, etc.

It is also possible to substitute the value of variables or, in fact, the result of evaluation, into a new expression, allowing on the fly construction of expressions. Such expressions are frequently used as labels in plots. This is achieved through use of *quoting* and *substitution*.

Function `bquote()` can be used to substitute variables or expressions enclosed in `.( )` by their value. Be aware that the argument to `bquote()` needs to be written as an expression; in this example, a tilde, ~, inserts a space between words. Furthermore, if the expressions include variables, these will be searched for in the environment rather than within `data`, except within calls to `aes()` or `vars()`.

```
ggplot(cars, aes(speed, dist)) +
  geom_point() +
  labs(title = bquote(Time~zone: .(Sys.timezone())),
       subtitle = bquote(Date: .(as.character(today()))))
```

```
)
```

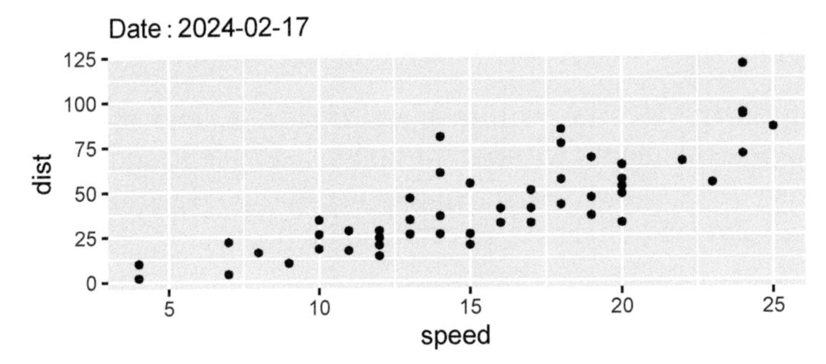

In the case of `substitute()` a named list can be passed as argument.

```
ggplot(cars, aes(speed, dist)) +
  geom_point() +
  labs(title = substitute(Time~zone: tz, list(tz = Sys.timezone())),
       subtitle = substitute(Date: date, list(date = as.character(today())))
       )
```

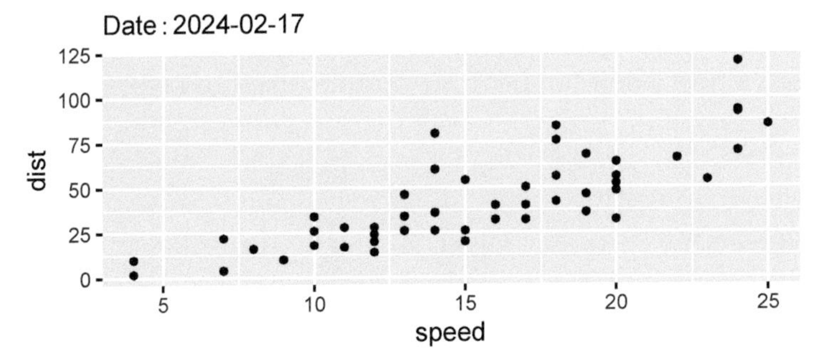

For example, substitution can be used to assemble an expression within a function based on the arguments passed. One case of interest is to retrieve the name of the object passed as an argument, from within a function.

```
deparse_test <- function(x) {
  print(deparse(substitute(x)))
}

a <- "saved in variable"

deparse_test("constant")
## [1] "\"constant\""
deparse_test(1 + 2)
## [1] "1 + 2"
deparse_test(a)
## [1] "a"
```

9.16 Creating Complex Data Displays

The grammar of graphics allows one to build and test plots incrementally. In daily use, when creating a completely new plot, it is best to start with a simple design for a plot, `print()` this plot, checking that the output is as expected and the code error-free. Afterwards, one can map additional *aesthetics* and add *geometries* and *statistics* gradually. The final steps are then to add *annotations* and the text or expressions used for titles, and axis and key labels. Another approach is to start with an existing plot and modify it, e.g., by using the same plotting code with different `data` or mapping different variables. When reusing code for a different data set, scale `limits` and `names` are likely to need to be edited.

9.33 Build a graphically complex data plot of your interest, step by step, taking advantage of the layered structure to test intermediate versions in an iterative design process, first by building up the complex plot in stages as a tool in debugging, and later using iteration in the processes of improving the graphic design of the plot, its readability, and effectiveness in conveying information.

9.17 Creating Sets of Plots

Plots to be presented at a given occasion or published as part of the same work need to be consistent in various respects: themes, scales and palettes, annotations, titles, and captions. To guarantee this consistency, we need to build plots modularly and avoid repetition by assigning names to the "modules" that need to be used multiple times.

A simple version of this approach was used in many examples above, where a base plot was modified by addition of different layers or scales.

9.17.1 Saving plot layers and scales in variables

When creating plots with 'ggplot2', objects are composed using operator + to assemble together the individual components. The functions that create plot layers, scales, etc. are constructors of objects and the objects they return can be stored in variables, and once saved, added to multiple plots at a later time.

A plot can be saved to a variable, here `p.base`, and, e.g., the value returned by a call to function `labs()`, into a different variable, here, `p.labels`.

```
p.base <- ggplot(data = mtcars,
                 aes(x = disp, y = mpg,
                     colour = factor(cyl))) +
          geom_point()

p.labels <- labs(x = "Engine displacement)",
                 y = "Gross horsepower",
                 colour = "Number of\ncylinders",
```

```
                      shape = "Number of\ncylinders")
```

When composing plots with the + operator, the left-hand-side operand must be a "gg" object. The right-hand-side operand is added to the "gg" plot object and the result returned as a new "gg" plot object.

The final plot can be assembled from the objects saved to variables. This is useful when creating several plots that should have consistent labels. The same approach can be used with other components. Below, the objects are combined with additional components to create different versions of the same plot.

```
p.base
p.base + p.labels + theme_bw(16)
p.base + p.labels + theme_bw(16) + ylim(0, NA)
```

We can also save intermediate results.

```
p.log <- p.base + scale_y_log10(limits=c(8,55))
p.log + p.labels + theme_bw(16)
```

9.17.2 Saving plot layers and scales in lists

If the pieces to be put together do not include a "gg" object, they can be collected into a list and saved. When the list is added to a "gg" plot object, the members of the list are added one by one to the plot respecting their order.

```
p.parts <- list(p.labels, theme_bw(16))
p1 + p.parts
```

9.34 Revise the code you wrote for the "playground" exercise in section 9.16, but this time, pre-building and saving groups of elements that you expect to be useful unchanged when composing a different plot of the same type, or a plot of a different type from the same data.

9.17.3 Using functions as building blocks

The "packaged" plots parts sometimes should adjust their behaviour at the time they are added to a plot. In this case a function that accepts the necessary arguments can be written, rather similarly as in the example for creating a new theme by wrapping function `theme_grey()` (see section 9.13.3 on page 368). These functions can return a "gg" object, a list of plot components, or a single plot component. The simplest use is to alter some defaults in existing constructor functions returning "gg" objects or layers. The ellipsis (`...`) allows passing named arguments to a nested function. In this case, every single argument passed by name to `bw_ggplot()` will be copied as an argument to the nested call to `ggplot()`. Be aware that supplying arguments by position, is possible only for parameters explicitly included in the definition of the wrapper function, thus, not supported with a function like this, with `...` as its only formal parameter.

```
bw_ggplot <- function(...) {
  ggplot(...) +
  theme_bw()
}
```

which could be used as follows.

```
bw_ggplot(data = mtcars,
          mapping = aes(x = disp, y = mpg,
          colour = factor(cyl))) +
          geom_point()
```

9.18 Generating Output Files

It is possible, when using RStudio, to directly export the displayed plot to a file using a menu. However, if the file will have to be generated again at a later time, or a series of plots need to be produced with consistent format, it is best to include the commands to export the plot in the script.

In R, files are created by printing to different devices. Printing is directed to a currently open device such a window in RStudio. Some devices produce screen output, while others write files. Devices depend on drivers. There are both devices that are part of R and additional ones defined in contributed packages.

Creating a file involves opening a device, printing and closing the device in sequence. In most cases, the file remains locked until the device is close.

For example, when rendering a plot to PDF, Encapsulated Postscript, SVG or other vector graphics formats, arguments passed to `width` and `height` are expressed in inches.

```
fig1 <- ggplot(data.frame(x = -3:3), aes(x = x)) +
  stat_function(fun = dnorm)
pdf(file = "fig1.pdf", width = 8, height = 6)
print(fig1)
dev.off()
```

For Encapsulated Postscript and SVG output, we only need to substitute `pdf()` with `postscript()` or `svg()`, respectively.

```
postscript(file = "fig1.eps", width = 8, height = 6)
print(fig1)
dev.off()
```

In the case of graphics devices for file output in BMP, JPEG, PNG, and TIFF bitmap formats, arguments passed to `width` and `height` are expressed in pixels.

```
tiff(file = "fig1.tiff", width = 1000, height = 800)
print(fig1)
dev.off()
```

Some graphics devices are part of base-R and others are implemented in contributed packages. In some cases, there are multiple graphic devices available for rendering graphics in a given file format. These devices usually use different libraries, or have been designed with different aims. These alternative graphic devices can also differ in their function signature, i.e., have differences in the parameters and their names.

Differences also exist in their limitations and supported features, so in cases when rendering fails inexplicably, it is worthwhile to switch to an alternative graphics device to find out if the problem is in the plot or in the rendering engine. Several of the new features added to 'grid' in R versions 4.1.0, 4.2.0, and 4.3.0 are currently supported only by some of the graphics devices.

9.19 Debugging Ggplots

R package 'gginnards' provides methods `str()` (enhanced), `num_layers()`, `top_layer()`, `bottom_layer()`, and `mapped_vars()`. It also defines `geoms` and `stats` that instead of creating a layer, pass to a function such as `print()` the data frame they receive through parameter `data`. These are simple functions that even if dependent on 'ggplot2' internals are not prone to easily break with 'ggplot2' updates.

Package 'ggtrace' provides much more detailed and sophisticated approaches to explore the internals of "gg" plot objects. Package 'ggplot2' itself gives access to some object components.

Of these tools, `geom_debug()` is probably the most intuitive to use, both on its own and as an argument to `stats`.

```
ggplot(data = iris, mapping = aes(x = Petal.Length, y = Species)) +
  stat_summary(geom = "debug")

## No summary function supplied, defaulting to `mean_se()`
## [1] "Summary of input 'data' to 'draw_panel()':"
##   y group     x     xmin     xmax PANEL flipped_aes
## 1 1     1 1.462 1.437440 1.486560     1        TRUE
## 2 2     2 4.260 4.193545 4.326455     1        TRUE
## 3 3     3 5.552 5.473950 5.630050     1        TRUE
```

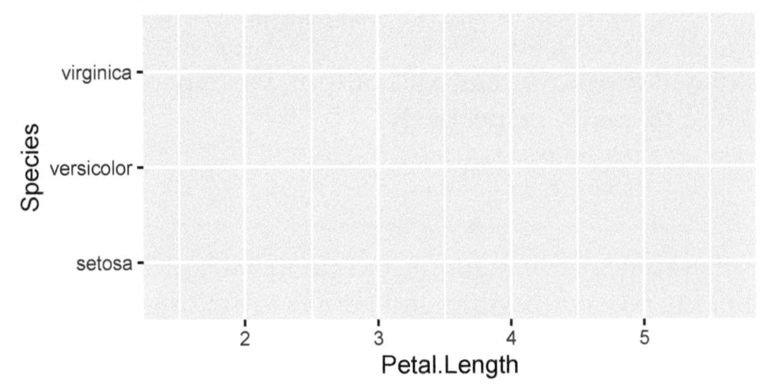


```
ggplot(data = iris, mapping = aes(x = Petal.Length)) +
  stat_bin(geom = "debug")

## `stat_bin()` using `bins = 30`. Pick better value with `binwidth`.
## [1] "Summary of input 'data' to 'draw_panel()':"
##    y count        x      xmin     xmax     width    density     ncount
## 1  2     2 1.017241 0.9155172 1.118966 0.2034483 0.06553672 0.07692308
## 2  9     9 1.220690 1.1189655 1.322414 0.2034483 0.29491525 0.34615385
## 3 26    26 1.424138 1.3224138 1.525862 0.2034483 0.85197740 1.00000000
## 4 11    11 1.627586 1.5258621 1.729310 0.2034483 0.36045198 0.42307692
## 5  2     2 1.831034 1.7293103 1.932759 0.2034483 0.06553672 0.07692308
```

```
## 6  0      0 2.034483 1.9327586 2.136207 0.2034483 0.00000000 0.00000000
##     ndensity flipped_aes PANEL group ymax ymin
## 1 0.07692308       FALSE     1    -1    2    0
## 2 0.34615385       FALSE     1    -1    9    0
## 3 1.00000000       FALSE     1    -1   26    0
## 4 0.42307692       FALSE     1    -1   11    0
## 5 0.07692308       FALSE     1    -1    2    0
## 6 0.00000000       FALSE     1    -1    0    0
```

count

20 -

10 -

0 -

2 4 6

Petal.Length

9.20 Further Reading

An in-depth discussion of the many extensions to package ‘ggplot2’ is outside the scope of this book. Several books describe in detail the use of ‘ggplot2’, being *ggplot2: Elegant Graphics for Data Analysis* (Wickham and Sievert 2016) the one written by the main author of the package. For inspiration or worked out examples, the book *R Graphics Cookbook* (Chang 2018) is an excellent reference. In depth explanations of the technical aspects of R graphics are available in the book *R Graphics* (Murrell 2019).

10

Base R and Extensions: Data Sharing

> Most programmers have seen them, and most good programmers realise they've written at least one. They are huge, messy, ugly programs that should have been short, clean, beautiful programs.
>
> John Bentley
> *Programming Pearls*, 1986

10.1 Aims of This Chapter

In this chapter, you will learn how to exchange data between R and some other applications. Base R and the recommended packages (installed by default) include several functions for importing and exporting data. Contributed packages provide both replacements for some of these functions and support for several additional file formats. In the present chapter, I aim at describing both data input and output covering in detail only the most common "foreign" data formats (those not native to R). The function pairs `save()` and `load()`, and `saveRDS()` and `readRDS()`, which save and read data in R's native formats, are described in chapter 4, sections 4.7.2 and 4.7.3 starting on page 118.

Data file formats that are foreign to R are not always well defined, making it necessary to reverse-engineer the algorithms needed to read them. These formats, even when clearly defined, may be updated by the developers of the foreign software that writes the files. Consequently, developing software to read and write files using foreign formats can easily result in long, messy, and ugly R scripts. We can also unwillingly write code that usually works but occasionally fails with specific files, or even worse, occasionally silently corrupts the imported data. The aim of this chapter is to provide guidance for finding functions for reading data encoded using foreign formats, covering both base R, including the 'foreign' package, and independently contributed packages. Such functions are well tested or validated and should be used whenever possible when importing data stored in foreign file formats.

DOI: 10.1201/9781003404187-10

10.2 Introduction

The first step in any data analysis with R is to input or read-in the data. Available sources of data are many and data can be stored or transmitted using various formats, both based on text or binary encodings. It is crucial that data are not altered (corrupted) when read and that in the eventual case of an error, errors are clearly reported. Most dangerous are silent non-catastrophic errors.

The very welcome increase of awareness of the need for open availability of data, makes the output of data from R into well-defined data-exchange formats another crucial step. Consequently, in many cases, an important step in data analysis is to export the data for submission to a repository, in addition to publication of the results of the analysis.

Faster internet access to data sources and cheaper random-access memory (RAM) has made it possible to efficiently work with relatively large data sets in R. That R keeps all data in memory (RAM), imposes limits to the size of data R functions can operate on. For data sets that do not fit in computer RAM, one can read selected lines from text files, use file formats like NetCDF that natively implement selective reading, or use queries to access data stored in local or remote databases.

Some contributed R packages support import of data saved in the same formats already supported by base R, but using different compromises between reliability, easy of use and performance. Functions in base R tend to prioritise reliability and protection from data corruption while some contributed packages prioritise performance. Other contributed packages make it possible to import and export data stored in file formats not supported by base R functions. Some of these formats are subject-area specific while others are in widespread use. Packages supporting download and upload of data sets from specific public repositories are also available (see `https://ropensci.org/packages/`).

10.3 Packages Used in This Chapter

```
install.packages(learnrbook::pkgs_ch10_2ed)
```

To run the examples included in this chapter, you need first to load some packages from the library (see section 6.4 on page 179 for details on the use of packages).

> ⚠ Several examples in this chapter make use of functions from the 'tidyverse' for data wrangling because some of the packages used to import data in "foreign" formats are themselves part of the 'tidyverse'.

```
library(learnrbook)
library(tibble)
library(purrr)
library(stringr)
```

```
library(dplyr)
library(tidyr)
library(readr)
library(readxl)
library(xlsx)
library(readODS)
library(pdftools)
library(foreign)
library(haven)
library(xml2)
library(XML)
library(ncdf4)
library(tidync)
library(lubridate)
library(jsonlite)
```

Some data sets used in this and other chapters are available in package 'learnrbook'. In addition to the R data objects, the package includes files saved in *foreign* formats used in examples of importing data. The files can be either read from the R library, or from a copy in a local folder. In this chapter, the code examples assume the user has copied the contents of folder "extdata" of the package to the current working folder.

The files can be copied by running the two statements below, assuming the current folder is the one that will be used to run the code examples in this chapter.

```
pkg.path <- system.file("extdata", package = "learnrbook")
file.copy(pkg.path, ".", overwrite = TRUE, recursive = TRUE)
## [1] TRUE
```

Some examples write files to disk, and the statements below ensure that the folder used in these examples exists, creating it if not found.

```
save.path = "./data"
if (!dir.exists(save.path)) {
  dir.create(save.path)
}
```

10.4 File Names and Operations

The naming of files affects data sharing irrespective of the format used for its encoding. The main difficulty is that different operating systems have different rules governing the syntax used for file names and file paths. In many cases, like when depositing data files in a public repository, we need to ensure that file names are valid across multiple operating systems (OSs). If the script used to create the files is itself expected to be OS agnostic, queries for file names and paths in R code should not make assumptions on the naming rules or available OS commands. This is especially important when developing R packages.

For maximum portability, file names should never contain white-space characters and contain at most one dot. For the widest possible portability, underscores should be avoided using dashes instead. As an example, instead of `my data.2019.csv`, use `my-data-2019.csv`.

R provides functions which help with portability, by hiding the idiosyncrasies of the different OSs from R code. In scripts, these functions should be preferred over direct call to OS commands (i.e., avoid calls to functions `shell()` or `system()` with OS commands as arguments) whenever possible. As the algorithm needed to extract a file name from a file path is OS specific, R provides functions such as `basename()`, whose implementation is OS specific but from the side of R code behave identically—these functions hide the differences among OSs from the user of R. The chunk below can be expected to work correctly under any OS for which R is available.

```
basename("extdata/my-file.txt")
## [1] "my-file.txt"
```

While in Unix and Linux folder nesting in file paths is marked with a forward slash character (/), under MS-Windows it is marked with a backslash character (\). Backslash (\) is an escape character in R and interpreted as the start of an embedded special character (see section 3.4 on page 41), while in R a forward slash (/) can be used for file paths under any OS, and escaped backslash (\\) is valid only under MS-Windows. Consequently, / should be always preferred to \\ to ensure portability, and is the approach used in this book.

```
basename("extdata/my-file.txt")
## [1] "my-file.txt"
basename("extdata\\my-file.txt")
## [1] "my-file.txt"
```

The complementary function to `basename()` is `dirname()` and extracts from a full file path the bare path to the containing folder.

```
dirname("extdata/my-file.txt")
## [1] "extdata"
```

Functions `getwd()` and `setwd()` can be used to get the path to the current working directory and to set a directory as current, respectively.

```
# not run
getwd()
```

Function `setwd()` returns the path to the current working directory, allowing to portably restore the working directory to the previous one. Both relative paths (relative to the current working directory), as in the example, or absolute paths (given in full) are accepted as an argument. In mainstream OSs "`.`" indicates the current directory and "`..`" the directory above the current one.

```
# not run
oldwd <- setwd("..")
getwd()
```

The returned value is always an absolute full path, so it remains valid even if the path to the working directory changes more than once before being restored.

```
# not run
oldwd
setwd(oldwd)
```

```
getwd()
```

Function `list.files()` returns a list of names of files and/or directories (= disk folders) portably across OSs. Function `list.dirs()` returns only the names of directories.

```
head(list.files())
## [1] "abbrev.sty"
## [2] "anscombe.svg"
## [3] "Aphalo-CR-9781032518435-Learn-R-proofs-2024-01-26.pdf"
## [4] "Aphalo-CR-9781032518435-Learn-R.pdf"
## [5] "aphalo-Learn-R-2ed-crc-2023-06-14.pdf"
## [6] "aphalo-learn-R-2ed-draft-2022-02-01.pdf"
head(list.dirs())
## [1] "."                   "./.git"              "./.git/hooks"        "./.git/info"
## [5] "./.git/logs"         "./.git/logs/refs"
```

10.1 In these functions, the default argument for parameter `path` is the current working directory, under Windows, Unix, and Linux indicated by `"."`. Convince yourself that this is indeed the default by calling the functions with an explicit argument. After this, play with the functions passing as argument to `path` other existing and non-existent file and directory paths.

10.2 Pass different arguments to parameter `full.names` of `list.files()` to obtain either a list of file paths or bare file names. Similarly, investigate how the returned list of files is affected by the argument passed to `all.names`.

Base R provides several functions for portably working with files, and they are listed in the help page for `files` and in individual help pages. Use `help("files")` to access the help for this "family" of functions. The chunk below exercises some of these functions.

```
if (!file.exists("xxx.txt")) {
  file.create("xxx.txt")
}
## [1] TRUE
file.size("xxx.txt")
## [1] 0
file.info("xxx.txt")
##         size isdir mode               mtime               ctime
## xxx.txt    0 FALSE  666 2024-02-17 22:36:11 2024-02-17 22:36:11
##                       atime exe
## xxx.txt 2024-02-17 22:36:11  no
file.rename("xxx.txt", "zzz.txt")
## [1] TRUE
file.exists("xxx.txt")
## [1] FALSE
file.exists("zzz.txt")
## [1] TRUE
file.remove("zzz.txt")
## [1] TRUE
```

10.3 Function `file.path()` can be used to construct a file path from its components in a way that is portable across OSs. Look at the help page and play with the function to assemble some paths that exist in the computer you are using.

10.5 Opening and Closing File Connections

Examples in the rest of this chapter use as an argument for the `file` formal parameter literal paths or URLs, and complete the reading or writing operations within the call to a function. Sometimes it is necessary to read or write a text file sequentially, one row or record at a time. In such cases, it is most efficient to keep the file open between reads and close the connection only when it is no longer needed. See `help(connections)` for details about the various functions available and their behaviour in different OSs. The code below opens a file connection, reads two lines, first the top one with column headers, then in a separate call to `readLines()`, the two lines or records with data, and finally closes the connection.

```
f1 <- file("extdata/not-aligned-ASCII-UK.csv", open = "r") # open for reading
readLines(f1, n = 1)
## [1] "col1,col2,col3,col4"

readLines(f1, n = 2)
## [1] "1.0,24.5,346,ABC" "23.4,45.6,78,Z Y"
close(f1)
```

When R is used in batch mode, the “files” `stdin`, `stdout` and `stderror` can be opened, and data read from, or written to. These *standard* sources and sinks, so familiar to C programmers, allow the use of R scripts as tools in data pipes coded as shell scripts under Unix and other OSs.

10.6 Plain-Text Files

In general, text files are the most portable approach to data storage but usually also the least efficient with respect to the size of the file. Text files are composed of encoded characters. This makes them easy to edit with text editors and easy to read from programs written in most programming languages. On the other hand, how the data encoded as characters is arranged can be based on two different approaches: positional or using a specific character as a separator.

The positional approach is more concise but almost unreadable to humans as the values run into each other. Reading of data stored using a positional approach requires access to a format definition and was common in FORTRAN and COBOL at the time when punch cards were used to store data. In the case of separators, different separators are in common use. Comma-separated values (CSV) encodings use either a comma or semicolon to separate the fields or columns. Tab-separated values (TSV) use the tab, or tabulator, character as a column separator. Sometimes, whitespace is used as a separator, most commonly when all values are to be converted into `numeric`.

Not all text files are born equal. When reading text files, and *foreign* binary files which may contain embedded text strings, there is potential for their misinterpretation during the import operation. One common source of problems, is that column headers are to be read as R names. As earlier discussed, there are strict rules, such as avoiding spaces or special characters if the names are to be used with the normal R syntax. On import, some functions will attempt to sanitise the names, but others not. Most such names are still accessible in R statements, but a special syntax is needed to protect them from triggering syntax errors through their interpretation as something different than variable or function names—in R jargon we say that they need to be quoted.

Some of the things we need to be on the watch for are: 1) Mismatches between the character encoding expected by the function used to read the file, and the encoding used for saving the file—usually because of different locales, i.e., language and country settings. 2) Leading or trailing (invisible) spaces present in the character values or column names—which are almost invisible when data frames are printed. 3) Wrongly guessed column classes—a typing mistake affecting a single value in a column, e.g., the wrong kind of decimal marker, can prevent the column from being recognised as numeric. 4) Mismatched decimal marker in CSV files—the marker depends on the locale (language and country settings).

If you encounter problems after import, such as failure of extraction of data frame columns by name, use function `names()` to get the names printed to the console as a character vector. This is useful because character vectors are always printed with each string delimited by quotation marks making leading and trailing spaces clearly visible. The same applies to use of `levels()` with factors created with data that might have contained mistakes or whitespace.

To demonstrate some of these problems, I create a data frame with name sanitation disabled, and in the second statement with sanitation enabled. The first statement is equivalent to the default behaviour of functions in package 'readr' and the second is equivalent to the behaviour of base R functions. 'readr' prioritises the integrity of the original data while R prioritises compatibility with R's naming rules.

```
data.frame(a = 1, "a " = 2, " a" = 3, check.names = FALSE)
##   a a   a
## 1 1  2  3
data.frame(a = 1, "a " = 2, " a" = 3)
##   a a. X.a
## 1 1  2   3
```

An even more subtle case is when characters can be easily confused by the user reading the output, or typing in the data: zero and o (`a0` vs. `ao`) or el and one (`al` vs. `a1`) can be difficult to distinguish in some fonts. When using encodings capable of storing many character shapes, such as unicode, in some cases two characters with almost identical visual shape may be encoded as different characters.

```
data.frame(al = 1, a1 = 2, ao = 3, a0 = 4)
##   al a1 ao a0
## 1  1  2  3  4
```

Reading data from a text file can result in very odd-looking values stored in R variables because of a mismatch in encoding, e.g., when a CSV file saved with MS-Excel is silently encoded using 16-bit unicode format, but read as an 8-bit unicode encoded file.

The hardest part of all these problems is to diagnose their origin, as function arguments and working environment options can in most cases be used to force the correct decoding of text files with diverse characteristics, origins, and vintages once one knows what is required. Function `tools:::showNonASCIIfile()` from the R 'tools' package, which is not exported, but available in recent and current (4.4.0) versions of R, can be used to test files for the presence on non-ASCII characters. This function takes as an argument the path to a file, and its companion function `tools:::showNonASCII()` a `character` string.

10.6.1 Base R and 'utils'

Text files containing data in columns can be divided into two broad groups. Those with fixed-width fields and those with delimited fields. Fixed-width fields were especially common in the early days of FORTRAN and COBOL when data storage capacity was very limited. These formats are frequently capable of encoding information using fewer characters than when delimited fields are used. The best way of understanding the differences is with examples. Although in this section we exemplify the use of functions by passing a file name as an argument, URLs and open file descriptors are also accepted (see section 10.5 on page 388). The file will be uncompressed on the fly if its name ends in `.gz`.

Wether columns containing character strings that cannot be converted into numbers are converted into factors or remain as character strings in the returned data frame depends on the value passed to parameter `stringsAsFactors`. The default changed in R version 4.0.0 from `TRUE` into `FALSE`. If code is to work consistently in old and new versions of R `stringsAsFactors = FALSE` has to be passed explicitly in calls to `read.csv()` (the approach used in the book).

In the first example, a file with fields solely delimited by "," is read. This is what is called comma-separated values (CSV) format that can be read and written with `read.csv()` and `write.csv()`, respectively.

The contents of file `not-aligned-ASCII-UK.csv` are shown below.

```
col1,col2,col3,col4
1.0,24.5,346,ABC
23.4,45.6,78,Z Y
```

The file is read and the returned value stored in a variable named `from_csv_a.df`, and printed.

```
from_csv_a.df <-
  read.csv("extdata/not-aligned-ASCII-UK.csv", stringsAsFactors = FALSE)
```

```
from_csv_a.df
##   col1 col2 col3 col4
## 1  1.0 24.5  346  ABC
## 2 23.4 45.6   78  Z Y
from_csv_a.df[["col4"]]
## [1] "ABC" "Z Y"
sapply(from_csv_a.df, class)
##        col1        col2        col3        col4
##   "numeric"   "numeric"   "integer" "character"
```

10.4 Read the file `not-aligned-ASCII-UK.csv` with function `read.csv2()` instead of `read.csv()`. Although this may look like a waste of time, the point of the exercise is for you to get familiar with R behaviour in case of such a mistake. This will help you recognise similar errors when they happen accidentally, which is quite common when files are shared.

Example file `aligned-ASCII-UK.csv` contains comma-separated values with added whitespace to align the columns, to make it easier to read by humans.

The contents of file `aligned-ASCII-UK.csv` are shown below.

```
col1, col2, col3, col4
 1.0, 24.5,  346,  ABC
23.4, 45.6,   78,  Z Y
```

The file is read and the returned value stored in a variable named `from_csv_b.df`, and printed. Although space characters are read as part of the fields, they are ignored when conversion to numeric takes place.

```
from_csv_b.df <-
  read.csv("extdata/aligned-ASCII-UK.csv", stringsAsFactors = FALSE)

from_csv_b.df
##   col1 col2 col3  col4
## 1  1.0 24.5  346   ABC
## 2 23.4 45.6   78   Z Y
from_csv_b.df[["col4"]]
## [1] "  ABC" "  Z Y"
sapply(from_csv_b.df, class)
##        col1        col2        col3        col4
##   "numeric"   "numeric"   "integer" "character"
```

By default, column names are sanitised but whitespace in character strings kept. Passing an additional argument changes this default so that leading and trailing whitespace are discarded. Most likely the default has been chosen so that by default data integrity is maintained.

```
from_csv_c.df <-
  read.csv("extdata/aligned-ASCII-UK.csv",
           stringsAsFactors = FALSE, strip.white = TRUE)

from_csv_c.df
##   col1 col2 col3 col4
## 1  1.0 24.5  346  ABC
## 2 23.4 45.6   78  Z Y
from_csv_c.df[["col4"]]
## [1] "ABC" "Z Y"
```

```
sapply(from_csv_c.df, class)
##        col1        col2        col3        col4
##   "numeric"   "numeric"   "integer" "character"
```

When character strings are converted into factors, leading and trailing whitespace is retained in the labels of factor levels. Leading and trailing whitespace are difficult to see when data frames are printed, as shown below. This example shows what problems were frequently encountered in earlier versions of R, and can still occur when factors are created. The recommended approach is to use the default `stringsAsFactors = FALSE` and do the conversion into factors in a separate step.

```
from_csv_b.df <-
  read.csv("extdata/aligned-ASCII-UK.csv", stringsAsFactors = TRUE)
```

Using `levels()` it can be seen that the labels of the automatically created factor levels contain leading spaces.

```
sapply(from_csv_b.df, class)
##      col1      col2      col3      col4
## "numeric" "numeric" "integer"  "factor"
from_csv_b.df[["col4"]]
## [1]    ABC    Z Y
## Levels:    ABC    Z Y
levels(from_csv_b.df[["col4"]])
## [1] "   ABC" "   Z Y"
```

Decimal points and exponential notation are allowed for floating point values. In English-speaking locales, the decimal mark is a point, while in many other locales it is a comma. The behaviour of R functions does not change when run under different locales. When a comma is used as decimal marker, we can a semicolon (`;`) is used as field marker.

This handled by using functions `read.csv2()` and `write.csv2()`. Furthermore, parameters `dec` and `sep` allow setting the decimal marker and field separator to arbitrary character strings.

Function `read.table()` does the actual work and functions like `read.csv()` only differ in the default arguments for the different parameters. By default, `read.table()` expects fields to be separated by whitespace (one or more spaces, tabs, new lines, or carriage return).

The contents of file `aligned-ASCII.txt` are shown below.

```
col1 col2 col3 col4
 1.0 24.5  346 ABC
23.4 45.6   78 "Z Y"
```

The file is read and the returned value stored in a variable named `from_txt_b.df`, and printed. Leading and trailing whitespace are removed because they are recognised as part of the separators. For character strings containing embedded spaces to be decoded as a single value, they need to be quoted in the file as in `aligned-ASCII.txt` above.

```
from_txt_b.df <-
  read.table("extdata/aligned-ASCII.txt",
             stringsAsFactors = FALSE, header = TRUE)
```

```
from_txt_b.df
##   col1 col2 col3 col4
## 1  1.0 24.5  346  ABC
## 2 23.4 45.6   78  Z Y
from_txt_b.df[["col4"]]
## [1] "ABC" "Z Y"
sapply(from_txt_b.df, class)
##        col1        col2        col3        col4
##   "numeric"   "numeric"   "integer" "character"
```

With a fixed-width format, no delimiters are needed. Decoding is based solely on the position of the characters in the line or record. A file like this cannot be interpreted without a description of the format used for saving the data. Files containing data stored in *fixed width format* can be read with function `read.fwf()`. Records for a single observation can be stored in a single or multiple lines. In either case, each line has fields of different but fixed known widths.

Function `read.fortran()` is a wrapper on `read.fwf()` that accepts format definitions similar to those used in FORTRAN. One particularity of FORTRAN *formatted data transfer* is that the decimal marker can be omitted in the saved file and its position specified as part of the format definition, a trick used to make text files (or stacks of punch cards!) smaller. Modern versions of FORTRAN support reading from and writing to other formats like those using field delimiters described above.

The contents of file `aligned-ASCII.fwf` are shown below.

```
 10245346ABC
234456 78Z Y
```

The file is read and the returned value stored in a variable named `from_fwf_a.df`, and printed. The format definition is passed as a separate character vector argument, e.g., `"2F3.1"` describes the format of the first two columns, `"I3"` describes the third column and `"A3"` the fourth.

```
from_fwf_a.df <-
  read.fortran("extdata/aligned-ASCII.fwf",
               format = c("2F3.1", "I3", "A3"),
               col.names = c("col1", "col2", "col3", "col4"))

from_fwf_a.df
##   col1 col2 col3 col4
## 1  1.0 24.5  346  ABC
## 2 23.4 45.6   78  Z Y
from_fwf_a.df[["col4"]]
## [1] "ABC" "Z Y"
sapply(from_fwf_a.df, class)
##        col1        col2        col3        col4
##   "numeric"   "numeric"   "integer" "character"
```

The file reading functions described above share with `read.table()` the same parameters. In addition to those described above, other frequently useful parameters are `skip` and `n`, which can be used to skip lines at the top of a file and limit the number of lines (or records) to read; `header`, which accepts a logical argument indicating if the fields in the first text line read should be decoded as column names rather than data; `na.strings`, to which can be passed a character vector with strings to be interpreted as `NA`; and `colClasses`, which provides control of the conversion of the fields to R classes and possibly skipping some columns altogether. All these parameters are described in the corresponding help pages.

10.5 In reality `read.csv()`, `read.csv2()` and `read.table()` are the same function with different default arguments to several of their parameters. Study the help page, and by passing suitable arguments, make `read.csv()` behave like `read.table()`, then make `read.table()` behave like `read.csv2()`.

A text file can be read as character strings, without attempting to decode them. This is occasionally useful, such as when the decoding is done in a script, or when needs to print a file as is. In this case, the function used is `readLines()`. The returned value is a character vector in which each member string corresponds to one line or record in the file, with the end-of-line markers stripped (see example in section 10.5 on page 388).

The next example shows how a *write* function matching one of the *read* functions described above can be used to save a data frame to a text file. The `write.csv()` function takes as an argument a data frame, or an object that can be coerced into a data frame, converts it to character strings, and saves them to a text file. A data frame, `my.df` with five rows is enough for a demonstration.

```
my.df <- data.frame(x = 1:5, y = 5:1 / 10, z = letters[1:5])
```

We write `my.df` to a CSV file suitable for an English language locale, and then display its contents.

```
write.csv(my.df, file = "my-file1.csv", row.names = FALSE)
file.show("my-file1.csv", pager = "console")
```

```
"x","y","z"
1,0.5,"a"
2,0.4,"b"
3,0.3,"c"
4,0.2,"d"
5,0.1,"e"
```

In most cases setting, as above, `row.names = FALSE` when writing a CSV file will help when it is read. Of course, if row names do contain important information, such as gene tags, you cannot skip writing the row names to the file unless you first copy these data into a column in the data frame. (Row names are stored separately as an attribute in `data.frame` objects, see section 4.6 on page 114 for details.)

10.6 Write the data frame `my.df` into text files with functions `write.csv2()` and `write.table()` instead of `read.csv()` and display the files.

Function `cat()` takes R objects and writes them after conversion to character strings to the console or a file, inserting one or more characters as separators, by default, a space. This separator can be set through parameter `sep`. In our example, we set `sep` to a new line (entered as the escape sequence "\n").

```
my.lines <- c("abcd", "hello world", "123.45")
cat(my.lines, file = "my-file2.txt", sep = "\n")
file.show("my-file2.txt", pager = "console")
```

```
abcd
hello world
123.45
```

10.6.2 'readr'

Package 'readr' is part of the 'tidyverse' suite. It defines functions that have different default behaviour and that are designed to be faster under different situations than those native to R. The functions from package 'readr' can sometimes wrongly decode their input and rarely even do this silently. The 'readr' functions guess more properties of the text file format; in most cases they succeed, which is very handy, but occasionally they fail. Automatic guessing can be overridden by passing arguments, and this is recommended for scripts that will be reused to read different files in the future. Another important advantage is that these functions read character strings formatted as dates or times directly into columns of class `POSIXct`. All `write` functions defined in 'readr' have an `append` parameter, which can be used to change the default behaviour of overwriting an existing file with the same name, to appending the output at its end.

Although we exemplify the use of these functions by passing a file name as an argument, as is the case with R native functions, URLs, and file descriptors are also accepted (see section 10.5 on page 388). The files read are uncompressed, and those written are compressed on the fly if their name ends in `.gz`, `.bz2`, `.xz`, or `.zip`.

Functions "equivalent" to native R functions described in the previous section have names formed by replacing the dot with an underscore, e.g., `read_csv()` ≈ `read.csv()`. The similarity refers to the format of the files read, but not the order, names, or roles of their formal parameters. For example, function `read_table()` has a slightly different behaviour than `read.table()`, although they both read fields separated by whitespace. Row names are not set in the returned `tibble`, which inherits from `data.frame`, but is not fully compatible (see section 8.4.2 on page 247).

Package 'readr' is under active development, and functions with the same name from different major versions are not fully compatible. Code for some examples from the first edition of the book no longer work because the updated implementation fails to recognise escaped special characters. Function `read_table2()` has been renamed `read_table()`.

These functions report to the console the specifications of the columns, which is important when these are guessed from the file contents, or even only from rows near the top of the file.

```
read_csv(file = "extdata/aligned-ASCII-UK.csv", show_col_types = FALSE)
## # A tibble: 2 x 4
##    col1  col2  col3 col4
##   <dbl> <dbl> <dbl> <chr>
## 1   1    24.5   346 ABC
## 2  23.4  45.6    78 Z Y

read_csv(file = "extdata/not-aligned-ASCII-UK.csv", show_col_types = FALSE)
## # A tibble: 2 x 4
```

```
##     col1  col2  col3 col4
##    <dbl> <dbl> <dbl> <chr>
## 1    1    24.5   346 ABC
## 2  23.4  45.6    78 Z Y
```

Package 'readr' is under active development, and different major versions are not fully compatible with each other. Because of the misaligned fields in file "not-aligned-ASCII.txt" in the past we needed to use read_table2(), which allowed misalignment of fields, similarly to read.table(). This function has been renamed as read_table() and read_table2() deprecated. However, parsing of both files fails if they are read with read_table(), quoted strings containing whitespace are no longer recognised. See above example using read.table(). Examples below are not run, but kept as they may work again in the future.

```
read_table(file = "extdata/aligned-ASCII.txt")
```

```
read_table(file = "extdata/not-aligned-ASCII.txt")
```

Function read_delim() with space as the delimiter succeeds only with the not-aligned file as in this file the separator is in all cases a single space.

```
read_delim(file = "extdata/not-aligned-ASCII.txt",
           delim = " ", show_col_types = FALSE)
## # A tibble: 2 x 4
##     col1  col2  col3 col4
##    <dbl> <dbl> <dbl> <chr>
## 1    1    24.5   346 ABC
## 2  23.4  45.6    78 Z Y
```

Function read_tsv() reads files encoded with the tab character as the delimiter, and read_fwf() reads files with fixed width fields. There is, however, no equivalent to read.fortran(), supporting implicit decimal points.

10.7 Use the "wrong" read_ functions to read the example files used above and/or your own files. As mentioned earlier, forcing errors will help you learn how to diagnose when such errors are caused by coding or data entry mistakes. In this case, as wrongly read data are not always accompanied by error or warning messages, carefully check the returned tibbles for misread data values.

The functions from R's package 'utils' read the whole file before attempting to guess the class of the columns or their alignment. This is reliable but slow for text files with many lines. The functions from 'readr' read by default only the top 1000 lines (guess_max = 1000) when guessing the format and class, assuming that the guessed properties also apply to the remaining lines of the file. This is more efficient, but rather risky. However, the functions from R's package 'utils' are faster at reading files with many fields (or columns) per line and few lines.

In earlier versions of 'readr', a problem was the failure to correctly decode numeric values when increasingly large numbers resulted in wider fields in the lines below those used for guessing. However, at the time of writing, this case is correctly handled. A guess based on the top lines of a text file also means that when values in lines below guess_max lines cannot be converted to numeric, the numeric column returned contains NA values. In contrast, in this situation, functions from R's package 'utils', skip decoding and return a character column. Below, a very small value for `guess_max` is used to demonstrate this behaviour with a file only a few lines in length.

```
read_table(file = "extdata/miss-aligned-ASCII.txt", show_col_types = FALSE)

## # A tibble: 4 x 4
##    col1   col2  col3 col4
##    <chr> <dbl> <dbl> <chr>
## 1 1.0     24.5   346 ABC
## 2 2.4     45.6    78 XYZ
## 3 20.4    45.6    78 XYZ
## 4 a       20    2500 abc

read_table(file = "extdata/miss-aligned-ASCII.txt", show_col_types = FALSE,
           guess_max = 3L)

## Warning: 1 parsing failure.
## row  col expected actual                             file
##    4 col1 a double      a 'extdata/miss-aligned-ASCII.txt'
## # A tibble: 4 x 4
##     col1  col2  col3 col4
##    <dbl> <dbl> <dbl> <chr>
## 1    1    24.5   346 ABC
## 2    2.4  45.6    78 XYZ
## 3   20.4  45.6    78 XYZ
## 4   NA    20    2500 abc
```

The `write_` functions from 'readr' are the counterpart to `write.` functions from 'utils'. In addition to the expected `write_csv()`, `write_csv2()`, `write_tsv()` and `write_delim()`, 'readr' provides functions that write MS-Excel-friendly CSV files. Function `write_excel_csv()` saves a text file with comma-separated fields suitable for import into MS-Excel.

```
write_excel_csv(my.df, file = "my-file6.csv")
file.show("my-file6.csv", pager = "console")

"x","y","z"
1,0.5,"a"
2,0.4,"b"
3,0.3,"c"
4,0.2,"d"
5,0.1,"e"
```

10.8 Compare the output from `write_excel_csv()` and `write_csv()`. What is the difference? Does it matter when you import the written CSV file into Excel (in the version you are using, and with the locale settings of your computer)?

The pair of functions `read_lines()` and `write_lines()` read and write character vectors without conversion, similarly to base R `readLines()` and `writeLines()`. Functions `read_file()` and `write_file()` read and write the contents of a whole text file into, and from, a single character string. Functions `read_file()` and `write_file()` can also be used with raw vectors to read and write binary files or text files of unknown encoding.

The contents of the whole file are returned as a character vector of length one, with the embedded new line markers. We use `cat()` to print it so these new line characters force the start of a new print-out line.

```
one.str <- read_file(file = "extdata/miss-aligned-ASCII.txt")
length(one.str)
## [1] 1
cat(one.str)
## col1  col2 col3 col4

## 1.0   24.5  346 ABC

## 2.4   45.6   78 XYZ

## 20.4   45.6   78 XYZ

##  a    20     2500 abc
```

10.9 Use `write_file()` to write a file that can be read with `read_csv()`.

10.7 XML and HTML Files

XML files contain text with special markup. Several modern data exchange formats are based on the XML standard (see `https://www.w3.org/TR/xml/`) which uses schemas for flexibility. Schemas define specific formats, allowing reading of formats not specifically targeted during development of the read functions. Even the modern XHTML standard used for web pages is based on such schemas, while HTML only differs slightly in its syntax.

10.7.1 'xml2'

Package 'xml2' provides functions for reading and parsing XTML and HTML files. This is a vast subject, of which I will only give a brief example.

Function `read_html()` can be used to read an HTML document, either locally or from a URL as below.

```
web_page <- read_html("https://www.learnr-book.info/")
```

Function `html_structure()` displays the structure of an HTML document (long text output not shown).

```
html_structure(web_page)
```

Function `xml_text()` extracts the text content of a field. Function `xml_find_all()` returns a field searched by name. Here used to extract the text from the `title` attribute, using functions `xml_find_all()` and .

```
xml_text(xml_find_all(web_page, ".//title"))
## [1] "Learn R: As a Language"
```

The functions defined in this package can be used to "harvest" data from web pages, but also to read data from files using formats that are defined through XML schemas.

10.8 GPX Files

GPX (GPS Exchange Format) files use an XML scheme designed for saving and exchanging data from geographic positioning systems (GPS). There is some variation on the variables saved depending on the settings of the GPS receiver. The example data used here is from a Transmeta BT747 GPS logger. The example below reads the data into a `tibble` as character strings. For plotting, the character values representing numbers and dates would need to be converted to numeric and datetime (`POSIXct`) values, respectively. In the case of plotting tracks on a map, it is preferable to use package 'sf' to import the tracks directly from the `.gpx` file into a layer (use of R pipe operator is described in section 5.5 on page 134).

```
xmlTreeParse(file = "extdata/GPSDATA.gpx", useInternalNodes = TRUE) |>
xmlRoot(x = _) |>
xmlToList(node = _) |>
_[["trk"]] |>
assign(x = "temp", value = _) |>
_[names(x = temp) == "trkseg"] |>
unlist(x = _, recursive = FALSE) |>
map_df(.x = _, .f = function(x) as_tibble(x = t(x = unlist(x = x))))
## # A tibble: 199 x 7
##   time                     speed  name          type  fix   .attrs.lat .attrs.lon
##   <chr>                    <chr>  <chr>         <chr> <chr> <chr>      <chr>
## 1 2018-12-08T23:09:02.000Z 0.0366 trkpt-2018-~ T     3d    -34.912071 138.660595
## 2 2018-12-08T23:09:04.000Z 0.0884 trkpt-2018-~ T     3d    -34.912067 138.660543
## 3 2018-12-08T23:09:06.000Z 0.0147 trkpt-2018-~ T     3d    -34.912102 138.660554
## # i 196 more rows
rm(temp) # cleanup
```

10.10 To understand what data transformation takes place in each statement of this pipe, start by running the first statement by itself, excluding the pipe operator, and continue adding one statement at a time, and at each step check the returned value and look out for what has changed from the previous step. Optionally you can insert a line `print() |>` at the point where you wish to see the data being "piped".

10.9 Worksheets

Microsoft Office, Open Office, and Libre Office are the most frequently used suites containing programs based on the worksheet paradigm. There is available a standardised file format for exchange of worksheet data, but it does not support all the features present in native file formats. We will start by considering MS-Excel. The file format used by MS-Excel has changed significantly over the years, and old formats tend to be less well supported by available R packages and may require the file to be updated to a more modern format with MS-Excel itself before import into R. The current format is based on XML and relatively simple to decode, whereas older binary formats are more difficult. Worksheets contain code as equations in addition to the actual data. In all cases, only values entered as such or those computed by means of the embedded equations can be imported into R rather than the equations themselves.

When directly reading from a worksheet, a column of cells with mixed type, can introduce `NA` values. A wrongly selected cell range from the worksheet can result in missing columns or rows, if the area is too small, or in rows or columns filled with `NA` values, if the range includes empty cells in the worksheet. Depending on the function used, it may be possible to ignore empty cells, by passing an argument.

Many problems related to the import of data from worksheets and workbooks are due to translation between two different formats that impose different restrictions on what is allowed or not. While in a worksheet it is allowed to set the "format" (as called in Excel, and roughly equivalent to `mode` in R) of individual cells, a variable (column) in an R data frame is expected to be vector, and thus contain members belonging the same `mode` or type. For the import to work as expected, the "format" must be consistent, i.e., all cells in a column to be imported are marked as one of the `Number`, `Date`, `Time`, or `Text` formats, with the possible exception of a *single row* of column headers with the names of the variables as `Text`. The default format `General` also works but as it does not ensure consistency, it makes more difficult to see format inconsistencies at a glance in Excel.

When reading a CSV file, text representing numbers will be recognised and converted, but only if the decimal point is encoded as expected from the arguments passed to the function call. So a single number with a comma instead of a dot as decimal marker (or vice versa) will result in most cases in the column not being decoded as numbers and returned as a `character` vector (or column) in the data frame. In the case of package 'readr', a `numeric` vector containing `NA` values for the non-decoded text may be returned instead of a `character` vector depending on whether the wrong decimal marker appears near the top or near the end of the file.

When importing data from a worksheet or workbook, my recommendation is first to check it in the original software to ensure that the cells to be imported are encoded as expected. When using a CSV as an intermediate step, it is crucial to also open this file in a plain-text editor such as the editor pane in RStudio (or Notepad in Windows or Nano, Emacs, etc., in Unix and Linux). Based on what field separator, decimal mark, and possibly character encoding has been used, which depends on the locale settings in the operating system of the computer and in the worksheet program, select a suitable function to call and the necessary arguments to pass to it.

10.9.1 CSV files as middlemen

If we have access to the original software used for creating a worksheet or workbook, then exporting worksheets to text files in CSV format and importing them into R using the functions described in sections 10.6 and 10.6.2 starting on pages 388 and 395 provides a broadly compatible route for importing data—with the caveat that one must ensure that delimiters and decimal marks match the expectations of the functions used. This approach is not ideal from the perspective of having to create intermediate csv formatted text files. A better approach is, when feasible, to import the data directly from the workbook or worksheets into R.

10.9.2 'readxl'

Package 'readxl' supports reading of MS-Excel workbooks, and selecting worksheets and regions within worksheets specified in ways similar to those used by MS-Excel itself. The interface is simple, and the package easy to install. We will import a file that in MS-Excel looks like the screen capture below.

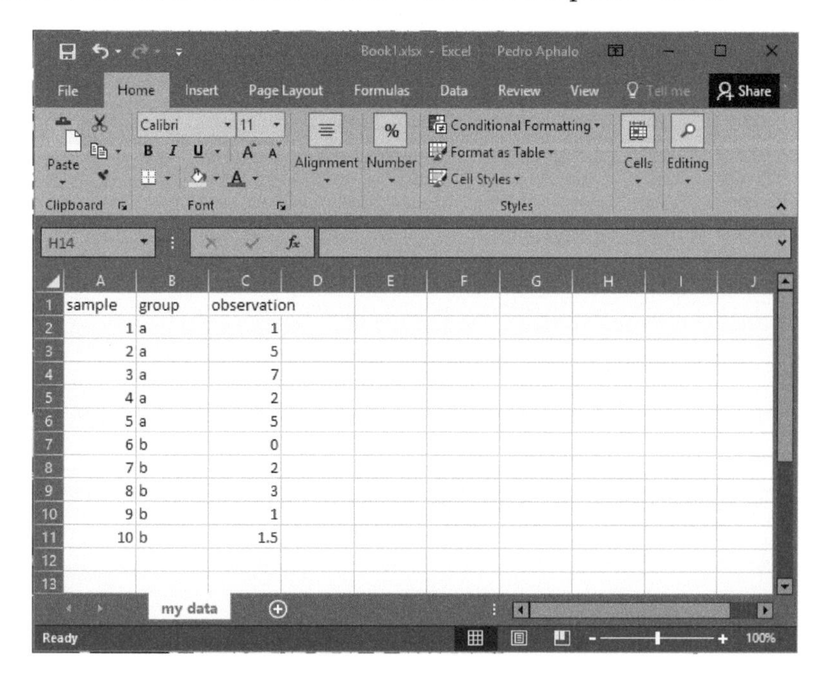

Function `excel_sheets()` lists the sheets contained in the workbook.

```
sheets <- excel_sheets("extdata/Book1.xlsx")
sheets
## [1] "my data"
```

In this case, the argument passed to `sheet` is redundant, as there is only a single worksheet in the file. It is possible to use either the name of the sheet or a positional index (in this case `1` would be equivalent to `"my data"`). Function `read_excel()` with no range specification imports the whole worksheet into a tibble, as can be expected from a package included in the 'tidyverse'.

```
Book1.df <- read_excel("extdata/Book1.xlsx",
                       sheet = "my data")
Book1.df
## # A tibble: 10 x 3
##   sample group observation
##    <dbl> <chr>       <dbl>
## 1      1 a             1
## 2      2 a             5
```

```
## 3          3 a                        7
## # i 7 more rows
```

It is also possible to read a region instead of the whole worksheet.

```
Book1_region.df <- read_excel("extdata/Book1.xlsx",
                              sheet = "my data",
                              range = "A1:B8")
Book1_region.df
## # A tibble: 7 x 2
##    sample group
##     <dbl> <chr>
## 1       1 a
## 2       2 a
## 3       3 a
## # i 4 more rows
```

Of the remaining arguments, the most useful ones have the same names and play similar roles as in 'readr' (see section 10.6.2 on page 395). For example, new names for the columns can be passed as an argument to override the names in the worksheet.

```
Book1_region.df <- read_excel("extdata/Book1.xlsx",
                              sheet = "my data",
                              range = "A2:B8",
                              col_names = c("A", "B"))
Book1_region.df
## # A tibble: 7 x 2
##         A B
##     <dbl> <chr>
## 1       1 a
## 2       2 a
## 3       3 a
## # i 4 more rows
```

10.9.3 'xlsx'

Package 'xlsx' can be more difficult to install as it uses Java functions to do the actual work. However, it is more comprehensive, with functions both for reading and writing MS-Excel worksheets and workbooks, in different formats including the older binary ones. Similarly to 'readr', it allows selected regions of a worksheet to be imported.

Function read.xlsx() can be used indexing the worksheet by name. The returned value is a data frame, and following the expectations of R package 'utils', character columns are *no longer* converted into factors by default.

```
Book1_xlsx.df <- read.xlsx("extdata/Book1.xlsx",
                           sheetName = "my data")
Book1_xlsx.df
##    sample group observation
## 1       1     a         1.0
## 2       2     a         5.0
## 3       3     a         7.0
## 4       4     a         2.0
## 5       5     a         5.0
## 6       6     b         0.0
```

```
## 7        7     b          2.0
## 8        8     b          3.0
## 9        9     b          1.0
## 10      10     b          1.5
sapply(Book1_xlsx.df, class)
##      sample       group observation
##   "numeric" "character"   "numeric"
```

With function `write.xlsx()`, we can write data frames out to Excel worksheets and even append new worksheets to an existing workbook.

```
set.seed(456321)
my.data <- data.frame(x = 1:10, y = letters[1:10])
write.xlsx(my.data,
           file = "extdata/my-data.xlsx",
           sheetName = "first copy")
write.xlsx(my.data,
           file = "extdata/my-data.xlsx",
           sheetName = "second copy",
           append = TRUE)
```

When opened in Excel, we get a workbook containing two worksheets, named using the arguments we passed through `sheetName` in the code chunk above.

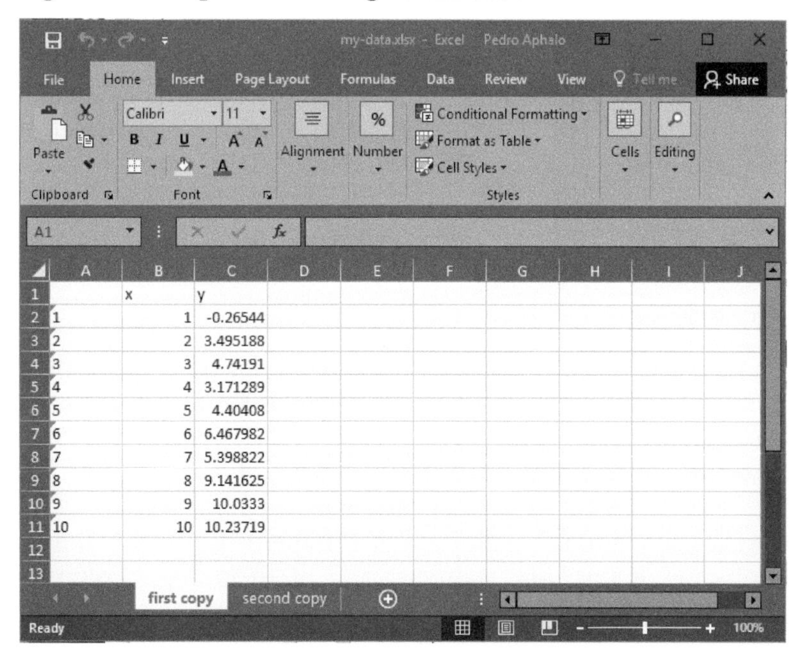

	A	B	C
1		x	y
2	1	1	-0.26544
3	2	2	3.495188
4	3	3	4.74191
5	4	4	3.171289
6	5	5	4.40408
7	6	6	6.467982
8	7	7	5.398822
9	8	8	9.141625
10	9	9	10.0333
11	10	10	10.23719

first copy | second copy

10.11 If you have some worksheet files available, import them into R to get a feel for how data is organised in the worksheets affects how easy or difficult it is to import them into R.

10.9.4 'readODS'

Package 'readODS' provides functions for reading data saved in files that follow the *Open Documents Standard*. Function `read_ods()` has a similar user interface to that of `read_excel()` and reads one worksheet at a time, with support only for skipping top rows and selecting ranges of columns and rows. The value returned is a tibble or, optionally, a data frame. Function `read_fods()` reads flat ODS files.

```
list_ods_sheets("extdata/Book1.ods")
## [1] "my_data"

ods.df <- read_ods("extdata/Book1.ods", sheet = 1)

ods.df
## # A tibble: 10 x 3
##   sample group observation
##    <dbl> <chr>       <dbl>
## 1      1 a               1
## 2      2 a               5
## 3      3 a               7
## # i 7 more rows
```

Functions `write_ods()` and `write_fods()` write a data frame into an ODS or FODS file.

10.10 Statistical Software

There are two different comprehensive packages for importing data saved from other statistical programs such as SAS, Statistica, SPSS, etc. The longtime "standard" is package 'foreign' included in base R, and package 'haven' is a newer contributed extension. In the case of files saved with old versions of statistical programs, functions from 'foreign' tend to be more robust than those from 'haven'.

10.10.1 'foreign'

Functions in package 'foreign' allow us to import data from files saved by several statistical analysis programs, including SAS, Stata, SPSS, Systat, Octave among others, and a function for writing data into files with formats native to SAS, Stata, and SPSS. R documents the use of these functions in detail in the *R Data Import/Export* manual. As a simple example, we use function `read.spss()` to read a `.sav` file, saved a few years ago with the then current version of SPSS. Only the first six rows and seven columns of the data frame are shown, including a column with dates, which appears as numeric.

```
my_spss.df <- read.spss(file = "extdata/my-data.sav", to.data.frame = TRUE)
my_spss.df[1:6, c(1:6, 17)]
##   block        treat mycotreat water1 pot harvest harvest_date
## 1     0 Watered, EM          1      1  14       1  13653705600
## 2     0 Watered, EM          1      1  52       1  13653705600
## 3     0 Watered, EM          1      1 111       1  13653705600
## 4     0 Watered, EM          1      1 127       1  13653705600
## 5     0 Watered, EM          1      1 230       1  13653705600
## 6     0 Watered, EM          1      1 258       1  13653705600
```

A second example, this time with a simple `.sav` file saved 15 years ago.

```
thiamin.df <- read.spss(file = "extdata/thiamin.sav", to.data.frame = TRUE)
head(thiamin.df)
##   THIAMIN CEREAL
## 1     5.2  wheat
## 2     4.5  wheat
## 3     6.0  wheat
## 4     6.1  wheat
## 5     6.7  wheat
## 6     5.8  wheat
```

Another example, for a Systat file saved on an PC more than 20 years ago, and read with `read.systat()`.

```
my_systat.df <- read.systat(file = "extdata/BIRCH1.SYS")
head(my_systat.df)
##   CONT DENS BLOCK SEEDL VITAL BASE ANGLE HEIGHT DIAM
## 1    1    1     1     2    44    2     0      1   53
## 2    1    1     1     2    41    2     1      2   70
## 3    1    1     1     2    21    2     0      1   65
## 4    1    1     1     2    15    3     0      1   79
## 5    1    1     1     2    37    3     0      1   71
## 6    1    1     1     2    29    2     1      1   43
```

Not all functions in 'foreign' return data frames by default, but all of them can be coerced to do so.

10.10.2 'haven'

Package 'haven' is less ambitious with respect to the number of formats supported, or their vintages, providing read and write functions for only three file formats: SAS, Stata, and SPSS. On the other hand, 'haven' provides flexible ways to convert the different labelled values that cannot be directly mapped to R modes. They also decode dates and times according to the idiosyncrasies of each of these file formats. In cases when the imported file contains labelled values, the returned `tibble` object needs some additional attention from the user. Labelled numeric columns in SPSS are not necessarily equivalent to factors, although they sometimes are. Consequently, conversion to factors cannot be automated and must be done manually in a separate step.

Function `read_sav()` can be used to import a `.sav` file saved by a recent version of SPSS. As in the previous section, we display below only the first six rows and seven columns of the data frame, including a column `treat` containing a labelled numeric vector and `harvest_date` with dates encoded as R date values.

```
my_spss.tb <- read_sav(file = "extdata/my-data.sav")
my_spss.tb[1:6, c(1:6, 17)]
## # A tibble: 6 x 7
##   block treat            mycotreat water1   pot harvest harvest_date
##   <dbl> <dbl+lbl>            <dbl>  <dbl> <dbl>   <dbl> <date>
## 1     0 1 [Watered, EM]          1      1    14       1 2015-06-15
## 2     0 1 [Watered, EM]          1      1    52       1 2015-06-15
## 3     0 1 [Watered, EM]          1      1   111       1 2015-06-15
## # i 3 more rows
```

In this case, the dates are correctly decoded.

Next, we import an SPSS's `.sav` file saved 20 years ago.

```
thiamin.tb <- read_sav(file = "extdata/thiamin.sav")
thiamin.tb
## # A tibble: 24 x 2
##   THIAMIN CEREAL
##     <dbl> <dbl+lbl>
## 1     5.2 1 [wheat]
## 2     4.5 1 [wheat]
## 3     6   1 [wheat]
## # i 21 more rows
```

```
thiamin.tb <- as_factor(thiamin.tb)
thiamin.tb
## # A tibble: 24 x 2
##   THIAMIN CEREAL
##     <dbl> <fct>
## 1     5.2 wheat
## 2     4.5 wheat
## 3     6   wheat
## # i 21 more rows
```

10.12 Compare the values returned by different `read` functions when applied to the same file on disk. Use `names()`, `str()`, and `class()` as tools in your exploration. If you are brave, also use `attributes()`, `mode()`, `dim()`, `dimnames()`, `nrow()`, and `ncol()`.

10.13 If you use or have in the past used other statistical software or a general-purpose language like Python, look for some old files and import them into R.

10.11 NetCDF Files

In some fields, including geophysics and meteorology, NetCDF is a very common format for the exchange of data. It is also used in other contexts in which data are referenced to a grid of locations, like with data read from Affymetrix microarrays used to study gene expression. NetCDF files are binary but use a format that allows the storage of metadata describing each variable together with the data itself in a well-organised and standardised format, which is ideal for exchange of moderately large data sets measured on a spatial or spatio-temporal grid.

Officially described as follows:

> NetCDF is a set of software libraries [from Unidata] and self-describing, machine-independent data formats that support the creation, access, and sharing of array-oriented scientific data.

That NetCDF files be selectively read, extracting the data from individual variables, is important as it allows computations in R with data sets too big to fit in a computer's RAM. Selective reading is possible using functions from packages 'ncdf4' or 'RNetCDF'. As a consequence of this flexibility, contrary to other data file reading operations, reading a NetCDF file is done in multiple steps—i.e., opening the file, reading metadata describing the variables and spatial grid, and finally selectively reading the data of interest.

10.11.1 'ncdf4'

Package 'ncdf4' supports reading of files using NetCDF version 4 or earlier formats. Functions in 'ncdf4' not only allow reading and writing of these files, but also their modification.

Below, first file `pevpr.sfc.mon.ltm.nc`, containing meteorological data, is opened with function `nc_open()`. The object returned is saved to `meteo_data.nc`. This object contains only an index to the file contents, whose structure is displayed with a call to `str()`, it plays the role of a file connection.

```
meteo_data.nc <- nc_open("extdata/pevpr.sfc.mon.ltm.nc")
str(meteo_data.nc, max.level = 1)
## List of 15
##  $ filename   : chr "extdata/pevpr.sfc.mon.ltm.nc"
##  $ writable   : logi FALSE
##  $ id         : int 65536
##  $ error      : logi FALSE
##  $ safemode   : logi FALSE
##  $ format     : chr "NC_FORMAT_NETCDF4_CLASSIC"
##  $ is_GMT     : logi FALSE
##  $ groups     :List of 1
##  $ fqgn2Rindex:List of 1
##  $ ndims      : num 4
##  $ natts      : num 8
##  $ dim        :List of 4
##  $ unlimdimid : num -1
##  $ nvars      : num 3
##  $ var        :List of 3
##  - attr(*, "class")= chr "ncdf4"
```

10.14 Increase `max.level` in the call to `str()` above and study how the connection object stores information on the dimensions and for each data variable. You can also `print(meteo_data.nc)` for a more complete printout once you have understood the structure of the object.

The dimensions of the data array are stored as metadata, in the file used mapping indexes to a grid of latitudes and longitudes and into a time vector as a third dimension. The dates are returned as character strings. The variables describing the grid are read one at a time with function `ncvar_get()`.

```
time.vec <- ncvar_get(meteo_data.nc, "time")
head(time.vec)
## [1] -657073 -657042 -657014 -656983 -656953 -656922
longitude <-  ncvar_get(meteo_data.nc, "lon")
head(longitude)
## [1] 0.000 1.875 3.750 5.625 7.500 9.375
latitude <- ncvar_get(meteo_data.nc, "lat")
head(latitude)
## [1] 88.5420 86.6531 84.7532 82.8508 80.9473 79.0435
```

The `time` vector contains only monthly values as the file contains a long-term series of monthly averages, expressed as days from 1800-01-01 corresponding to the first day of each month of year "1". We use package 'lubridate' for the conversion. To find the indexes for the grid point of interest, it is necessary to study the vectors `longitude` and `latitude` saved above.

Next, the potential evapotranspiration is read for one grid point, and used to construct a data frame, with some values recycled.

```
pet.tb <-
    tibble(time = time.vec,
           month = month(ymd("1800-01-01") + days(time)),
           lon = longitude[6],
           lat = latitude[2],
           pet = ncvar_get(meteo_data.nc, "pevpr")[6, 2, ]
           )
pet.tb
## # A tibble: 12 x 5
##          time month   lon   lat   pet
##     <dbl[1d]> <dbl> <dbl> <dbl> <dbl>
## 1    -657073    12  9.38  86.7  4.28
## 2    -657042     1  9.38  86.7  5.72
## 3    -657014     2  9.38  86.7  4.38
## # i 9 more rows
```

To read data for several grid points, different approaches are available. However, the order of nesting of dimensions can make adding the dimensions as columns error prone. It is much simpler to use package 'tidync' described next.

10.11.2 'tidync'

Package 'tidync' provides functions that make it easy to extract subsets of the data from an NetCDF file. The initial steps are the same operations as in the examples for 'ncdf4'.

Function `tidync()` is used to open the file and simultaneously activate the first grid. The returned object is saved as `meteo_data.tnc`. This object is subsequently used to access the file, and when printed displays a summary of the file structure and data encoding.

```
meteo_data.tnc <- tidync("extdata/pevpr.sfc.mon.ltm.nc")
meteo_data.tnc
##
## Data Source (1): pevpr.sfc.mon.ltm.nc ...
##
## Grids (5) <dimension family> : <associated variables>
##
## [1]  D0,D1,D2 : pevpr, valid_yr_count    **ACTIVE GRID** ( 216576  values per variable
## [2]    D3,D2     : climatology_bounds
## [3]    D0        : lon
## [4]    D1        : lat
## [5]    D2        : time
##
## Dimensions 4 (3 active):
##
##   dim   name  length     min     max start count    dmin    dmax unlim coord_dim
##   <chr> <chr>  <dbl>   <dbl>   <dbl> <int> <int>   <dbl>   <dbl> <lgl> <lgl>
```

```
## 1 D0    lon      192  0        3.58e2     1    192  0        3.58e2 FALSE TRUE
## 2 D1    lat       94 -8.85e1  8.85e1     1     94 -8.85e1  8.85e1 FALSE TRUE
## 3 D2    time      12 -6.57e5 -6.57e5     1     12 -6.57e5 -6.57e5 FALSE TRUE
##
## Inactive dimensions:
##
##   dim   name  length   min   max unlim coord_dim
##   <chr> <chr>  <dbl> <dbl> <dbl> <lgl> <lgl>
## 1 D3    nbnds      2     1     2 FALSE FALSE
```

Function `hyper_dims()` returns a description of the grid for which observations are available.

```
hyper_dims(meteo_data.tnc)
## # A tibble: 3 x 7
##   name  length start count    id unlim coord_dim
##   <chr>  <dbl> <int> <int> <int> <lgl> <lgl>
## 1 lon      192     1   192     0 FALSE TRUE
## 2 lat       94     1    94     1 FALSE TRUE
## 3 time      12     1    12     2 FALSE TRUE
```

Function `hyper_vars()` returns a description of the observations or variables available at each grid point.

```
hyper_vars(meteo_data.tnc)
## # A tibble: 2 x 6
##      id name            type     ndims natts dim_coord
##   <int> <chr>           <chr>    <int> <int> <lgl>
## 1     4 pevpr           NC_FLOAT     3    14 FALSE
## 2     5 valid_yr_count NC_FLOAT     3     4 FALSE
```

Function `hyper_tibble()` extracts a subset of the data into a tibble in long (or tidy) format. The selection of the grid point is done in the same operation and in this case using `signif()` to test for an approximate match to actual longitude and latitude values. A pipe is used to add the decoded dates, using the pipe operator (`|>`) and methods from 'dplyr' (see section 8.7.2 on page 262). The decoding of dates is done using functions from package 'lubridate' (see section 8.8 on page 267).

```
hyper_tibble(meteo_data.tnc,
             lon = signif(lon, 1) == 9,
             lat = signif(lat, 2) == 87) |>
  mutate(.data = _, month = month(ymd("1800-01-01") + days(time))) |>
  select(.data = _, -time)
## # A tibble: 12 x 5
##   pevpr valid_yr_count   lon   lat month
##   <dbl>          <dbl> <dbl> <dbl> <dbl>
## 1  4.28       1.19e-39  9.38  86.7    12
## 2  5.72       1.19e-39  9.38  86.7     1
## 3  4.38       1.29e-39  9.38  86.7     2
## # i 9 more rows
```

In this second example, data are extracted for all grid points along latitudes by omitting the test for `lat` from the chunk above. The tibble is assembled automatically and columns for the active dimensions added. The decoding of the months remains the same as above.

```
hyper_tibble(meteo_data.tnc,
             lon = signif(lon, 1) == 9) |>
  mutate(.data = _, month = month(ymd("1800-01-01") + days(time))) |>
  select(.data = _, -time)
## # A tibble: 1,128 x 5
##    pevpr valid_yr_count   lon   lat month
##    <dbl>          <dbl> <dbl> <dbl> <dbl>
## 1  1.02        1.19e-39  9.38  88.5    12
## 2  4.28        1.19e-39  9.38  86.7    12
## 3  3.03        9.18e-40  9.38  84.8    12
## # i 1,125 more rows
```

10.15 Instead of extracting data for one longitude across latitudes, extract data across longitudes for one latitude near the Equator.

10.12 Remotely Located Data

Many of the functions described above accept a URL address in place of a file name. Consequently, files can be read remotely without having to first download and save a copy in the local file system. This can be useful, especially when file names are generated within a script. However, one should avoid, especially in the case of servers open to public access, repeatedly downloading the same file as this unnecessarily increases network traffic and workload on the remote server. Because of this, our first example reads a small file from my own web site. See section 10.6 on page 388 for details on the use of these and other functions for reading text files.

```
logger.df <-
      read.csv2(file = "http://r4photobiology.info/learnr/logger_1.txt",
                header = FALSE,
                col.names = c("time", "temperature"))
sapply(logger.df, class)
##        time temperature
## "character"   "numeric"
```

While functions in package 'readr' support the use of URLs, those in packages 'readxl' and 'xlsx' do not. Consequently, the file has to be first downloaded and saved locally, and subsequently imported as described in section 10.9.2 on page 401. Function `download.file()` in the R 'utils' package can be used to download files using URLs. It supports different modes such as binary or text for the contents, and write or append for the local file, and different methods such as `"internal"`, `"wget"`, and `"libcurl"`.

For portability, MS-Excel files should be downloaded in binary mode, setting `mode = "wb"`, which is required under MS-Windows.

```
download.file("http://r4photobiology.info/learnr/my-data.xlsx",
              "data/my-data-dwn.xlsx",
              mode = "wb")
```

Functions from packages 'foreign' and 'haven', useful for reading files saved by other statistical software, support URLs. See section 10.10 on page 404 for more information about importing this kind of data into R. The two examples below read a file saved by SPSS located in a remote server, using these two packages.

```
remote_thiamin.df <-
  read.spss(file = "http://r4photobiology.info/learnr/thiamin.sav",
            to.data.frame = TRUE)
head(remote_thiamin.df)
##   THIAMIN CEREAL
## 1     5.2  wheat
## 2     4.5  wheat
## 3     6.0  wheat
## 4     6.1  wheat
## 5     6.7  wheat
## 6     5.8  wheat
```

```
remote_my_spss.tb <-
    read_sav(file = "http://r4photobiology.info/learnr/thiamin.sav")
remote_my_spss.tb
## # A tibble: 24 x 2
##   THIAMIN CEREAL
##     <dbl> <dbl+lbl>
## 1     5.2 1 [wheat]
## 2     4.5 1 [wheat]
## 3     6   1 [wheat]
## # i 21 more rows
```

Next, we download from NOAA's server a NetCDF file with long-term means for potential evapotranspiration, the same file used above in the 'ncdf4' example. This is a moderately large file at 834 KB. In this case, it is not possible to directly open the connection to the NetCDF file and it has to be downloaded. The `if` statement ensures that the file is downloaded only if the local copy is missing (to refresh the local copy simply delete the existing one). Once downloaded, the file can be opened as shown in section 10.11 on page 406.

```
if (!file.exists("extdata/pevpr.sfc.mon.ltm.nc")) {
  my.url <- paste("ftp://ftp.cdc.noaa.gov/Datasets/ncep.reanalysis.derived/",
                  "surface_gauss/pevpr.sfc.mon.ltm.nc",
                  sep = "")
  download.file(my.url,
                mode = "wb",
                destfile = "extdata/pevpr.sfc.mon.ltm.nc")
}
pet_ltm.nc <- nc_open("extdata/pevpr.sfc.mon.ltm.nc")
```

For portability, NetCDF files should be downloaded in binary mode, setting `mode = "wb"`, which is required under MS-Windows.

Some NetCDF file servers support the OPeNDAP protocol. In these servers, it is possible to open the files remotely and only download a part of the file. Function `open.nc()` from package 'RNetCDF' transparently supports OPeNDAP URLs.

10.13 Databases

One of the advantages of using databases is that subsets of cases and variables can be retrieved, even remotely, making it possible to work in R both locally and remotely with huge data sets. One should remember that R natively keeps whole objects in RAM, and consequently, available machine memory limits the size of data sets with which it is possible to work. Package 'dbplyr' provides the tools to work with data in databases using the same verbs as when using 'dplyr' with data stored in memory (RAM) (see chapter 8). This is an important subject, but extensive enough to be outside the scope of this book. We provide a few simple examples to show the very basics but interested readers should consult *R for Data Science* (Wickham et al. 2023).

The additional steps compared to using 'dplyr' start with the need to establish a connection to a local or remote database. We will use R package 'RSQLite' to create a local temporary SQLite database. 'dbplyr' backends supporting other database systems are also available. We will use meteorological data from 'learnrbook' for this example.

```
library(dplyr)
con <- DBI::dbConnect(RSQLite::SQLite(), dbname = ":memory:")
copy_to(con, weather_wk_25_2019.tb, "weather",
        temporary = FALSE,
        indexes = list(
          c("month_name", "calendar_year", "solar_time"),
          "time",
          "sun_elevation",
          "was_sunny",
          "day_of_year",
          "month_of_year"
        )
)
weather.db <- tbl(con, "weather")
colnames(weather.db)
##  [1] "time"           "PAR_umol"       "PAR_diff_fr"    "global_watt"
##  [5] "day_of_year"    "month_of_year"  "month_name"     "calendar_year"
##  [9] "solar_time"     "sun_elevation"  "sun_azimuth"    "was_sunny"
## [13] "wind_speed"     "wind_direction" "air_temp_C"     "air_RH"
## [17] "air_DP"         "air_pressure"   "red_umol"       "far_red_umol"
## [21] "red_far_red"
weather.db |>
  filter(.data = _, sun_elevation > 5) |>
  group_by(.data = _, day_of_year) |>
  summarise(.data = _, energy_Wh = sum(global_watt, na.rm = TRUE) * 60 / 3600)
## # Source:   SQL [?? x 2]
## # Database: sqlite 3.45.0 [:memory:]
##   day_of_year energy_Wh
##         <dbl>     <dbl>
## 1         162     7500.
## 2         163     6660.
## 3         164     3958.
## # i more rows
```

Package 'dbplyr' translates data pipes that use 'dplyr' syntax into SQL queries to databases, either local or remote. As long as there are no problems with the backend, the use of a database is almost transparent to the R user.

It is always good to clean up, and in the case of the book, the best way to test that the examples can be run in a "clean" system.

```
unlink("./data", recursive = TRUE)
unlink("./extdata", recursive = TRUE)
```

10.14 Data Acquisition from Physical Devices

Numerous modern data acquisition devices based on microcontrollers, including internet-of-things (IoT) devices, have servers (or daemons) that can be queried over a network connection to retrieve either real-time or logged data. Formats based on XML schemas or the JSON format are commonly used for data.

10.14.1 'jsonlite'

The next example retrieves data from USB module from *YoctoPuce* (`http://www.yoctopuce.com/`) using a software hub running locally. The module used in this example is a YoctoMeteo capable of storing measured data in its own memory.

Before running this example, data recording needs to be anabled in the YoctoPuce module and allow some time for the module to collect some data. In the call to function `fromJSON()`, `"C1-Meteo"` needs to be replaced by the ID assigned to the module used. The example uses an instance of Yocto VirtualHub running locally and listening at port 4444, the default. The same code can be used over the network by editing the string saved in `hub.url`.

Function `fromJSON()` from package 'jsonlite' can be used to retrieve logged data from one sensor module.

```
hub.url <- "http://localhost:4444/"
Meteo01.df <-
    fromJSON(paste(hub.url, "byName/C1-Meteo/dataLogger.json",
                   sep = ""), flatten = TRUE)
str(Meteo01.df, max.level = 2)
## 'data.frame': 3 obs. of  4 variables:
##  $ id     : chr  "humidity" "pressure" "temperature"
##  $ unit   : chr  "g/m3" "mbar" "'C"
##  $ calib  : chr  "0," "0," "0,"
##  $ streams:List of 3
##   ..$ :'data.frame': 447 obs. of  5 variables:
##   ..$ :'data.frame': 444 obs. of  5 variables:
##   ..$ :'data.frame': 447 obs. of  5 variables:
```

The minimum, mean, and maximum values for each logging interval need to be split from a single vector. We do this by indexing with a logical vector (recycled).

The data returned are in long form, including measured values, quantity names, units, and the date and time when each value was acquired.

```
Meteo01.df[["streams"]][[which(Meteo01.df$id == "temperature")]] |>
  as_tibble(x = _) |>
  dplyr::transmute(.data = _,
                   utc.time = as.POSIXct(utc, origin = "1970-01-01", tz = "UTC"),
                     t_min = unlist(val)[c(TRUE, FALSE, FALSE)],
                     t_mean = unlist(val)[c(FALSE, TRUE, FALSE)],
                    t_max = unlist(val)[c(FALSE, FALSE, TRUE)]) -> temperature.df

Meteo01.df[["streams"]][[which(Meteo01.df$id == "humidity")]] |>
  as_tibble(x = _) |>
  dplyr::transmute(.data = _,
                   utc.time = as.POSIXct(utc, origin = "1970-01-01", tz = "UTC"),
                     hr_min = unlist(val)[c(TRUE, FALSE, FALSE)],
                     hr_mean = unlist(val)[c(FALSE, TRUE, FALSE)],
                     hr_max = unlist(val)[c(FALSE, FALSE, TRUE)]) -> humidity.df

full_join(temperature.df, humidity.df)

## Joining with `by = join_by(utc.time)`
## # A tibble: 114 x 7
##   utc.time            t_min t_mean t_max hr_min hr_mean hr_max
##   <dttm>              <dbl>  <dbl> <dbl>  <dbl>   <dbl>  <dbl>
## 1 2023-10-15 18:06:00  26.1   26.1  26.3   10.4    10.5   10.6
## 2 2023-10-15 19:00:00  25.8   26.0  26.1   10.5    10.7   10.8
## 3 2023-10-15 20:00:00  25.7   26.0  26.0   10.6    10.8   10.9
## # i 111 more rows
```

Most YoctoPuce input modules have a built-in datalogger, and the stored data can also be downloaded as a csv file through a physical or virtual hub. As shown above, it is possible to control them through the HTML server in the physical or virtual hubs. Alternatively the R package 'reticulate' can be used to control YoctoPuce modules by means of the Python library giving access to their full API.

10.15 Further Reading

R includes the manual "R Data Import/Export", a very useful reference.

Since this is the end of the book, I recommend as further reading the writings of Burns as they are full of insight. Having arrived at the end of *Learn R: As a Language*, you should read *S Poetry* (Burns 1998) and *Tao Te Programming* (Burns 2012). If you want to never get caught unaware by R's idiosyncrasies, read also *The R Inferno* (Burns 2011).

Bibliography

Aho, A. V. and J. D. Ullman (1992). *Foundations of Computer Science*. Computer Science Press. ISBN: 0716782332.

Aiken, H., A. G. Oettinger and T. C. Bartee (1964). 'Proposed automatic calculating machine'. In: *IEEE Spectrum* 1.8, pp. 62–69. DOI: `10.1109/mspec.1964.6500770`.

Bates, D., M. Mächler, B. Bolker and S. Walker (2015). 'Fitting Linear Mixed-Effects Models Using lme4'. In: *Journal of Statistical Software* 67.1, pp. 1–48. DOI: `10.18637/jss.v067.i01` (cit. on p. 230).

Becker, R. A. and J. M. Chambers (1984). *S: An Interactive Environment for Data Analysis and Graphics*. Chapman and Hall/CRC. ISBN: 0-534-03313-X (cit. on p. 10).

Becker, R. A., J. M. Chambers and A. R. Wilks (1988). *The New S Language: A Programming Environment for Data Analysis and Graphics*. Chapman & Hall. ISBN: 0-534-09192-X (cit. on p. 10).

Boas, R. P. (1981). 'Can we make mathematics intelligible?' In: *The American Mathematical Monthly* 88.10, pp. 727–731.

Burns, P. (1998). *S Poetry* (cit. on pp. 177, 414).

— (2011). *The R Inferno*. URL: `http://www.burns-stat.com/pages/Tutor/R_inferno.pdf` (visited on 27/07/2017) (cit. on p. 414).

— (2012). *Tao Te Programming*. Lulu. ISBN: 9781291130454 (cit. on pp. 8, 414).

Chambers, J. M. (2016). *Extending R*. The R Series. Chapman and Hall/CRC. ISBN: 1498775713 (cit. on pp. 10, 22, 185, 186).

Chang, W. (2018). *R Graphics Cookbook*. 2nd ed. O'Reilly UK Ltd. ISBN: 1491978600 (cit. on pp. 367, 381).

Cleveland, W. S. (1985). *The Elements of Graphing Data*. Wadsworth, Inc. ISBN: 978-0534037291 (cit. on p. 273).

Crawley, M. J. (2012). *The R Book*. Wiley, p. 1076. ISBN: 0470973927 (cit. on pp. 212, 241).

Dalgaard, P. (2008). *Introductory Statistics with R*. Springer, p. 380. ISBN: 0387790543 (cit. on p. 241).

Diez, D., M. Cetinkaya-Rundel and C. D. Barr (2019). *OpenIntro Statistics*. 4th ed. 422 pp. URL: `https://www.openintro.org/stat/os4.php` (visited on 20/11/2022) (cit. on p. 241).

Everitt, B. S. and T. Hothorn (2010). *A Handbook of Statistical Analyses Using R*. 2nd ed. Chapman & Hall/CRC, p. 376. ISBN: 1420079336 (cit. on p. 241).

— (2011). *An Introduction to Applied Multivariate Analysis with R*. Springer, p. 288. ISBN: 1441996494 (cit. on p. 241).

Faraway, J. J. (2004). *Linear Models with R*. Boca Raton, FL: Chapman & Hall/CRC, p. 240 (cit. on p. 241).

Faraway, J. J. (2006). *Extending the Linear Model with R: Generalized Linear, Mixed Effects and Nonparametric Regression Models.* Chapman & Hall/CRC, p. 345. ISBN: 158488424X (cit. on pp. 241, 242).

Gagolewski, M. (2023). *Deep R Programming.* Zenodo. 438 pp. ISBN: 978-0-6455719-2-9. DOI: `10.5281/ZENODO.7490464`. URL: `https://deepr.gagolewski.com/` (visited on 05/01/2024) (cit. on p. 124).

Gandrud, C. (2015). *Reproducible Research with R and R Studio.* 2nd ed. Chapman & Hall/CRC The R Series). Chapman and Hall/CRC. 323 pp. ISBN: 1498715370 (cit. on pp. 19, 20).

Hall, J. N. and R. L. Schwartz (1997). *Effective Perl Programming. Writing Better Programs with Perl.* Addison-Wesley. 288 pp. ISBN: 9780201419757 (cit. on p. 4).

Hamming, R. W. (1987). *Numerical Methods for Scientists and Engineers.* Dover Publications Inc. 752 pp. ISBN: 0486652416.

Holmes, S. and W. Huber (2019). *Modern Statistics for Modern Biology.* Cambridge University Press. 382 pp. ISBN: 1108705294 (cit. on pp. 241, 332).

Hughes, T. P. (2004). *American Genesis.* The University of Chicago Press. 530 pp. ISBN: 0226359271 (cit. on p. 130).

Hyndman, R. and G. Athanasopoulos (2021). *Forecasting: Principles and Practice.* 3rd ed. Melbourne, Australia: OTexts (cit. on pp. 235, 242).

Ihaka, R. and R. Gentleman (1996). 'R: A Language for Data Analysis and Graphics'. In: *J. Comput. Graph. Stat.* 5, pp. 299–314 (cit. on p. 10).

Ihaka, R. (1998). *R : Past and Future History. A Draft of a Paper for Interface 98.* Interface Symposium on Computer Science and Statistics. The University of Auckland. URL: `https://www.stat.auckland.ac.nz/~ihaka/downloads/Interface98.pdf`. Draft (cit. on p. 22).

James, G., D. Witten, T. Hastie and R. Tibshirani (2013). *An Introduction to Statistical Learning: with Applications in R.* Springer, p. 426. ISBN: 978-1461471370 (cit. on p. 242).

Kernigham, B. W. and P. J. Plauger (1981). *Software Tools in Pascal.* Reading, Massachusetts: Addison-Wesley. 366 pp. (cit. on pp. 134, 244).

Kernighan, B. W. and R. Pike (1999). *The Practice of Programming.* Addison Wesley. 288 pp. ISBN: 020161586X (cit. on p. 21).

Knuth, D. E. (1987). *The TeXbook.* Reading, Massachusetts: Addison-Wesley Publishing Company, p. 483. ISBN: 0-201-13448-9 (cit. on p. 20).

Knuth, D. E. (1984). 'Literate programming'. In: *The Computer Journal* 27.2, pp. 97–111 (cit. on pp. 19, 130).

Koponen, J. and J. Hildén (2019). *Data Visualization Handbook.* Espoo, Finland: Aalto University. ISBN: 9789526074498 (cit. on p. 272).

Lamport, L. (1994). *LaTeX: A Document Preparation System.* English. 2nd ed. Reading: Addison-Wesley, p. 272. ISBN: 0-201-52983-1 (cit. on pp. 20, 130).

Leisch, F. (2002). 'Dynamic generation of statistical reports using literate data analysis'. In: *Proceedings in Computational Statistics.* Compstat 2002. Ed. by W. Härdle and B. Rönz. Heidelberg, Germany: Physika Verlag, pp. 575–580. ISBN: 3-7908-1517-9 (cit. on p. 19).

Lemon, J. (2020). *Kickstarting R.* URL: `https://cran.r-project.org/doc/contrib/Lemon-kickstart/kr_intro.html` (visited on 07/02/2020).

Matloff, N. (2011). *The Art of R Programming: A Tour of Statistical Software Design.* No Starch Press, p. 400. ISBN: 1593273843 (cit. on pp. 84, 124, 168, 244).

Mehtätalo, L. and J. Lappi (2020). *Biometry for Forestry and Environmental Data with Examples in R.* Boca Raton: Taylor & Francis Group. 411 pp. ISBN: 9781498711487 (cit. on p. 241).

Murrell, P. (2019). *R Graphics.* 3rd ed. Portland: Chapman and Hall/CRC. 423 pp. ISBN: 1498789056 (cit. on pp. 121, 124, 271, 381).

Newham, C. and B. Rosenblatt (2005). *Learning the* `bash` *Shell.* O'Reilly UK Ltd. 352 pp. ISBN: 0596009658 (cit. on p. 21).

Peng, R. D. (2022). *R Programming for Data Science.* Leanpub. 182 pp. URL: `https://leanpub.com/rprogramming` (visited on 27/07/2023) (cit. on p. 270).

Pinheiro, J. C. and D. M. Bates (2000). *Mixed-Effects Models in S and S-Plus.* New York: Springer (cit. on pp. 180, 230, 242).

Ram, K., C. Boettiger, S. Chamberlain, N. Ross, M. Salmon and S. Butland (2019). 'A Community of Practice Around Peer Review for Long-Term Research Software Sustainability'. In: *Computing in Science & Engineering* 21.2, pp. 59–65. DOI: `10.1109/mcse.2018.2882753` (cit. on p. 183).

Ramsay, J. (2009). *Functional Data Analysis with R and MATLAB.* Springer-Verlag New York, p. 214. ISBN: 9780387981840 (cit. on p. 226).

Sarkar, D. (2008). *Lattice: Multivariate Data Visualization with R.* 1st ed. Springer, p. 268. ISBN: 0387759689 (cit. on p. 271).

Smith, H. F. (1957). 'Interpretation of adjusted treatment means and regressions in analysis of covariance'. In: *Biometrics* 13, pp. 281–308 (cit. on p. 213).

Tufte, E. R. (1983). *The Visual Display of Quantitative Information.* Cheshire, CT: Graphics Press. 197 pp. ISBN: 0-9613921-0-X (cit. on p. 320).

Venables, W. N. and B. D. Ripley (2002). *Modern Applied Statistics with S.* 4th ed. New York: Springer. ISBN: 0-387-95457-0 (cit. on pp. 212, 241).

Wickham, H. (2010). 'A Layered Grammar of Graphics'. In: *Journal of Computational and Graphical Statistics* 19.1, pp. 3–28. DOI: `10.1198/jcgs.2009.07098` (cit. on p. 273).

— (2019). *Advanced R.* 2nd ed. Chapman and Hall/CRC. 588 pp. ISBN: 0815384572 (cit. on pp. 168, 185).

Wickham, H. and J. Bryan (2023). *R Packages. Organize, Test, Document, and Share Your Code.* O'Reilly Media, Incorporated. ISBN: 9781098134945 (cit. on p. 186).

Wickham, H., M. Cetinkaya-Rundel and G. Grolemund (2023). *R for Data Science. Import, Tidy, Transform, Visualize, and Model Data.* O'Reilly Media. ISBN: 9781492097402 (cit. on pp. 270, 412).

Wickham, H. and C. Sievert (2016). *ggplot2: Elegant Graphics for Data Analysis.* 2nd ed. Springer. XVI + 260. ISBN: 978-3-319-24277-4 (cit. on pp. 180, 271, 367, 381).

Wirth, N. (1976). *Algorithms + Data Structures = Programs.* Englewood Cliffs: Prentice-Hall, p. 366.

Wood, S. N. (2017). *Generalized Additive Models.* Chapman and Hall/CRC. 476 pp. ISBN: 1498728332 (cit. on pp. 226, 230, 242).

Xie, Y. (2013). *Dynamic Documents with R and knitr.* The R Series. Chapman and Hall/CRC, p. 216. ISBN: 1482203537 (cit. on pp. 19, 20, 130).

Xie, Y. (2016). *bookdown: Authoring Books and Technical Documents with R Markdown.* Chapman and Hall/CRC. ISBN: 9781138700109 (cit. on p. 130).

Xie, Y., J. J. Allaire and G. Grolemund (2018). *R Markdown. The Definitive Guide.* Chapman and Hall/CRC. 304 pp. ISBN: 1138359335 (cit. on p. 130).

Zachry, M. and C. Thralls (2004). 'An Interview with Edward R. Tufte'. In: *Technical Communication Quarterly* 13.4, pp. 447–462. DOI: 10.1207/s15427625tcq1304_5.

Zuur, A. F. (2012). *A Beginner's Guide to Generalized Additive Models with R.* 1st ed. Newburgh: Highland Statistics. 194 pp. ISBN: 9780957174122 (cit. on p. 230).

Zuur, A. F., E. N. Ieno and E. Meesters (2009). *A Beginner's Guide to R.* 1st ed. Springer, p. 236. ISBN: 0387938362 (cit. on p. 241).

General Index

Alphabetic Index of R Names

Index of R Names by Category

R names and symbols grouped into the categories 'classes and modes', 'constant and special values', 'control of execution', 'data objects', 'functions and methods', 'names and their scope', and 'operators'.

Frequently Asked Questions

Frequently asked questions and their answers appear in the body of the book preceded by the icon ? and highlighted by a marginal bar of the same colour as the icon.